Stochastic Approximation and Its Applications

Nonconvex Optimization and Its Applications

Volume 64

The titles published in this series are listed at the end of this volume.

Stochastic Approximation and Its Applications

by

Han-Fu Chen

Institute of Systems Science,
Academy of Mathematics and System Science,
Chinese Academy of Sciences,
Beijing, P.R. China

KLUWER ACADEMIC PUBLISHERS

DORDRECHT / BOSTON / LONDON

A C.I.P. Catalogue record for this book is available from the Library of Congress.

ISBN 1-4020-0806-6

Published by Kluwer Academic Publishers,
P.O. Box 17, 3300 AA Dordrecht, The Netherlands.

Sold and distributed in North, Central and South America
by Kluwer Academic Publishers,
101 Philip Drive, Norwell, MA 02061, U.S.A.

In all other countries, sold and distributed
by Kluwer Academic Publishers,
P.O. Box 322, 3300 AH Dordrecht, The Netherlands.

Printed on acid-free paper

Printed in the Netherlands.

Contents

Preface

Estimating unknown parameters based on observation data containing information about the parameters is ubiquitous in diverse areas of both theory and application. For example, in system identification the unknown system coefficients are estimated on the basis of input-output data of the control system; in adaptive control systems the adaptive control gain should be defined based on observation data in such a way that the gain asymptotically tends to the optimal one; in blind channel identification the channel coefficients are estimated using the output data obtained at the receiver; in signal processing the optimal weighting matrix is estimated on the basis of observations; in pattern classification the parameters specifying the partition hyperplane are searched by learning, and more examples may be added to this list.

All these parameter estimation problems can be transformed to a root-seeking problem for an unknown function. To see this, let y_k denote the observation at time k, i.e., the information available about the unknown parameters at time k. It can be assumed that the parameter under estimation denoted by x^0 is a root of some unknown function $f(\cdot)$: $f(x^0) = 0$. This is not a restriction, because, for example, $\|x - x^0\|^2$ may serve as such a function. Let x_k be the estimate for x^0 at time k. Then the available information y_{k+1} at time $k+1$ can formally be written as

$$y_{k+1} = f(x_k) + \epsilon_{k+1},$$

where

$$\epsilon_{k+1} = y_{k+1} - f(x_k).$$

Therefore, by considering y_{k+1} as an observation on $f(\cdot)$ at x_k with observation error ϵ_{k+1}, the problem has been reduced to seeking the root x^0 of $f(\cdot)$ based on $\{y_k\}$.

It is clear that for each problem to specify $f(\cdot)$ is of crucial importance. The parameter estimation problem is possible to be solved only if $f(\cdot)$

is appropriately selected so that the observation error $\{\epsilon_k\}$ meets the requirements figured in convergence theorems.

If $f(\cdot)$ and its gradient can be observed without error at any desired values, then numerical methods such as Newton-Raphson method among others can be applied to solving the problem. However, this kind of methods cannot be used here, because in addition to the obvious problem concerning the existence and availability of the gradient, the observations are corrupted by errors which may contain not only the purely random component but also the structural error caused by inadequacy of the selected $f(\cdot)$.

Aiming at solving the stated problem, Robbins and Monro proposed the following recursive algorithm

$$x_{k+1} = x_k + a_k y_{k+1}, \qquad a_k > 0,$$

to approximate the sought-for root x^0, where a_k is the step size. This algorithm is now called the Robbins-Monro (RM) algorithm. Following this pioneer work of stochastic approximation, there have been a large amount of applications to practical problems and research works on theoretical issues.

At beginning, the probabilistic method was the main tool in convergence analysis for stochastic approximation algorithms, and rather restrictive conditions were imposed on both $f(\cdot)$ and $\{\epsilon_k\}$. For example, it is required that the growth rate of $f(x)$ is not faster than linear as $\|x\|$ tends to infinity and $\{\epsilon_k\}$ is a martingale difference sequence [78]. Though the linear growth rate condition is restrictive, as shown by simulation it can hardly be simply removed without violating convergence for RM algorithms.

To weaken the noise conditions guaranteeing convergence of the algorithm, the ODE (ordinary differential equation) method was introduced in [72, 73] and further developed in [65]. Since the conditions on noise required by the ODE method may be satisfied by a large class of $\{\epsilon_k\}$ including both random and structural errors, the ODE method has been widely applied for convergence analysis in different areas. However, in this approach one has to *a priori* assume that the sequence of estimates $\{x_k\}$ is bounded. It is hard to say that the boundedness assumption is more desirable than a growth rate restriction on $f(\cdot)$.

The stochastic approximation algorithm with expanding truncations was introduced in [27], and the analysis method has then been improved in [14]. In fact, this is an RM algorithm truncated at expanding bounds, and for its convergence the growth rate restriction on $f(\cdot)$ is not required. The convergence analysis method for the proposed algorithm is called the trajectory-subsequence (TS) method, because the analysis is carried out at trajectories where the noise condition is satisfied and in contrast to the ODE method the noise condition need not be verified on the whole sequence $\{x_k\}$ but is verified only along convergent subsequences $\{x_{n_k}\}$. This makes a great difference when dealing with the state-dependent noise $\{\epsilon_{k+1}(x_k)\}$, because a convergent subsequence $\{x_{n_k}\}$ is always bounded while the boundedness of the whole sequence $\{x_k\}$ is not guaranteed before establishing its convergence. As shown in Chapters 4, 5, and 6 for most of parameter estimation problems after transforming them to a root-seeking problem, the structural errors are unavoidable, and they are state-dependent.

The expanding truncation technique equipped with TS method appears a powerful tool in dealing with various parameter estimation problems: it not only has succeeded in essentially weakening conditions for convergence of the general stochastic approximation algorithm but also has made stochastic approximation possible to be successfully applied in diverse areas. However, there is a lack of a reference that systematically describes the theoretical part of the method and concretely shows the way how to apply the method to problems coming from different areas. To fill in the gap is the purpose of the book.

The book summarizes results on the topic mostly distributed over journal papers and partly contained in unpublished material. The book is written in a systematical way: it starts with a general introduction to stochastic approximation and then describes the basic method used in the book, proves the general convergence theorems and demonstrates various applications of the general theory.

In Chapter 1 the problem of stochastic approximation is stated and the basic methods for convergence analysis such as probabilistic method, ODE method, TS method, and the weak convergence method are introduced.

Chapter 2 presents the theoretical foundation of the algorithm with expanding truncations: the basic convergence theorems are proved by TS method; various types of noises are discussed; the necessity of the imposed noise condition is shown; the connection between stability of the equilibrium and convergence of the algorithm is discussed; the robustness of stochastic approximation algorithms is considered when the commonly used conditions deviate from the exact satisfaction, and the moving root tracking is also investigated. The basic convergence theorems are presented in Section 2.2, and their proof is elementary and purely deterministic.

Chapter 3 describes asymptotic properties of the algorithms: convergence rates for both cases whether or not the gradient of $f(\cdot)$ is degenerate; asymptotic normality of $\{x_k\}$ and asymptotic efficiency by averaging method.

Starting from Chapter 4 the general theory developed so far is applied to different fields. Chapter 4 deals with optimization by using stochastic approximation methods. Convergence and convergence rates of the Kiefer-Wolfowitz (KW) algorithm with expanding truncations and randomized differences are established. A global optimization method consisting in combination of the KW algorithms with search methods is defined, and its a.s. convergence as well as asymptotic behaviors are established. Finally, the global optimization method is applied to solving the model reduction problem.

In Chapter 5 the general theory is applied to the problems arising from signal processing. Applying the stochastic approximation method to blind channel identification leads to a recursive algorithm estimating the channel coefficients and continuously improving the estimates while receiving new signal in contrast to the existing "block" algorithms. Applying TS method to principal component analysis results in improving conditions for convergence. Stochastic approximation algorithms with expanding truncations with TS method are also applied to adaptive filters with and without constraints. As a result, conditions required for convergence have been considerably improved in comparison with the existing results. Finally, the expanding truncation technique and TS method are applied to the asynchronous stochastic approximation.

In the last chapter, the general theory is applied to problems arising from systems and control. The ideal parameter for operation is identified for stochastic systems by using the methods developed in this book. Then the obtained results are applied to the adaptive quadratic control problem. Adaptive regulation for a nonlinear nonparametric system and learning pole assignment are also solved by the stochastic approximation method.

The book is self-contained in the sense that there are only a few points using knowledge for which we refer to other sources, and these points can be ignored when reading the main body of the book. The basic mathematical tools used in the book are calculus and linear algebra based on which one will have no difficulty to read the fundamental convergence Theorems 2.2.1 and 2.2.2 and their applications described in the subsequent chapters. To understand other material, probability concept, especially the convergence theorems for martingale difference sequences are needed. Necessary concept of probability theory is given in Appendix A. Some facts from probability that are used at a few specific points are listed in Appendix A but without proof, because omitting the corresponding parts still makes the rest of the book readable. However, the proof of convergence theorems for martingales and martingale difference sequences is provided in detail in Appendix B.

The book is written for students, engineers and researchers working in the areas of systems and control, communication and signal processing, optimization and operation research, and mathematical statistics.

HAN-FU CHEN

Acknowledgments

The support of the National Key Project of China and the National Natural Science Foundation of China is gratefully acknowledged. The author would like to express his gratitude to Dr. Haitao Fang for his helpful suggestions and useful discussions. The author would also like to thank Ms. Jinling Chang for her skilled typing and to thank my wife Shujun Wang for her constant support.

Chapter 1

ROBBINS-MONRO ALGORITHM

Optimization is ubiquitous in various research and application fields. It is quite often that an optimization problem can be reduced to finding zeros (roots) of an unknown function $f(\cdot)$, which can be observed but the observation may be corrupted by errors. This is the topic of stochastic approximation (SA). The error source may be observation noise, but may also come from structural inaccuracy of the observed function. For example, one wants to find zeros of $f(\cdot)$, but he actually observes functions $f_k(\cdot)$ which are different from $f(\cdot)$. Let us denote by y_{k+1} the observation at time k, ϵ_{k+1} the observation noise:

$$y_{k+1} = f_k(x) + \epsilon_{k+1} = f(x) + \epsilon_{k+1} + (f_k(x) - f(x)).$$

Here, $f_k(x) - f(x)$ is the additional error caused by the structural inaccuracy. It is worth noting that the structural error normally depends on x, and it is hard to require it to have a certain probabilistic property such as independence, stationarity or martingale property. We call this kind of noises as state-dependent noise.

The basic recursive algorithm for finding roots of an unknown function on the basis of noisy observations is the Robbins-Monro (RM) algorithm, which is characterized by its simplicity in computation. This chapter serves as an introduction to SA, describing various methods for analyzing convergence of the RM algorithm.

In Section 1.1 the motivation of RM algorithm is explained, and its limitation is pointed out by an example. In Section 1.2 the classical approach to analyzing convergence of RM algorithm is presented, which is based on probabilistic assumptions on the observation noise. To relax restrictions made on the noise, a convergence analysis method connecting convergence of the RM algorithm with stability of an ordinary differential equation (ODE) was introduced in nineteen seventies. The ODE method is demonstrated in Section 1.3. In Section 1.4 the convergence analysis is carried out at a sample path by considering convergent subsequences. So, we call this method as Trajectory-Subsequence (TS) method, which is the basic tool used in the subsequent chapters.

In this book our main concern is the path-wise convergence of the algorithm. However, there is another approach to convergence analysis called the weak convergence method, which is briefly introduced in Section 1.5. Notes and references are given in the last section.

This chapter introduces main methods used in literature for convergence analysis, but restricted to the single root case. Extension to more general cases in various aspects is given in later chapters.

1.1. Finding Zeros of a Function.

Many theoretical and practical problems in diverse areas can be reduced to finding zeros of a function. To see this it suffices to notice that solving many problems finally consists in optimizing some function $L(\cdot)$, i.e., finding its minimum (or maximum). If $L(\cdot)$ is differentiable, then the optimization problem reduces to finding the roots of $f(x)$, where $f(x) \triangleq \frac{dL(x)}{dx}$, the derivative of $L(\cdot)$.

In the case where the function $L(\cdot)$ or its derivatives can be observed without errors, there are many numerical methods for solving the problem. For example, the gradient method, by which the estimate x_k for the root x^0 of $f(\cdot)$ is recursively generated by the following algorithm

$$x_{k+1} = x_k - \left(f'(x_k)\right)^{-1} f(x_k) \qquad (1.1.1)$$

where $f'(\cdot)$ denotes the derivative of $f(\cdot)$. This kind of problems belongs to the topics of optimization theory, which considers general cases where $L(\cdot)$ may be nonconvex, nonsmooth, and with constraints.

In contrast to the optimization theory, SA is devoted to finding zeros of an unknown function $f(\cdot)$ which can be observed, but the observations are corrupted by errors.

Since $f(x_k)$ is not exactly known and $f'(\cdot)$ even may not exist, (1.1.1)-like algorithms are no longer applicable. Consider the following simple example. Let $f(\cdot)$ be a linear function

$$f(x) = x - b.$$

If the derivative of $f(\cdot)$ is available, i.e., if we know $f'(x) \equiv 1$ and if $f(x_k)$ can precisely be observed, then according to (1.1.1)

$$x_{k+1} = x_k - f(x_k) = b.$$

This means that the gradient algorithm leads to the zero ($= b$) of $f(\cdot)$ by one step.

Assume the derivative of $f(\cdot)$ is unavailable but $f(\cdot)$ can exactly be observed.

Let us replace $(f'(x_k))^{-1}$ by $\frac{1}{k+1}$ in (1.1.1). Then we derive

$$x_{k+1} = x_k - \frac{1}{k+1} f(x_k), \tag{1.1.2}$$

or

$$x_{k+1} = (1 - \frac{1}{k+1})x_k + \frac{b}{k+1}. \tag{1.1.3}$$

This is a linear difference equation, which can inductively be solved, and the solution of (1.1.3) can be expressed as follows

$$\begin{aligned} x_{k+1} =& \frac{1}{k+1} x_1 + \sum_{i=1}^{k} (\frac{i+1}{k+1}) \frac{b}{i+1} \\ =& \frac{1}{k+1} x_1 + \frac{k}{k+1} b. \end{aligned}$$

Clearly, x_{k+1} tends to the root b of $f(\cdot)$ as $k \longrightarrow \infty$ for any initial value x_1. This is an attractive property: although the gradient of $f(\cdot)$ is unavailable, we can still approach the sought-for root if the inverse of the gradient is replaced by a sequence of positive real numbers decreasingly tending to zero.

Let us consider the case where $f(\cdot)$ is observed with errors:

$$y_{k+1} = f(x_k) + \epsilon_{k+1}, \tag{1.1.4}$$

where y_{k+1} denotes the observation at time $k+1$, ϵ_{k+1} the corresponding observation error and x_k the estimate for the root of $f(\cdot)$ at time k.

It is natural to ask, how will $\{x_k\}$ behave if the exact value of $f(x_k)$ in (1.1.2) is replaced by its error-corrupted observation y_{k+1}, i.e., if $\{x_k\}$ is recursively derived according to the following algorithm:

$$x_{k+1} = x_k - \frac{1}{k+1} y_{k+1}. \tag{1.1.5}$$

In our example, $f(x) = x - b$, and (1.1.5) turns to be

$$\begin{aligned} x_{k+1} &= x_k - \frac{1}{k+1}(f(x_k) + \epsilon_{k+1}) \\ &= (1 - \frac{1}{k+1})x_k + \frac{b}{k+1} - \frac{\epsilon_{k+1}}{k+1} \end{aligned}$$

Similar to (1.1.3), the solution of this difference equation is

$$\begin{aligned} x_{k+1} &= \frac{1}{k+1}x_1 + \sum_{i=1}^{k}(\frac{i+1}{k+1})\frac{b}{i+1} - \sum_{i=1}^{k}(\frac{i+1}{k+1})\frac{\epsilon_{i+1}}{i+1} \\ &= \frac{1}{k+1}x_1 + \frac{k}{k+1}b - \sum_{i=1}^{k}\frac{\epsilon_{i+1}}{k+1}. \end{aligned} \tag{1.1.6}$$

Therefore, x_{k+1} converges to b, the root of $f(\cdot)$, if $\frac{1}{k+1}\sum_{i=1}^{k}\epsilon_{i+1}$ tends to zero as $k \longrightarrow \infty$. This means that replacement of gradient by a sequence of $f(\cdot)$-independent numbers still works even in the case of error-corrupted observations, if the observation errors can be averaged out. It is worth noting that in lieu of (1.1.5) we have to take the positive sign before $\frac{1}{k+1}$, i.e., to consider

$$x_{k+1} = x_k + \frac{1}{k+1}y_{k+1}, \tag{1.1.7}$$

if $f(x) = -x - b$ rather than $x - b$, or more general, if $f(x)$ is decreasing as x increases.

This simple example demonstrates the basic features of the algorithm (1.1.5) or (1.1.7): 1) The algorithm may converge to a root of $f(\cdot)$; 2) The limit of the algorithm, if exists, should not depend on the initial value; 3) The convergence rate is defined by that how fast the observation errors are averaged out.

From (1.1.6) it is seen that the convergence rate is defined by $\frac{1}{n}\sum_{i=1}^{n}\epsilon_i$ for linear functions. In the case where $\{\epsilon_i\}$ is a sequence of independent and identically distributed random variables with zero mean and bounded variance, then

$$\frac{1}{n}\sum_{i=1}^{n}\epsilon_i = O((\frac{\log\log n}{n})^{\frac{1}{2}})$$

by the iterated logarithm law.

This means that convergence rate for algorithms (1.1.5) or (1.1.7) with error-corrupted observations should not be faster than $O(\frac{1}{\sqrt{n}})$.

1.2. Probabilistic Method

We have just shown how to find the root of an unknown linear function based on noisy observations. We now formulate the general problem.

Let $f(\cdot)$ be an unknown $\mathbb{R}^l \longrightarrow \mathbb{R}^l$ function with unknown root x^0 : $f(x^0) = 0$. Assume $f(\cdot)$ can be observed at each point $x \in \mathbb{R}^l$ with noise

$$y_{k+1} = f(x_k) + \epsilon_{k+1}, \tag{1.2.1}$$

where y_{k+1} is the observation at time $k+1$, ϵ_{k+1} is the observation noise, and x_k is the estimate for x^0 at time k.

Stochastic approximation algorithms recursively generate $\{x_k\}$ to approximate x^0 based on the past observations. In the pioneer work of this area Robbins and Monro proposed the following algorithm

$$x_{k+1} = x_k + a_k y_{k+1} \tag{1.2.2}$$

to estimate x^0, where step size $\{a_k\}$ is decreasing and satisfies the following conditions $a_k > 0$, $a_k \xrightarrow[k\longrightarrow\infty]{} 0$ and $\sum\limits_{i=1}^{\infty} a_i = \infty$. They proved $E\|x_k - x^0\|^2 \xrightarrow[k\longrightarrow\infty]{} 0$.

We explain the meaning of conditions required for step size $\{a_k\}$. Condition "$a_k \longrightarrow 0$" aims at reducing the effect of observation noises. To see this, consider the case where x_k is close to x^0 and $f(x_k)$ is close to zero, say, $\|x_k - x^0 + a_k f(x_k)\| < \epsilon$ with ϵ small.

Throughout the book, $\|x\|$ always means the Euclidean norm of a vector x, and $\|A\|$ denotes the square root of the maximum eigenvalue of the matrix $A^T A$, where A^T means the transpose of the matrix A.

By (1.2.2) $x_{k+1} - x^0 = x_k - x^0 + a_k f(x_k) + a_k \epsilon_{k+1}$, and $\|x_{k+1} - x^0\| \geq a_k\|\epsilon_{k+1}\| - \epsilon$. Even in the Gaussian noise case, $a_k\|\epsilon_{k+1}\|$ may be large if a_k has a positive lower bound. Therefore, in order to have the desired consistency, i.e., $x_k \xrightarrow[k\longrightarrow\infty]{} x^0$, it is necessary to use decreasing gains such that $a_k \longrightarrow 0$. On the other hand, consistency can neither be achieved, if a_k decreases too fast as $k \longrightarrow \infty$. To see this, let $\sum\limits_{i=0}^{\infty} a_i < \infty$. Then even for the noise-free case, i.e., $\epsilon_k \equiv 0$, from (1.2.2) we have $\sum\limits_{i=0}^{\infty} \|x_{i+1} - x_i\| \leq \sum\limits_{i=0}^{\infty} a_i\|f(x_i)\| < \infty$ if $f(\cdot)$ is a bounded function. Therefore, in this case

$$\|x_k - x^0\| = \|x_k - x_0 + x_0 - x^0\| \geq \|x_0 - x^0\| - \sum_{i=0}^{\infty} \|x_{i+1} - x_i\| > 0,$$

if the initial value x_0 is far from the true root x^0, and hence x_k will never converge to x^0.

The algorithm (1.2.2) is now called Robbins-Monro (RM) algorithm.

The classical approach to convergence analysis of SA algorithms is based on the probabilistic analysis for trajectories. We now present a typical convergence theorem by this approach. Related concept and results from probability theory are given in Appendices A and B.

In fact, we will use the martingale convergence theorem to prove the path-wise convergence of x_k, i.e., to show $x_k \xrightarrow[k\to\infty]{} x^0$ a.s. For this, the following set of conditions will be used.

A1.2.1 *The step size is such that*

$$a_i > 0, \quad \sum_{i=1}^{\infty} a_i = \infty, \ \textit{and} \ \sum_{i=1}^{\infty} a_i^2 < \infty.$$

A1.2.2 *There exists a continuously twice differentiable Lyapunov function* $v(\cdot)$ $\mathbb{R}^l \longrightarrow R$ *satisfying the following conditions.*

i) Its second derivative is bounded;

ii) $v(x) > 0$, $\forall x \neq x^0$, $v(x^0) = 0$ *and* $v(x) \longrightarrow \infty$ *as* $\|x\| \longrightarrow \infty$*;*

iii) For any $\epsilon > 0$ *there is a* $\beta_\epsilon > 0$ *such that*

$$\sup_{\|x-x^0\|>\epsilon} v_x^T(x) f(x) = -\beta_\epsilon < 0$$

where $v_x(x)$ *denotes the gradient of* $v(\cdot)$.

A1.2.3 *The observation noise* $(\epsilon_k, \mathcal{F}_k)$ *is a martingale difference sequence with*

$$E(\epsilon_k|\mathcal{F}_{k-1}) = 0, \quad E\|\epsilon_k\|^2 < \infty, \tag{1.2.3}$$

where $\{\mathcal{F}_k\}$ *is a family of nondecreasing* σ*-algebras.*

A1.2.4 *The function* $f(\cdot)$ *and the conditional second moment of the observation noise have the following upper bound*

$$\|f(x)\|^2 + E(\|\epsilon_k\|^2|\mathcal{F}_{k-1}) < c(1 + v(x)), \quad \forall k \geq 0, \tag{1.2.4}$$

where c *is a positive constant.*

Prior to formulating the theorem we need some auxiliary results.

Let $(x_k, \mathcal{F}_k)$ be an adapted sequence, i.e., x_k is $\mathcal{F}_k$-measurable $\forall k$.

Define σ, the first exist time of $\{x_k\}$ from a Borel set $G \in \mathbb{R}^l$: $[\sigma \leq n] = [x_0 \notin G] \cup [x_0 \in G, x_1 \notin G] \cup \cdots \cup [x_0 \in G, \cdots, x_{n-1} \in G, x_n \notin G]$.

It is clear that $[\sigma \leq n] \in \mathcal{F}_n$, i.e., σ is a Markov time.

Lemma 1.2.1 *Assume* $v(\cdot) \geq 0$ *and* $v(x_k)$ *is a nonnegative supermartingale, i.e.,*

$$F(v(x_k)|\mathcal{F}_{k-1}) \leq v(x_{k-1}) \ a.s.$$

Then $(v(x_{\sigma\wedge k}), \mathcal{F}_k)$ *is also a nonnegative supermartingale, where* $\sigma \wedge k = \min(\sigma, k)$.

The proof is given in Appendix B, Lemma B-2-1.

The following lemma concerning convergence of an adapted sequence will be used in the proof for convergence of the RM algorithm, but the lemma is of interest by itself.

Lemma 1.2.2 *Let* $(v_k, \mathcal{F}_k)$, $(\alpha_k, \mathcal{F}_k)$ *be two nonnegative adapted sequences.*

i) If $E(v_{k+1}|\mathcal{F}_k) \le v_k + \alpha_k$ *and* $E\sum_{i=1}^{\infty} \alpha_i < \infty$, *then* v_k *converges a.s. to a finite limit.*

ii) If $E(v_{k+1}|\mathcal{F}_k) \le v_k - \alpha_k$, *then* $\sum_{i=1}^{\infty} \alpha_i < \infty$ *a.s.*

Proof. For proving i) set

$$u_{k+1} = v_{k+1} + E\Big(\sum_{i=k+1}^{\infty} \alpha_i|\mathcal{F}_{k+1}\Big). \tag{1.2.5}$$

Then we have

$$E(u_{k+1}|\mathcal{F}_k) \le v_k + \alpha_k + E\Big(\sum_{i=k+1}^{\infty} \alpha_i|\mathcal{F}_k\Big) = u_k.$$

By the convergence theorem for nonnegative supermartingales, u_k converges a.s. as $k \longrightarrow \infty$.

Since $E\sum_{i=1}^{\infty} \alpha_i < \infty$, by the convergence theorem for martingales it follows that $E(\sum_{i=1}^{\infty} \alpha_i|\mathcal{F}_k)$ converges a.s. as $k \longrightarrow \infty$. Since α_k is $\mathcal{F}_k$-measurable and $\mathcal{F}_k$ is nondecreasing, we have $E(\sum_{i=1}^{\infty} \alpha_i|\mathcal{F}_k) = \sum_{i=1}^{k} \alpha_i + E(\sum_{i=k+1}^{\infty} \alpha_i|\mathcal{F}_k)$. Noticing that both $E(\sum_{i=1}^{\infty} \alpha_i|\mathcal{F}_k)$ and $\sum_{i=1}^{k} \alpha_i$ converge a.s. as $k \longrightarrow \infty$, we conclude that $E(\sum_{i=k+1}^{\infty} \alpha_i|\mathcal{F}_k)$ is also convergent a.s. as $k \longrightarrow \infty$. Consequently, from (1.2.5) it follows that v_k converges a.s. as $k \longrightarrow \infty$.

For proving ii) set

$$u_{k+1} = v_{k+1} + \sum_{i=1}^{k} \alpha_i.$$

Taking conditional expectation leads to

$$E(u_{k+1}|\mathcal{F}_k) \leq v_k - \alpha_k + \sum_{i=1}^{k} \alpha_i = u_k.$$

Again, by the convergence theorem for nonnegative supermartingales, u_k converges a.s. as $k \longrightarrow \infty$. Since by the same theorem v_k also converges a.s. as $k \longrightarrow \infty$, it directly follows that $\sum_{i=1}^{\infty} \alpha_i < \infty$ a.s. □

Theorem 1.2.1 *Assume Conditions A1.2.1–A1.2.4 hold. Then for any initial value, x_k given by the RM algorithm (1.2.2) converges to the root x^0 of $f(\cdot)$ a.s. as $k \longrightarrow \infty$.*

Proof. Let $v(\cdot)$ be the Lyapunov function given in A1.2.2. Expanding $v(x_{k+1})$ to the Taylor series, we obtain

$$\begin{aligned} v(x_{k+1}) =& v(x_k) + a_k v_x^T(x_k)\big(f(x_k) + \epsilon_{k+1}\big) \\ &+ \frac{1}{2} a_k^2 (f(x_k) + \epsilon_{k+1})^T v_{xx}(\xi_k)\big(f(x_k) + \epsilon_{k+1}\big) \\ \leq& v(x_k) + a_k v_k^T(x_k)\epsilon_{k+1} + a_k v_k^T(x_k) f(x_k) \\ &+ c_1 a_k^2 \big(\|f(x_k)\|^2 + \|\epsilon_{k+1}\|^2\big), \end{aligned} \tag{1.2.6}$$

where $v_x(\cdot)$ and v_{xx} denote the gradient and Hessian of $v(\cdot)$, respectively, ξ_k is a vector with components located in-between the corresponding components of x_k and x_{k+1}, and c_1 denotes the constant such that $\|v_{xx}(\cdot)\| \leq c_1$ (by A1.2.2).

Noticing that x_k is $\mathcal{F}_k$-measurable, and $E(\epsilon_{k+1}|\mathcal{F}_k) = 0$, taking conditional expectation for (1.2.6), by (1.2.4) we derive

$$E(v(x_{k+1})|\mathcal{F}_k) \leq v(x_k) + a_k v_k^T(x_k) f(x_k) + c c_1 a_k^2 (1 + v(x_k)) \tag{1.2.7}$$

Since $\sum_{i=1}^{\infty} a_k^2 < \infty$ by (A1.2.1), we have $\prod_{i=1}^{\infty}(1 + c c_1 a_i^2) < \infty$.
Denoting

$$v_{k+1} = (1 + v(x_{k+1})) \prod_{i=k+1}^{\infty} (1 + c c_1 a_i^2) \tag{1.2.8}$$

and noticing $v_x^T(x_k)f(x_k) \le 0$ by A1.2.2, iii) from (1.2.7) and (1.2.8) it follows that

$$\begin{aligned} &E(v_{k+1}|\mathcal{F}_k) \\ &\le \Big(1 + v(x_k) + cc_1a_k^2v(x_k) + cc_1a_k^2 + a_kv_x^T(x_k)f(x_k)\Big) \prod_{i=k+1}^{\infty}(1+cc_1a_k^2) \\ &= (1+v(x_k))\prod_{i=k}^{\infty}(1+cc_1a_i^2) + a_kv_x^T(x_k)f(x_k)\prod_{i=k+1}^{\infty}(1+cc_1a_i^2) \\ &= v_k + a_kv_k^T(x_k)f(x_k)\prod_{i=k+1}^{\infty}(1+cc_1a_i^2). \end{aligned} \tag{1.2.9}$$

Therefore, $E(v_{k+1}|\mathcal{F}_k) \le v_k$ and v_k converges a.s. by the convergence theorem for nonnegative supermartingales.

Since $\prod_{i=k}^{\infty}(1+cc_1a_i^2) \xrightarrow[k\to\infty]{} 1$, $v(x_k)$ also converges a.s. $k \longrightarrow \infty$.

For any $\epsilon > 0$, denote

$$G_\epsilon = \{x : \|x - x^0\| > \epsilon\}.$$

Let σ_ϵ^0 be the first exit time of $\{x_k\}$ from G_ϵ, and let

$$\sigma_\epsilon^i \triangleq \min\{t : t > \sigma_\epsilon^{i-1}, x_t \in G_\epsilon^c\},$$

where G_ϵ^c denotes the complement to G_ϵ. This means that σ_ϵ^i is the first exit time from G_ϵ after σ_ϵ^{i-1}.

Since $v_k^T(x_k)f(x_k)$ is nonpositive, from (1.2.9) it follows that

$$E(v_{k+1}|\mathcal{F}_k) \le v_k + a_kI_{[\sigma_\epsilon^i>k]}v_k^T(x_k)f(x_k)$$

for any i.

Then by (1.2.2), this implies that

$$E(v_{k+1}|\mathcal{F}_k) \le v_k - \beta_\epsilon a_kI_{[\sigma_k^i>k]}.$$

By Lemma 1.2.2, ii), the above inequality implies

$$\sum_{k=1}^{\infty} a_kI_{[\sigma_\epsilon^i>k]} < \infty \text{ a.s.}$$

which means that σ_ϵ^i must be finite a.s. Otherwise, we would have $\sum_{k=1}^{\infty} a_k < \infty$, a contradiction to A1.2.1. Therefore, after σ_ϵ^i, $\forall i$ with possible exception of a set with probability zero the trajectory of $\{x_k\}$ must enter G_ϵ^c.

Consequently, there is a subsequence $\{x_{k_i}\}$ such that $\|x_{k_i} - x^0\| \le \epsilon$, where $k_i \longrightarrow \infty$ as $i \longrightarrow \infty$.

By the arbitrariness of ϵ, we then conclude that there is a subsequence, denoted still by $\{x_{k_i}\}$, such that $x_{k_i} \xrightarrow[i\to\infty]{} x^0$. Hence $v(x_{k_i}) \xrightarrow[i\to\infty]{} 0$.

However, we have shown that $v(x_k)$ converges a.s. Therefore, $v(x_k) \xrightarrow[k\to\infty]{} 0$ a.s. By A1.2.2, ii) we then conclude that $x_k \xrightarrow[k\to\infty]{} x^0$ a.s. □

Remark 1.2.1 If Condition A1.2.2 iii) changes to

$$\sup_{\|x-x^0\|>\epsilon} v_x^T(x)f(x) = \beta_\epsilon > 0,$$

then the algorithm (1.2.2) should accordingly change to

$$x_{k+1} = x_k - a_k y_{k+1}. \tag{1.2.10}$$

We now explain conditions required in Theorem 1.2.1. As noted in Section 1.1, the step size should satisfy $\sum_{i=1}^{\infty} a_i = \infty$, but the condition $\sum_{i=1}^{\infty} a_i^2 < \infty$ may be weakened to $a_k \xrightarrow[k\to\infty]{} 0$.

Condition A1.2.2 requires existence of a Lyapunov function $v(\cdot)$. This kind of conditions is normally necessary to be imposed for convergence of the algorithms, but the analytic properties of $v(\cdot)$ may be weakened. The noise condition A1.2.3 is rather restrictive. As to be shown in the subsequent chapters, ϵ_k may be composed of not only the random noise but also structural errors which hardly have nice probabilistic properties such as martingale difference, stationarity or with bounded variances etc.

As in many cases, one can take $\|x - x^0\|^2$ to serve as $v(x)$. Then from (1.2.4) it follows that the growth rate of $f(x)$ as $\|x\| \longrightarrow \infty$ should not be faster than linear. This is a major restriction to apply Theorem 1.2.1. However, if we *a priori* assume that $\{x_k\}$ generated by the algorithm (1.2.2) is bounded, then $\|f(x_k)\|$ is bounded provided $f(\cdot)$ is locally bounded, and then the linear growth is not a restriction for $\{\|f(x_k)\|, k = 1, 2, \dots\}$.

1.3. ODE Method

As mentioned in Section 1.2, the classical probabilistic approach to analyzing SA algorithms requires rather restrictive conditions on the observation noise. In nineteen seventies a so-called ordinary differential equation (ODE) method was proposed for analyzing convergence of SA algorithms. We explain the idea of the method. The estimate $\{x_k\}$ generated by the RM algorithm is interpolated to a continuous function with interpolating length equal to $\{a_k\}$, the step size used in the algorithm. The tail part x_t of the interpolating function is shown to satisfy an ordinary differential equation $\dot{x} = f(x)$. The sought-for root x^0 is the equilibrium of the ODE. By stability of this equation, or by assuming existence of a Lyapunov function, it is proved that $x_t \xrightarrow[t \to \infty]{} x^0$. From this, it can be deduced that $x_k \longrightarrow x^0$.

For demonstrating the ODE method we need two facts from analysis, which are formulated below as propositions.

Proposition 1.3.1 *(Arzelà-Ascoli) Let $\{f_\lambda(t), t \in [0, \infty)\}$ be a set of equi-continuous and uniformly bounded functions, $\lambda \in \Lambda$, where by equi-continuity we mean that for any $t \in [0, \infty)$ and any $\epsilon > 0$, there exists $\delta > 0$ such that*

$$|f_\lambda(t) - f_\lambda(s)| \leq \epsilon, \quad \forall \lambda \in \Lambda, \ \textit{whenever}\ |t - s| < \delta.$$

Then there are a continuous function $f(t)$ and a subsequence $f_{\lambda_k}(t)$ of functions which converge to $f(t)$ uniformly in any finite interval of t, i.e.,

$$|f_{\lambda_k}(t) - f(t)| \xrightarrow[k \to \infty]{} 0$$

uniformly with respect to t belonging to any finite interval.

Proposition 1.3.2 *For the following ODE*

$$\dot{x}_t = -f(x_t), \quad t \geq 0 \ \textit{with}\ f(x^0) = 0 \tag{1.3.1}$$

if there exists a continuously differentiable function such that $v(x) > 0$, $\forall x \neq x^0$, $v(x^0) = 0$, $v(x) \longrightarrow \infty$ as $\|x\| \longrightarrow \infty$ and

$$v_x^T(x) f(x) < 0, \quad \forall x \neq x^0,$$

then the solution x_t to (1.3.1), starting from any initial value, tends to x^0 as $t \longrightarrow \infty$, i.e., x^0 is the global asymptotically stable solution to (1.3.1).

Let us introduce the following conditions.

A1.3.1 $a_k > 0$, $a_k \xrightarrow[k \to \infty]{} 0$ *and* $\sum\limits_{k=1}^{\infty} a_k = \infty$.

A1.3.2 *There exists a twice continuously differentiable Lyapunov function $v(\cdot)$ such that $v(x) > 0$, $\forall x \neq x^0$, $v(x^0) = 0$, $v(x) \longrightarrow \infty$ as $\|x\| \longrightarrow \infty$, and*

$$v_x^T(x) f(x) < 0, \quad \forall x \neq x^0.$$

In order to describe conditions on noise, we introduce an integer-valued function $m(k,T)$ for any $T>0$ and any integer k.

For $T>0$ define

$$m(k,T) \triangleq \max\left\{m : \sum_{i=k}^{m} a_i \leq T\right\}. \tag{1.3.2}$$

Noticing that a_i tends to zero, for any fixed $T>0$, $m(k,T)$ diverges to infinity as $k \longrightarrow \infty$. In fact, $m(k,T)$ counts the number of iterations starting from time k as long as the sum of step sizes does not exceed T. The integer-valued function $m(k,T)$ will be used throughout the book.

The following conditions will be used:

A1.3.3 $\{\epsilon_k\}$ *satisfies the following conditions*

$$\lim_{T\longrightarrow 0} \limsup_{n\longrightarrow\infty} \frac{1}{T}\left\| \sum_{k=n}^{m(n,T)} a_i\epsilon_{i+1}\right\| = 0, \quad a_k\epsilon_{k+1} \xrightarrow[k\longrightarrow\infty]{} 0; \tag{1.3.3}$$

A1.3.4 $f(\cdot)$ *is continuous.*

Theorem 1.3.1 *Assume that A1.3.1, A1.3.2, and A1.3.4 hold. If for a fixed sample* (ω) *A1.3.3 holds and* $\{x_k\}$ *generated by the RM algorithm (1.2.2) is bounded, then for this* ω x_k *tends to* x^0 *as* $k \longrightarrow \infty$.

Proof. Set

$$t_k = \sum_{i=0}^{k} a_i, \quad t_0 = 0.$$

Define the linear interpolating function x_t^0:

$$x_t^0 \triangleq \frac{t_k - t}{a_k} x_{k-1} + \frac{t - t_{k-1}}{a_k} x_k, \quad t \in [t_{k-1}, t_k]. \tag{1.3.4}$$

It is clear that x_t^0 is continuous and $x_{t_k}^0 = x_k$.

Further, define $q_k \triangleq \sum_{i=0}^{k} a_i\epsilon_{i+1}$ and the corresponding linear interpolating function q_t^0, which is defined by (1.3.4) with x_k replaced by q_k.

Since we will deal with the tail part of x_t^0, we define $x_n(t)$ by shifting time in x_t^0:

$$x_k(t) \triangleq x_{t+k}^0. \tag{1.3.5}$$

Thus, we derive a family of continuous functions $\{x_k(t)\}$.

Let us define the constant interpolating function

$$\overline{x}_t = x_k, \quad t \in [t_{k-1}, t_k). \tag{1.3.6}$$

Then summing up both sides of (1.2.2) yields

$$x_{k+1} = x_0 + \sum_{i=0}^{k} a_i (f(x_i) + \epsilon_{i+1}), \tag{1.3.7}$$

and hence

$$x_k(t) = x_0 + \int_0^{t+k} f(\overline{x}_s) ds + q^0_{t+k}. \tag{1.3.8}$$

By the boundedness assumption on $\{x_k\}$, the family $\{x_k(t)\}$ is uniformly bounded. We now prove it is equi-continuous.

By definition,

$$t_{m(0,t)} = \sum_{i=0}^{m(0,t)} a_i \le t < \sum_{i=0}^{m(0,t)+1} a_i = t_{m(0,t)+1}. \tag{1.3.9}$$

Hence, we have

$$q^0_{t+k+\Delta} = \sum_{i=0}^{m(0,t+k+\Delta)} a_i \epsilon_{i+1} + o(1), \tag{1.3.10}$$

where $o(1) \xrightarrow[k \to \infty]{} 0$ since $a_k \epsilon_{k+1} \xrightarrow[k \to \infty]{} 0$.

From this it follows that

$$\begin{aligned} \frac{1}{\Delta} \left\| q^0_{t+k+\Delta} - q^0_{t+k} \right\| &= \frac{1}{\Delta} \left\| \sum_{i=m(0,t+k)+1}^{m(0,t+k+\Delta)} a_i \epsilon_{i+1} \right\| + o(1) \\ &= \frac{1}{\Delta} \left\| \sum_{i=m(0,t+k)+1}^{m(m(0,t+k)+1,\Delta)} a_i \epsilon_{i+1} \right\| + o(1), \end{aligned} \tag{1.3.11}$$

which tends to zero as $k \longrightarrow \infty$ and then $\Delta \longrightarrow 0$ by A1.3.3.

For any $\Delta > 0$ we have

$$\| x_k(t+\Delta) - x_k(t) \| \le \left\| \int_{t+k}^{t+k+\Delta} f(\overline{x}_s) ds \right\| + \left\| q^0_{t+k+\Delta} - q^0_{t+k} \right\|.$$

By boundedness of $f(\overline{x}_s)$ and (1.3.11) we see that $\{x_k(t)\}$ is equi-continuous.

By Proposition 1.3.1, we can select from $\{x_k(t)\}$ a convergent subsequence $\{x_{n_k}(t)\}$ which tends to a continuous function $x(t)$.

Consider the following difference with $\Delta > 0$:

$$\frac{x(t+\Delta)-x(t)}{\Delta} = \frac{1}{\Delta}\Bigg[\lim_{k\longrightarrow\infty}\sum_{i=m(0,t+n_k)}^{m(m(0,t+n_k),\Delta)} a_i\big[f(x_i)-f(x(t))\big] + \Delta f(x(t))\Bigg], \tag{1.3.12}$$

which is derived by using (1.3.11).

By (1.3.9) it is clear that for $i \in \{m(0,t+n_k), m(0,t+n_k)+1, \ldots, m(m(0,t+n_k),\Delta)\}$

$$t_{m(0,t+n_k)+i} = \sum_{i=0}^{m(0,t+n_k)+i} a_i = t + n_k + \delta_{ki}, \text{ where } |\delta_{ki}| \le \Delta.$$

Then from (1.3.12) we obtain

$$\begin{aligned}\frac{x(t+\delta)-x(t)}{\Delta} &= f(x_t) + \frac{1}{\Delta}\lim_{k\longrightarrow\infty}\sum_{i=0}^{m(m(0,t+n_k),\Delta)-m(0,t+n_k)} a_{m(0,t+n_k)+i} \\ &\quad\cdot\big[f(x_{m(0,t+n_k)+i})-f(x(t))\big] \\ &= f(x_t) + \frac{1}{\Delta}\lim_{k\longrightarrow\infty}\sum_{i=0}^{m(m(0,t+n_k),\Delta)-m(0,t+n_k)} a_{m(0,t+n_k)+i} \\ &\quad\cdot\big[f(x^0_{t_{m(0,t+n_k)+i}})-f(x(t))\big] \\ &= f(x_t) + \frac{1}{\Delta}\lim_{k\longrightarrow\infty}\sum_{i=0}^{m(m(0,t+n_k),\Delta)-m(0,t+n_k)} a_{m(0,t+n_k)+i} \\ &\quad\cdot\big[f(x^0_{t+n_k+\delta_{ki}})-f(x(t))\big] \\ &= f(x_t) + \frac{1}{\Delta}\lim_{k\longrightarrow\infty}\sum_{i=0}^{m(m(0,t+n_k),\Delta)-m(0,t+n_k)} a_{m(0,t+n_k)+i} \\ &\quad\cdot\big[f(x_{n_k}(t+\delta_{ki}))-f(x(t))\big]. \end{aligned} \tag{1.3.13}$$

Tending Δ to zero in (1.3.13), by continuity of $f(\cdot)$ and uniform convergence of $x_{n_k}(t)$ to $x(t)$, we conclude that the last term in (1.3.13) converges to zero, and

$$\frac{dx(t)}{dt} = f(x(t)). \tag{1.3.14}$$

By A1.3.2 and Proposition 1.3.2 we see $x(t) \longrightarrow x^0$ as $t \longrightarrow \infty$.

We now prove that $x_k \longrightarrow x^0$. Assume the converse: there is a subsequence $x_{s_k} \xrightarrow[k\longrightarrow\infty]{} \overline{x} \neq x^0$.

Then for $k \geq k_1$, $\|x_{s_k} - \overline{x}\| < \|\overline{x} - x^0\|/4$. There is a t_1 such that $\|x(t) - x^0\| < \|\overline{x} - x^0\|/4$, $\forall t \geq t_1$.

By (1.3.4) we have

$$x_{s_k} = x^0_{t_{s_k}} = x^0_{[t_{s_k}-t_1]+t_1+\delta_k} = x_{[t_{s_k}-t_1]}(t_1 + \delta_k), \tag{1.3.15}$$

where $t_{s_k} - t_1 > 0$ and $[t_{s_k} - t_1]$ denotes the integer part of $t_{s_k} - t_1$, so $0 \leq \delta_k < 1$.

It is clear that the family of functions $\{x_{m_k}(t)\}$ indexed by $[t_{s_k} - t_1] \triangleq m_k$ is uniformly bounded and equi-continuous. Hence, we can select a convergent subsequence, denoted still by $\{x_{m_k}(t)\}$. The limit satisfies the ODE (1.3.14) and coincides with $x(t)$ being the limit of $x_{n_k}(t)$ by the uniqueness of the solution to (1.3.14).

By the uniform convergence we have

$$\|x_{m_k}(t) - x(t)\| < \|\overline{x} - x^0\|/4, \quad \forall t \in [t_1, t_1 + 1] \text{ for } k \geq k_2,$$

which implies that $\|x_{m_k}(t) - x^0\| < \|\overline{x} - x^0\|/2$, $\forall t \in [t_1, t_1 + 1]$, $\forall k \geq k_2$. From here by (1.3.15) it follows that

$$\|x_{s_k} - x^0\| < \|\overline{x} - x^0\|/2 \text{ for } [t_{s_k} - t_1] \geq k_2.$$

Then we obtain a contradictory inequality:

$$\begin{aligned}\|\overline{x} - x^0\| &\leq \|\overline{x} - x_{s_k}\| + \|x_{s_k} - x^0\| < \|\overline{x} - x^0\|/4 + \|\overline{x} - x^0\|/2 \\ &= \frac{3}{4}\|\overline{x} - x^0\|\end{aligned}$$

for k large enough such that $k \geq k_1$ and $[t_{s_k} - t_1] \geq k_2$. This completes the proof of $x_k \longrightarrow x^0$. □

We now compare conditions used in Theorem 1.3.1 with those in Theorem 1.2.1.

Conditions A1.3.1 and A1.3.2 are slightly weaker than A1.2.1 and A1.2.2, but they are almost the same. The noise condition A1.3.3 is significantly weaker than those used in Theorem 1.2.1, because under the conditions of Theorem 1.2.1 we have

$$\sum_{i=1}^{\infty} a_i \epsilon_{i+1} < \infty \text{ a.s.},$$

which certainly implies A1.3.3.

As a matter of fact, Condition A1.3.3 may be satisfied by sequences much more general than martingale difference sequences.

Example 1.3.1 Assume $\epsilon_k \longrightarrow 0$, but ϵ_k may be any random or deterministic sequence. Then $\{\epsilon_k\}$ satisfies A1.3.3.
This is because

$$\lim_{T\longrightarrow 0}\limsup_{k\longrightarrow\infty}\frac{1}{T}\Big\|\sum_{i=k}^{m(k,T)} a_i\epsilon_i\Big\| \le \limsup_{k\longrightarrow\infty}\max_{i\ge k}\|\epsilon_{i+1}\| = 0.$$

Example 1.3.2 Let ϵ_k be an MA process, i.e.,

$$\epsilon_k = w_k + c_1 w_{n-1} + \cdots + c_r w_{k-r},$$

where $(w_k, \mathcal{F}_k)$ is a martingale difference sequence with $\sup_k E(\|w_{k+1}\|^2 | \mathcal{F}_k) < \infty$.
Then under condition A1.2.1, $\sum_1^\infty a_i w_{i+1} < \infty$ a.s., and hence $\sum_{i=1}^\infty a_i \epsilon_{i+1} < \infty$ a.s. Consequently, A1.3.3 is satisfied for almost all sample paths ω.

Condition A1.3.4 requires continuity of $f(\cdot)$ which is not required in A1.2.4. At first glance, unlike A1.2.4, Condition A1.3.4 does not impose any growth rate condition on $f(\cdot)$, but Theorem 1.3.1 *a priori* requires the boundedness of $\{x_k\}$, which is an implicit requirement for the growth rate of $f(x)$ as $\|x\| \longrightarrow \infty$.

The ODE method is widely used in convergence analysis for algorithms arising from various application areas, because from the noise it requires no probabilistic property which would be difficult to verify. Concerning the weakness of the ODE method, we have mentioned that it *a priori* assumes that $\{x_k\}$ is bounded. This condition is difficult to be verified in general case. The other point should be mentioned that Condition A1.3.3 is also difficult to be verified in the case where ϵ_{i+1} depends on the past $\{x_i, i \le k\}$, which often occurs when $\{\epsilon_k\}$ contains structural errors of $f(\cdot)$. This is because A1.3.3 may be verifiable if x_k is convergent, but ϵ_{k+1} may badly behave depending upon the behavior of $\{x_i, i \le k\}$. So we are somehow in a cyclic situation: with A1.3.3 we can prove convergence of $\{x_k\}$, on the other hand, with convergent $\{x_k\}$ we can verify A1.3.3. This difficulty will be overcome by using Trajectory-Subsequence (TS) method to be introduced in the next section and used in subsequent chapters.

1.4. Truncated RM Algorithm and TS Method

In Section 1.2 we considered the root-seeking problem where the sought-for root x^0 may be any point in $\mathbb{R}^l$. If the region x^0 belongs to is known, then we may use the truncated algorithm and the growth rate restriction on $f(\cdot)$ can be removed.

Let us assume that $\|x^0\| < c_1$ and c_1 is known. In lieu of (1.2.2) we now consider the following truncated RM algorithm:

$$\begin{aligned} x_{k+1} =& (x_k + a_k y_{k+1}) I_{[\|x_k+a_k y_{k+1}\|<b]} \\ & + x^* I_{[\|x_k+a_k y_{k+1}\| \geq b]}, \end{aligned} \tag{1.4.1}$$

where the observation y_{k+1} is given by (1.2.1), $x^* \in \mathbb{R}^l$ is a given point, and $I_{[\|x\|<b]} \triangleq \begin{cases} 0, & \text{if } \|x\| \geq b, \\ 1, & \text{if } \|x\| < b. \end{cases}$

The constant b used in (1.4.1) will be specified later on.

The algorithm (1.4.1) means that it coincides with the RM algorithm when it evolves in the sphere $[x : \|x\| < b]$, but if $x_k + a_k y_{k+1}$ exits the sphere $[x : \|x\| < b]$, then the algorithm is pulled back to the fixed point x^*.

We will use the following set of conditions:

A1.4.1 *The step size* $\{a_k\}$ *satisfies the following conditions*

$$a_k > 0, \quad a_k \longrightarrow 0, \text{ and } \sum_{i=1}^{\infty} a_i = \infty;$$

A1.4.2 *There exists a continuously differentiable Lyapunov function* $v(\cdot)$ *(not necessarily being nonnegative) such that* $v(x^0) = 0$, $v(x) \neq 0$, $\forall x \neq x^0$, $\sup_{\Delta \geq \|x-x^0\| \geq \delta} f^T(x) v_x(x) < 0$, *and for* $x^* \in \mathbb{R}^l$ *(which is used in (1.4.1)) there is* $x_0 > 0$ *such that* $v(x^*) < \inf_{\|x\|=c_0} v(x)$;

A1.4.3 *For any convergent subsequence* $\{x_{n_k}\}$ *of* $\{x_k\}$

$$\lim_{T \longrightarrow 0} \limsup_{k \longrightarrow \infty} \frac{1}{T} \| \sum_{i=n_k}^{m(n_k,t)} a_i \epsilon_{i+1} \| = 0, \quad \forall t \in [0, T], \tag{1.4.2}$$

where $m(k, T)$ *is given by (1.3.2);*

A1.4.4 $f(\cdot)$ *is measurable and locally bounded.*

We first compare these conditions with A1.3.1–A1.3.4. We note that A1.4.1 is the same as A1.3.1, while A1.4.2 is weaker than A1.2.2.

The difference between A1.3.3 and A1.4.3 consists in that Condition (1.4.2) is required to be verified only along convergent subsequences, while (1.3.3) in A1.3.3 has to be verified along the whole sequence $\{x_k\}$.

It will be seen that A1.4.3 in many problems can be verified while A1.3.3 is difficult to verify.

Comparing A1.4.4 with A1.3.4 we find that the conditions on $f(\cdot)$ have now been weakened. The growth rate restriction used in Theorem 1.2.1 and the boundedness assumption on $\{x_k\}$ imposed in Theorem 1.3.1 have been removed in the following theorem.

Theorem 1.4.1 *Assume Conditions A1.4.1, A1.4.2, and A1.4.4 hold and the constant c_0 in A1.4.2 is available. Set $b = c_0 \vee c_1$ for (1.4.1). If for some sample path ω, A1.4.3 holds, then x_k given by (1.4.1) converges to x^0 for this ω.*

Proof. We say that $\{v(x_{n_k}, v(x_{n_k+1}), \dots, v(x_{m_k})\}$ crosses an interval $[\delta_1, \delta_2]$, if $n_k < m_k$, $v(x_{n_k}) \le \delta_1$, $v(x_{m_k}) \ge \delta_2$, and

$$\delta_1 < v(x_i) < \delta_2, \quad \forall i : n_k < i < m_k.$$

We first prove that the number of truncations in (1.4.1) may happen at most for a finite number of steps. Assume the converse: there are infinitely many truncations occurring in (1.4.1). Since $v(x^*) < \inf_{\|x\| \le b} v(x)$ by A1.4.2, there is an interval $[\delta_1, \delta_2]$ such that

$$[\delta_1, \delta_2] \subset \left(v(x^*), \inf_{\|x\| \le b} v(x) \right), \quad \delta_1 \delta_2 > 0, \tag{1.4.3}$$

and there are infinitely many $\{v(x_{n_k}), \dots, v(x_{m_k})\}$, $k = 1, 2, \dots$, that cross $[\delta_1, \delta_2]$.

Since $\{x_k\}$ is bounded, wc may extract a convergent subsequence from $\{x_{n_k}\}$. Let us denote the extracted convergent subsequence still by $\{x_{n_k}\}$: $x_{n_k} \xrightarrow[k \to \infty]{} \overline{x} \in \{x : \|x\| < b\}$. It is clear that

$$v(\overline{x}) = \lim_{k \to \infty} v(x_{n_k}) = \delta_1. \tag{1.4.4}$$

Since the limit of $\{x_{n_k}\}$ is located in the open sphere $\{x : \|x\| < b\}$, there is an $\epsilon < 0$ such that

$$\|x_{n_k}\| < b - \epsilon \tag{1.4.5}$$

for all sufficiently large k.

Since $\{f(x_k)\}$ is bounded by A1.4.4 and the boundedness of $\{x_k\}$, using (1.4.2) we have

$$\| \sum_{i=n_k}^{m+1} a_i f(x_i) + \sum_{i=n_k}^{m+1} a_i \epsilon_{i+1} \| \le \frac{\epsilon}{2}, \quad \forall m : \{n_k \le m \le m(n_k, T)\} \tag{1.4.6}$$

if T is small enough and k is large enough.

This incorporating with (1.4.5) implies that

$$x_{n_k} + \left\| \sum_{i=n_k}^{m+1} a_i(f(x_i) + \epsilon_{i+1}) \right\| < b - \frac{\epsilon}{2}, \quad \forall m : \{n_k \le m \le m(n_k, T)\}.$$

Therefore, the norm of x_{m+1}

$$x_{m+1} = x_m + a_m y_{m+1}, \quad m \ge n_k, \quad m : \{n_k \le m \le m(n_k, T)\} \quad (1.4.7)$$

cannot reach the truncation bound b. In other words, the algorithm (1.4.1) turns to be an untruncated RM algorithm (1.4.7) for $m : \{n_k \le m \le m(n_k, T)\}$ for small T and large k.

By the mean theorem there exists a vector ξ with components located in-between the corresponding components of x_{n_k} and $x_{m(n_k,T)+1}$ such that

$$v(x_{m(n_k,T)+1}) - v(x_{n_k}) = (x_{m(n_k,T)+1} - x_{n_k})^T v_x(\overline{x}) + \left(x_{m(n_k,T)+1} - x_{n_k}\right)^T (v_x(\xi) - v_x(\overline{x})). \quad (1.4.8)$$

Notice that by (1.4.2) the left-hand side of (1.4.6) is of $O(T)$ for all sufficiently large k since $\|f(x_i)\|$ is bounded. From this it follows that i) for small enough $T > 0$ and large enough k

$$|v(x_m) - v(x_{n_k})| < \delta_2 - \delta_1 \text{ and } v(x_m) < \delta_2, \quad \forall m : \{n_k \le m \le m(n_k, T)\},$$

and hence $m(n_k, T) + 1 < m_k$, and ii) the last term in (1.4.8) is of $o(T)$ since $v_x(\xi) - v_x(\overline{x}) \longrightarrow 0$ as $T \longrightarrow 0$. From (1.4.7) and (1.4.8) it then follows that

$$v\left(x_{m(n_k,T)+1}\right) - v(x_{n_k}) = \sum_{i=n_k}^{m(n_k,T)} a_i y_{i+1}^T v_x(\overline{x}) + o(T)$$

$$= \sum_{i=n_k}^{m(n_k,T)} a_i f^T(x_i) v_x(x_i) + \sum_{i=n_k}^{m(n_k,T)} a_i f^T(x_i)(v_x(\overline{x}) - v_x(x_i))$$

$$+ \sum_{i=n_k}^{m(n_k,T)} a_i v_x^T(\overline{x}) \epsilon_{i+1} + o(T). \quad (1.4.9)$$

Since $\delta_1 \delta_2 > 0$, the interval $[\delta_1, \delta_2]$ does not contain the origin. Noticing that $v(x_m) \in [\delta_1, \delta_2]$, $\forall m : n_k \le m \le m(n_k, T) + 1$, we find $v(x_m) \neq 0$, and that there is $\delta > 0$ such that

$$\|x_m - x^0\| > \delta, \quad \forall m : \{n_k, \le m \le m(n_k, T)\}$$

for sufficiently small $T > 0$ and all large enough k. Then by A1.4.2 there is $\alpha > 0$ such that

$$f^T(x_m)v_x(x_m) \le -\alpha, \quad \forall m : \{n_k, \le m \le m(n_k, T)\}$$

for all large k and small enough T. As mentioned above $|v_x(\overline{x}) - v_x(x_i)| \xrightarrow[T \to 0]{} 0$, from (1.4.9) we have

$$v(x_{m(n_k,T)+1}) - v(x_{n_k}) \le -\alpha T + o(T) + o(1) \le -\frac{\alpha}{2}T, \tag{1.4.10}$$

for sufficiently large k and small enough T, where $o(1)$ denotes a magnitude tending to zero as $k \longrightarrow \infty$.

Taking (1.4.4) into account, from (1.4.10) we find that

$$v(x_{m(n_k,T)+1}) < \delta_1$$

for large k. However, we have shown that

$$m(n_k, T) + 1 < m_k, \text{ i.e., } v(x_{m(n_k,T)+1}) > \delta_1.$$

The obtained contradiction shows that the number of truncations in (1.4.1) can only be finite.

We have proved that starting from some large k_0, the algorithm (1.4.1) develops as an RM algorithm

$$x_{k+1} = x_k + a_k y_{k+1}, \quad k \ge k_0,$$

and $\{x_k\}$ is bounded.

We are now in a position to show that $v(x_k)$ converges.

Assume it were not true. Then we would have

$$-\infty < \liminf_{k \to \infty} v(x_k) < \limsup_{k \longrightarrow \infty} v(x_k) < \infty.$$

Then there would exist an interval $[\delta_1, \delta_2]$ not containing the origin and $\{v(x_{n_k}), \dots, v(x_{m_k})\}$ would cross $[\delta_1, \delta_2]$ for infinitely many k.

Again, without loss of generality, assuming $x_{n_k} \xrightarrow[k \longrightarrow \infty]{} \overline{x}$, by the same argument as that used above, we will arrive at (1.4.9) and (1.4.10) for large k, and obtain a contradiction. Thus, $v(x_k)$ tends to a finite limit as $k \longrightarrow \infty$.

It remains to show that $x_k \xrightarrow[k \longrightarrow \infty]{} x^0$.

Assume the converse that there is a subsequence $x_{n_k} \xrightarrow[k \longrightarrow \infty]{} \overline{x} \ne x^0$. Then there is a $\delta > 0$, such that $\|x_{n_k} - x^0\| > \delta$ for all sufficiently large k. We still have (1.4.8), (1.4.9), and (1.4.10) for some $\alpha > 0$.

Tending $k \longrightarrow \infty$ in (1.4.10), by convergence of $v(x_k)$ we arrive at a contradictory inequality:

$$0 \leq -\frac{\alpha}{2}T.$$

This means $x_k \xrightarrow[k \longrightarrow \infty]{} x^0$.

$\square$

In this section we have demonstrated an analysis method which is different from those used in Sections 1.2 and 1.3. This method is based on analyzing the sample-path behavior, and conclusions on the whole sequence $\{x_k\}$ are deducted from the local behaviors of estimates x_m that are obtained immediately after x_{n_k}, which denotes a convergent subsequence of $\{x_k\}$. We call this method as Trajectory-Subsequence (TS) Method. The TS method is the main tool to be used in subsequent chapters for analyzing more general cases. It will be seen that the TS method is powerful in dealing with complicated errors including both random noise and structural inaccuracy of the function.

The obvious weakness of Theorem 1.4.1 is the assumption on the availability of the upper bound b for $\|x^0\|$. This limitation will be removed later on.

1.5. Weak Convergence Method

Up-to now we have worked with decreasing gains which are necessary for path-wise convergence when observations are corrupted by noise. However, in some applications people prefer to using constant gain:

$$x_{k+1} = x_k + \epsilon y_{k+1}, \tag{1.5.1}$$

$$y_{k+1} = f(x_k) + \epsilon_{k+1}, \tag{1.5.2}$$

where in contrast to (1.2.2) a constant $\epsilon > 0$ stands for a_k which tends to zero as $k \longrightarrow \infty$.

Define the piece-wise constant interpolating function $x^\epsilon(\cdot)$ as

$$x^\epsilon(t) = x_k \text{ for } t \in [\epsilon k, \epsilon(k+1)).$$

Then $x^\epsilon(\cdot) \in D^d[0,\infty)$, which is the space of real functions on $[0,\infty)$ that are right continuous and have left-hand limits, endowed with the Skorohod topology. Convergence of $f_n(\cdot) \in D^d[0,\infty)$ to a continuous function $f(\cdot)$ in the Skorohod topology is equivalent to the uniform convergence on any bounded interval.

Let P^ϵ and P be probability measures determined by stochastic processes $x^\epsilon(\cdot)$ and $x(\cdot)$, respectively on $D^d(0,\infty)$ with σ-algebra induced by the Skorohod topology.

If for any bounded continuous function $F(x)$ defined on $D^d[0,\infty)$

$$\int F(x)dP^\epsilon(x) \underset{\epsilon\longrightarrow 0}{\longrightarrow} \int F(x)dP(x),$$

then we say that $x^\epsilon(\cdot)$ weakly converges to $x(\cdot)$.

If for any $\delta > 0$, there is a compact measurable set B_δ in $D^d[0,\infty)$ such that

$$\sup_{\epsilon>0} P(x^\epsilon(\cdot) \notin B_\delta) \leq \delta,$$

then $x^\epsilon(\cdot)$ is called tight.

Further, $\{x^\delta(\cdot)\}$ is called relatively compact if each subsequence of $\{x^\epsilon(\cdot)\}$ contains a weakly convergent subsequence.

In the weak convergence analysis an important role is played by the Prohorov's Theorem, which says that on a complete and separable metric space, tightness is equivalent to relative compactness. The weak convergence method establishes the weak limit of $x^\epsilon(\cdot)$ as $\epsilon \longrightarrow 0$ and convergence of $x^\epsilon(t_\epsilon + \cdot)$ to x^0 in probability as $\epsilon \longrightarrow 0$ where $t_\epsilon \longrightarrow \infty$ as $\epsilon \longrightarrow 0$.

Theorem 1.5.1 *Assume the following conditions:*

A1.5.1 $\{x_k\}$ *is a.s. bounded;*

A1.5.2 $f(\cdot)$ *is continuous;*

A1.5.3 $(\epsilon_k, \mathcal{F}_k)$ *is adapted,* $\{\epsilon_k\}$ *is uniformly integrable in the sense that*

$$\lim_{\lambda\longrightarrow\infty} \sup_k E\left(\|\epsilon_k\| I_{\{\|\epsilon_k\|\geq\lambda\}}\right) = 0$$

and

$$\lim_{m,n\longrightarrow\infty} \frac{1}{m} \sum_{i=n}^{n+m-1} E(\epsilon_i|\mathcal{F}_n) = 0 \text{ in mean.}$$

Then $\{x^\epsilon(\cdot)\}$ *is tight in* $D^d[0,\infty)$, *and* $x^\epsilon(\cdot)$ *weakly converges to* $x(\cdot)$ *that is a solution to*

$$\dot{x} = f(x), \quad x(0) = x_0. \tag{1.5.3}$$

Further, if x^0 *is asymptotically stable for (1.5.3), then for any* $t_\epsilon \longrightarrow \infty$ *as* $\epsilon \longrightarrow 0$, *the distance between* $\{x^\epsilon(t_\epsilon + \cdot)\}$ *and* x^0, $dist(x^\epsilon(t_\epsilon + \cdot), x^0)$, *converges to zero in probability as* $\epsilon \longrightarrow 0$.

In stead of proof, we only outline its basic idea. First, it is shown that we can extract a subsequence of $\{x^\epsilon(\cdot)\}$ weakly converging to $x(\cdot)$.

For notational simplicity, denote the subsequence still by $\{x^\epsilon(\cdot)\}$. By the Skorohod representation, we may assume $x^\epsilon(\cdot) \xrightarrow[\epsilon\to 0]{} x(\cdot)$ a.s. For this we need only, if necessary, to change the probabilistic space and take $\widetilde{x}^\epsilon(\cdot)$ and $\widetilde{x}(\cdot)$ on this new space such that $\widetilde{x}^\epsilon(\cdot) \xrightarrow[\epsilon\to 0]{} \widetilde{x}(\cdot)$ a.s. and $\widetilde{x}^\epsilon(\cdot)$, $\widetilde{x}(\cdot)$ have the same distributions as those of $x^\epsilon(\cdot)$ and $x(\cdot)$, respectively. Then, it is proved that

$$\zeta(t) = x(t) - x_0 - \int_0^t f(s)ds$$

is a martingale. Since $\zeta(0) = 0$ and as can be shown, $\zeta(t)$ is Lipschitz continuous, it follows that $\zeta(t) \equiv 0$.

Since $\{x^\epsilon(\cdot)\}$ is relatively compact and the limit does not depend on the extracted subsequence, the whole family $\{x^\epsilon(t)\}$ weakly converges to $x(t)$ as $\epsilon \longrightarrow 0$ and $x(t)$ satisfies (1.5.3). By asymptotic stability of x^0, $\operatorname{dist}(x^\epsilon(t_\epsilon + \cdot), x^0) \xrightarrow[\epsilon\to 0]{} 0$.

Remark 1.5.1 The boundedness assumption on $\{x_k\}$ may be removed. For this a smooth function $q^m(x)$ is introduced such that

$$q^M(x) = \begin{cases} 1, & \|x\| \le M \\ 0, & \|x\| \ge M+1 \end{cases}$$

and the following truncated algorithm

$$x_{k+1}^M = x_k^M + \epsilon y_{k+1} q^M(x_k^M)$$

is considered in lieu of (1.5.1). Then $\{x_k^M\}$ is interpolated to a piece-wise constant function $x^{\epsilon,M}(t) = x_k^M$ for the $t \in [k\epsilon, \epsilon(k+1))$. It is shown that $\{x^{\epsilon,M}(\cdot)\}$ is tight, and weakly convergent as $\epsilon \longrightarrow 0$. The limit $x^M(\cdot)$ satisfies $\dot{x}^M = f(x^M)q^M(x^M)$.

Finally, by showing $\limsup_{M\to\infty} \limsup_{\epsilon\to 0} P\{x^{\epsilon,M}(t) \ne x^\epsilon(x)$, for some $t \le T\} = 0$ for each T, it is proved that $\{x^\epsilon(\cdot)\}$ itself is tight and weakly converges to $x(\cdot)$ satisfying (1.5.3).

1.6. Notes and References

The stochastic approximation algorithm was first proposed by Robbins and Monro in [82], where the mean square convergence of the algorithm was established under the independence assumption on the observation noise. Later, the noise was extended from independent sequence to martingale difference sequences (e.g. [7, 40, 53]).

The probabilistic approach to convergence analysis is well summarized in [78].

The ODE approach was proposed in [65, 72], and then it was widely used [4, 85]. For detailed presentation of the ODE method we refer to [65, 68].

The proof of Arzelá-Ascoli Theorem can be found in ([37], p.266).

Section 1.4 is an introduction to the method described in detail in coming chapters. For stability and Lyapunov functions we refer to [69].

The weak convergence method was developed by Kushner [64, 68]. The Skorohod topology and Prohorov's theorem can be found in [6, 41].

For probability concepts briefly presented in Appendix A, we refer to [30, 32, 70, 76, 84]. But the proof of the convergence theorem for martingale difference sequences, which are frequently used throughout the book, is given in Appendix B.

Chapter 2

STOCHASTIC APPROXIMATION ALGORITHMS WITH EXPANDING TRUNCATIONS

In Chapter 1 the RM algorithm, the basic algorithm used in stochastic approximation(SA), was introduced, and four different methods for analyzing its convergence were presented. However, conditions imposed for convergence are rather strong.

Comparing theorems derived by various methods in Chapter 1, we find that the TS method introduced in Section 1.4 requires the weakest condition on noise. The trouble is that the sought-for root has to be inside the truncation region. This motivates us to consider SA algorithms with expanding truncations with the purpose that the truncation region will finally cover the sought-for root whose location is unknown. This is described in Section 2.1.

General convergence theorems of the SA algorithm with expanding truncations are given in Section 2.2. The key point of the proof is to show that the number of truncations is finite. If this is done, then the estimate sequence is bounded and the algorithm turns to be the conventional RM algorithm in a finite number of steps. This is realized by using the TS method. It is worth noting that the fundamental convergence theorems given in this section are analyzed by a completely elementary method, which is deterministic and is limited to the knowledge of calculus. In Section 2.3 the state-independent conditions on noise are given to guarantee convergence of the algorithm when the noise itself is state-dependent. In Section 2.4 conditions on noise are discussed. It appears that the noise condition in the general convergence theorems in a certain sense is necessary. In Section 2.5 the convergence theorem is given for the case where the observation noise is non-additive.

In the multi-root (of $f(\cdot)$) case, up-to Section 2.6 we have only established that the distance between the estimate and the root set tends to zero. But, by no means this implies convergence of the estimate itself. This is briefly discussed in Section 2.4, and is considered in Section 2.6 in connection with properties of the equilibrium of $f(\cdot)$. Conditions are given to guarantee the trajectory convergence. It is also considered whether the limit of the estimate is a stable or unstable equilibrium of $f(\cdot)$. In Section 2.7 it is shown that a small distortion of conditions may cause only a small estimation error in limit, while Section 2.8 of this chapter considers the case where the sought-for root is moving during the estimation process. Convergence theorems are derived with the help of the general convergence theorem given in Section 2.2. Notes and references are given in the last section.

2.1. Motivation

In Chapter 1 we have presented four types of convergence theorems using different analysis methods for SA algorithms. However, none of these theorems is completely satisfactory in applications. Theorem 1.2.1 is proved by using the classical probabilistic method, which requires restrictive conditions on the noise and $f(\cdot)$. As mentioned before, the noise may contain component caused by the structural inaccuracy of the function, and it is hard to assume this kind of noise to be mutually independent or to be a martingale difference sequence etc. The growth rate restriction imposed on the function not only is sever, but also is unavoidable in a certain sense. To see this, let us consider the following example:

$$f(x) = -(x-10)^3, \quad \epsilon_k \equiv 0, \quad a_k = \frac{1}{k+1}.$$

It is clear that conditions A1.2.1, A1.2.2, and A1.2.3 are satisfied, where for A1.2.2 one may take $v(x) = (x-10)^2$. The only condition that is not satisfied is (1.2.4), since $(f(x))^2 = (x-10)^6$, while the right-hand side of (1.2.4) is a second order polynomial. Simple calculation shows that x_k given by RM algorithm rapidly diverges:

$$x^0 = 10, x_0 = 0, x_1 = 1000, x_2 = -485148500, x_3 \approx 3.8 \times 10^{25}.$$

From this one might conclude that the growth rate restriction would be necessary.

However, if we take the initial value x_0 with $|x_0 - 10| < 1$, then x_k given by the RM algorithm converges to x^0. To reduce initial value x_0, in a certain sense, it is equivalent to use step size not from a_0 but from a_{k_0} for some k_0. The difficulty consists in that from which a_{k_0} we should start the algorithm. This is one of the motivations to use expanding truncations to be introduced later.

Theorem 1.3.1 proved in Section 1.3 demonstrates the ODE method. By this approach, the condition imposed on the noise has significantly been weakened and it covers a class of noises much larger than that treated by the probabilistic method. However, it *a priori* requires $\{x_n\}$ be bounded. This is the case if x_k converges, but before establishing its convergence, this is an artificial condition, which is not satisfied even for the simple example given above. Further, although the noise condition (1.3.3) is much more general than that used in Theorem 1.2.1, it is still difficult to be verified for the state-dependent noise. For example, $\epsilon_{k+1} = x_k w_{k+1}$, where $(w_k, \mathcal{F}_k)$ is a martingale difference sequence with $\sup_k E(\|w_{k+1}\|^2|\mathcal{F}_k) < \infty$. If $\{\|x_k\|\}$ is bounded and $\sum_{k=1}^{\infty} a_k^2 < \infty$, then $\sum\limits_{k=1}^{\infty} a_k x_k w_{k+1} < \infty$ a.s. and (1.3.3) holds. However, in general, it is difficult to directly verify (1.3.3) because the behavior of $\{x_k\}$ is unknown. This is why we use Condition (1.4.2) which should be verified only along convergent subsequences. With convergent $\{x_{n_k}\}$, the noise $\epsilon_{n_k+1} = x_{n_k} w_{n_k+1}$ is easier to be dealt with.

Considering convergent subsequences, the path-wise convergence is proved for a truncated RM algorithm by using the TS method in Theorem 1.4.1. The weakness of algorithms with fixed truncation bounds is that the sought-for root of $f(\cdot)$ has to be located in the truncation region. But, in general, this cannot be ensured. This is another motivation to consider algorithms with expanding truncations.

The weak convergence method explained in Section 1.5 can avoid boundedness assumption on $\{x_k\}$, but it can ensure convergence in distribution only, while in practical computation one always deals with a sample path. Hence, people in applications are mainly interested in path-wise convergence.

The SA algorithm with expanding truncations was introduced in order to remove the growth rate restriction on $f(\cdot)$. It has been developed in two directions: weakening conditions imposed on noise and improving the analysis method. By the TS method we can show that the SA algorithm with expanding truncations converges under a truly weak condition on noise, which, in fact, is also necessary for a wide class of $f(\cdot)$.

In Chapter 1, the root x^0 of $f(\cdot) : \mathbb{R}^l \longrightarrow \mathbb{R}^l$ is a singleton. From now on we will consider the general case. Let J be the root set of $f(\cdot) : J = \{x \in \mathbb{R} : f(x) = 0\}$.

We now define the algorithm. Let $\{M_k\}$ be a sequence of positive numbers increasingly diverging to infinity, and let x^* be a fixed point in

$\mathbb{R}^l$. Fix an arbitrary initial value x_0, and denote by x_k the estimate at time k, serving as the kth approximation to J. Define x_k by the following recursion:

$$x_{k+1} = (x_k + a_k y_{k+1}) I_{[\|x_k + a_k y_{k+1}\| \le M_{\sigma_k}]} + x^* I_{[\|x_k + a_k y_{k+1}\| > M_{\sigma_k}]}, \tag{2.1.1}$$

$$\sigma_k = \sum_{i=1}^{k-1} I_{[\|x_i + a_i y_{i+1}\| > M_{\sigma_i}]}, \qquad \sigma_0 = 0, \tag{2.1.2}$$

$$y_{k+1} = f(x_k) + \epsilon_{k+1}, \tag{2.1.3}$$

where $I_{[\text{inequality}]}$ is an indicator function meaning that it equals 1 if the inequality indicated in the bracket is fulfilled, and 0 if the inequality does not hold.

We explain the algorithm. σ_k is the number of truncations up-to time k. M_{σ_k} serves as the truncation bound when the $(k+1)th$ estimate is generated. From (2.1.1) it is seen that if the estimate at time $k+1$ calculated by the RM algorithm remains in the truncation region, i.e., if $\|x_k + a_k y_{k+1}\| \le M_{\sigma_k}$, then the algorithm evolves as the RM algorithm. If $(x_k + a_k y_{k+1})$ exits from the sphere with radius M_{σ_k}, i.e., if $\|x_k + a_k y_{k+1}\| > M_{\sigma_k}$, then the estimate at time $k+1$ is pulled back to the pre-specified point x^*, and the truncation bound is enlarged from M_{σ_k} to $M_{\sigma_{k+1}}$.

Consequently, if it can be shown that the number of truncations is finite, or equivalently, $\{x_k\}$ generated by (2.1.1) and (2.1.2) is bounded, then the algorithm (2.1.1) and (2.1.2) turns to be the one without truncations, i.e., to be the RM algorithm after a finite number of steps. This actually is the key step when we prove convergence of (2.1.1) and (2.1.2).

The convergence analysis of (2.1.1) and (2.1.2) will be given in the next section, and the analysis is carried out in a deterministic way at a fixed sample without involving any interpolating function.

2.2. General Convergence Theorems by TS Method

In This section by TS method we establish convergence of the RM algorithm with expanding truncations defined by (2.1.1)–(2.1.3) under general conditions. Let us first list conditions to be used.

A2.2.1 $a_k > 0$, $a_k \xrightarrow[k\to\infty]{} 0$ *and* $\sum_{k=1}^{\infty} a_k = \infty$.

A2.2.2 *There is a continuously differentiable function (not necessarily being nonnegative)* $v(\cdot): \mathbb{R}^l \longrightarrow R$ *such that*

$$\sup_{\delta \le d(x,J) \le \Delta} f^T(x) v_x(x) < 0 \tag{2.2.1}$$

for any $\Delta > \delta > 0$, *and* $v(J) \triangleq \{v(x) : x \in J\}$ *is nowhere dense, where* J *is the zero set of* $f(\cdot)$, *i.e.,* $f(x) = 0, \forall x \in J$, $d(x, J) = \inf_y \{\|x - y\| : y \in J\}$ *and* $v_x(\cdot)$ *denotes the gradient of* $v(\cdot)$. *Further,* x^* *used in (2.1.1) is such that* $v(x^*) < \inf_{\|x\|=c_0} v(x)$ *for some* $c_0 > 0$ *and* $\|x^*\| < c_0$.

For introducing condition on noise let us denote by $(\Omega, \mathcal{F}, P)$ the probability space. Let $\epsilon_{k+1}(\cdot, \cdot) = (\mathbb{R}^l \times \Omega, \mathcal{B}^l \times \mathcal{F}) \longrightarrow (\mathbb{R}^l \times \mathcal{B}^l)$ be a measurable function defined on the product space. Fixing an $\omega \in \Omega$ means that a sample path is under consideration. Let the noise ϵ_{k+1} be given by

$$\epsilon_{k+1} = \epsilon_{k+1}(x_k, \omega), \quad \omega \in \Omega.$$

Thus, the state-dependent noise is considered, and for fixed x_k, $\epsilon_{k+1}(x_k, \omega)$ may be random.

A2.2.3 *For the sample path* ω *under consideration for any sufficiently large integer* $N (\ge N_0)$

$$\lim_{T \longrightarrow 0} \limsup_{k \longrightarrow \infty} \frac{1}{T} \| \sum_{i=n_k}^{m(n_k, T_k)} a_i \epsilon_{i+1}(x_i(\omega), \omega) I_{[\|x_i(\omega)\| \le N]} \| = 0, \quad \forall T_k \in [0, T] \tag{2.2.2}$$

for any $\{n_k\}$ such that $x_{n_k}(\omega)$ converges, where $m(k, T)$ is given by (1.3.2) and $x_i(\omega)$ denotes x_i given by (2.1.1)–(2.1.3) and valued at the sample path ω.

In the sequel, the algorithm (2.1.1)–(2.1.3) is considered for the fixed ω for which A2.2.3 holds, and ω in $x_i(\omega)$ will often be suppressed if no confusion is caused.

A2.2.4 $f(\cdot)$ *is measurable and locally bounded.*

Remark 2.2.1 Comparing A2.2.1–A2.2.4 with A1.4.1–A1.4.4, we find that if the root set J degenerates to a singleton x^0, then the only essential difference is that an indicator function $I_{[\|x_i(\omega)\| \le N]}$ is included in (2.2.2) while (1.4.2) stands without it. It is clear that if $\{x_k\}$ is bounded, then this makes no difference. However, before establishing the boundedness of $\{x_k\}$, condition (2.2.2) is easier to be verified. The key point here is that in contrast to Section 1.4 we do not assume availability of the upper bound for the roots of $f(\cdot)$.

Remark 2.2.2 It is worth noting that $a_{n_k}\epsilon_{n_k+1} \xrightarrow[k\to\infty]{} 0$ if $\{x_{n_k}\}$ converges. To see this it suffices to take $T_k = a_{n_k}$ in (2.2.2).

Theorem 2.2.1 *Let $\{x_k\}$ be given by (2.1.1)–(2.1.3) for a given initial value x_0. Assume A2.2.1–A2.2.4 hold. Then, $d(x_k, J) \xrightarrow[k\to\infty]{} 0$ for the sample path ω for which A2.2.3 holds.*

Proof. The proof is completed by six steps by considering convergent subsequences at the sample path. This is why we call the analysis method used here as TS method.

Step 1. We show that there are constants $M > 0, T > 0$ such that for any $t \in [0, T]$ there exists $k_t > 0$ such that for any $k > k_t$

$$\|\sum_{i=n_k}^{m+1} a_i y_{i+1}\| \le M, \quad \forall m : n_k - 1 \le m \le m(n_k, t) \tag{2.2.3}$$

if $\{x_{n_k}\}$ is a convergent subsequence of $\{x_k\} : x_{n_k} \xrightarrow[k\to\infty]{} \overline{x}$, where M is independent of t and k.

Since $a_{n_k} f(x_{n_k}) \xrightarrow[k\to\infty]{} 0$, $a_{n_k}\epsilon_{n_{k+1}} \xrightarrow[k\to\infty]{} 0$, we need only to prove (2.2.3) for $m : n_k \le m \le m(n_k, t)$.

If the number of truncations in (2.1.1)–(2.1.3) is finite, then there is an N such that $\sigma_k = \sigma_N, \forall k \ge N$, i.e., there is no more truncation for $k \ge N$. Hence, $\|\sum_{i=n_k}^{m+1} a_i y_{i+1}\| = \|x_{m+2} - x_{n_k}\| \le 2M_{\sigma_N}$, whenever $n_k \ge N$. In this case, we may take $M = 2M_{\sigma_N}$ in (2.2.3).

We now prove (2.2.3) for the case where $\sigma_k \longrightarrow \infty$ as $k \longrightarrow \infty$.

Assume the converse that (2.2.3) is not true. Take $c > \|\overline{x}\|$. There is k_c such that

$$\|x_{n_k}\| \le (c + \|\overline{x}\|)/2, \quad \forall k \ge k_c. \tag{2.2.4}$$

Take a sequence of positive real numbers $t_j > 0$ and $t_j \longrightarrow 0$ as $j \longrightarrow \infty$. Since (2.2.3) is not true, for $j = 1$ there are $k_1 > k_c$ and $m_1 : n_{k_1} \le m_1 \le m(n_{k_1}, t_1)$ such that

$$\|\sum_{i=n_{k_1}}^{m_1+1} a_i y_{i+1}\| > (c - \|\overline{x}\|)/2,$$

and for any $j > 1$ there are $k_j > k_{j-1}$ and $m_j : n_{k_j} \le m_j \le m(n_{k_j}, t_j)$ such that

$$\|\sum_{i=n_{k_j}}^{m_j+1} a_i y_{i+1}\| > (c - \|\overline{x}\|)/2. \tag{2.2.5}$$

Without loss of generality we may assume

$$m_j = \inf\Big\{m : \|\sum_{i=n_{k_j}}^{m+1} a_i y_{i+1}\| > (c - \|\overline{x}\|)/2\Big\}. \tag{2.2.6}$$

Then for any $m : n_{k_j} \le m \le m_j$, from (2.2.4) and (2.2.6) it follows that

$$\|x_{n_{k_j}} + \sum_{i=n_{k_j}}^{m} a_i y_{i+1}\| < c. \tag{2.2.7}$$

Since $\sigma_k \longrightarrow \infty$, there is j_0 such that $M_{\sigma_{n_j}} > c$, $\forall j \ge j_0$. Then from (2.2.7) it follows that

$$x_{m+1} = x_m + a_m y_{m+1}, \quad \forall m : n_{k_j} \le m \le m_j, \tag{2.2.8}$$

and by (2.2.4), (2.2.7), and (2.2.8)

$$\|x_m\| \le c, \text{ and hence } \|f(x_m)\| \le c_1, \quad \forall m : n_{k_j} \le m \le m_j + 1, \tag{2.2.9}$$

by A2.2.4, where c_1 is a constant.

Let $N > c \vee N_0$, where N_0 is specified in A2.2.3. Then from A2.2.3 for any $T_j : 0 \le T_j < T$,

$$\limsup_{j\longrightarrow\infty} \|\sum_{i=n_{k_j}}^{m(n_{k_j},T_j)} a_i \epsilon_{i+1} I_{[\|x_i\| \le N]}\| = 0. \tag{2.2.10}$$

For any fixed $T > 0$, if j is large enough, then $t_j < T$ and $m_j + 1 < m(n_{k_j}, T)$, and by (2.2.10)

$$\limsup_{j\longrightarrow\infty} \|\sum_{i=n_{k_j}}^{m_j+1} a_i \epsilon_{i+1} I_{[\|x_i\| \le N]}\| = 0. \tag{2.2.11}$$

Since $\|x_m\| \le c < N, \forall m : n_{k_j} \le m \le m_j + 1$, from (2.2.11) it follows that

$$\limsup_{j\longrightarrow\infty} \|\sum_{i=n_{k_j}}^{m_j+1} a_i \epsilon_{i+1}\| = 0. \tag{2.2.12}$$

Taking $T_j = \sum_{i=n_{k_j}}^{m_j} a_i$ and $T_j = \sum_{i=n_{k_j}}^{m_j+1} a_i$ respectively in (2.2.10) and noticing from (2.2.9) $\|x_i\| \leq c < N$, $\forall i : n_{k_j} \leq i \leq m_j + 1$, we then have

$$\limsup_{j \longrightarrow \infty} \| \sum_{i=n_{k_j}}^{m_j} a_i \epsilon_{i+1} \| = 0 \quad \text{and} \quad \limsup_{j \longrightarrow \infty} \| \sum_{i=n_k}^{m_j+1} a_i \epsilon_{i+1} \| = 0,$$

and hence

$$\lim_{j \longrightarrow \infty} a_{m_j+1} \epsilon_{m_j+2} = 0. \tag{2.2.13}$$

From (2.2.8), it follows that

$$\|x_{m_j+1} - x_{n_{k_j}}\| \leq \sum_{i=n_{k_j}}^{m_j} a_i \|f(x_i)\| + \| \sum_{i=n_{k_j}}^{m_j} a_i \epsilon_{i+1} \| \xrightarrow[j \longrightarrow \infty]{} 0, \tag{2.2.14}$$

where the second term on the right-hand of the inequality tends to zero by (2.2.12) and (2.2.13), while the first term tends to zero because

$$\sum_{i=n_{k_j}}^{m_j} a_i \|f(x_i)\| \leq c_1 \sum_{i=n_{k_j}}^{m_j} a_i \leq c_1 t_j \xrightarrow[j \longrightarrow \infty]{} 0.$$

Noticing that $a_{m+j+1} y_{m_j+2} = a_{m_j+1} \big(f(x_{m_j+1}) + \epsilon_{m_j+2} \big) \xrightarrow[j \longrightarrow \infty]{} 0$ by (2.2.9) and (2.2.13), we then by (2.2.14) have

$$\|x_{m_j+1} - x_{n_{k_j}} + a_{m_j+1} y_{m_j+2}\| \leq \|x_{m_j+1} - x_{n_{k_j}}\| + \|a_{m_j+1} y_{m_j+2}\| \xrightarrow[j \longrightarrow \infty]{} 0.$$

On the other hand, by (2.2.6) we have

$$\|x_{m_j+1} - x_{n_{k_j}} + a_{m_j+1} y_{m_j+2}\| = \| \sum_{i=n_{k_j}}^{m_j+1} a_i y_{i+1} \| > (c - \|\bar{x}\|)/2, \quad \forall j.$$

The obtained contradiction proves (2.2.3).

Step 2. We now show that for all k large enough

$$\|x_{m+1} - x_{n_k}\| \leq 2c_2 t, \quad \forall m : n_k \leq m \leq m(n_k, t), \quad \forall t \in [0, T], \tag{2.2.15}$$

if T is small enough, where c_2 is a constant.

If the number of truncations in (2.1.1)–(2.1.3) is finite, then $\{x_k\}$ is bounded and hence $\{f(x_k)\}$ is also bounded.

Then for large enough k there is no truncation, and by (2.2.2) for $\forall m : n_k \leq m \leq m(n_k, t)$

$$\|x_{m+1} - x_{n_k}\| \leq \|\sum_{i=n_k}^{m} a_i f(x_i)\| + \|\sum_{i=n_k}^{m} a_i \epsilon_{i+1}\|$$

$$\leq \sup_i \|f(x_i)\| \sum_{i=n_k}^{m} a_i + o(t) \leq c_3 t, \tag{2.2.16}$$

if T is small enough. In (2.2.16), for the last inequality the boundedness of $\{x_k\}$ is invoked, and c_3 is a constant.

Thus, it suffices to prove (2.2.15) for the case where $\sigma_k \xrightarrow[k \to \infty]{} \infty$.

From (2.2.3) it follows that for any $t \in [0, T]$

$$\|x_{n_k} + \sum_{i=n_k}^{m} a_i y_{i+1}\| \leq M + \|\bar{x}\| + 1 \leq M_{\sigma_k} \quad \forall m : n_k \leq m \leq m(n_k, t) \tag{2.2.17}$$

if k is large enough.

This implies that for $\forall m : n_k \leq m \leq m(n_k, t)$

$$x_{m+1} = x_m + a_m y_{m+1}, \quad \|x_{m+1}\| \leq M + 1 + \|\bar{x}\|, \quad \|f(x_m)\| \leq c_2, \tag{2.2.18}$$

where c_2 is a constant. The last inequality of (2.2.18) yields

$$\|\sum_{i=n_k}^{m} a_i f(x_i)\| \leq c_2 t. \tag{2.2.19}$$

With $N \geq (M + 1 + \|\bar{x}\|) \vee N_0$ in A2.2.3, from (2.2.2) we have

$$\|\sum_{i=n_k}^{m(n_k,t)} a_i \epsilon_{i+1} I_{[\|x_i\| \leq N]}\| = \|\sum_{i=n_k}^{m(n_k,t)} a_i \epsilon_{i+1}\| \leq c_2 t \tag{2.2.20}$$

for large enough k and small enough T.

Combining (2.2.18), (2.2.19), and (2.2.20) leads to

$$\|x_{m+1} - x_{n_k}\| \leq 2c_2 t, \quad \forall m : n_k \leq m \leq m(n_k, t)$$

for all large enough k. This together with (2.2.16) verifies (2.2.15).

Step 3. We now show the following assertion:

For any interval $[\delta_1, \delta_2]$ with $\delta_1 < \delta_2$ and $d([\delta_1, \delta_2], v(J)) > 0$, the sequence $\{v(x_k)\}$ cannot cross $[\delta_1, \delta_2]$ infinitely many times with $\{\|x_{n_k}\|\}$

bounded, where by "crossing $[\delta_1, \delta_2]$ by $v(x_{n_k}, \ldots, v(x_{m_k})$" we mean that $v(x_{n_k}) \le \delta_1, v(x_{m_k}) \ge \delta_2$, and $\delta_1 < v(x_i) < \delta_2, \forall i : n_k < i < m_k$.

Assume the converse: there are infinitely many crossings $v(x_{n_k}), \ldots, v(x_{m_k})$, $k = 1, 2, \ldots$, and $\{\|x_{n_k}\|\}$ is bounded.

By boundedness of $\{\|x_{n_k}\|\}$, without loss of generality, we may assume $x_{n_k} \xrightarrow[k \to \infty]{} \overline{x}$.

By setting $t = a_{n_k}$ in (2.2.15), we have

$$\|x_{n_k+1} - x_{n_k}\| \le 2c_2 a_{n_k} \xrightarrow[k \to \infty]{} 0. \tag{2.2.21}$$

But by definition $v(x_{n_k+1}) > \delta_1 \ge v(x_{n_k})$, so we have

$$v(x_{n_k}) \xrightarrow[k \to \infty]{} \delta_1 = v(\overline{x}) \quad \text{and} \quad d(\bar{x}, J) \triangleq \delta > 0. \tag{2.2.22}$$

From (2.2.15) we see that if take t sufficiently small, then

$$d(x_m, J) \ge \frac{\delta}{2}, \quad \forall m : n_k \le m \le m(n_k, t) \tag{2.2.23}$$

for sufficiently large k.

By (2.2.18) and (2.2.15), for large k we then have

$$v(x_{m(n_k,t)+1}) - v(x_{n_k}) = \sum_{i=n_k}^{m(n_k,t)} a_i y_{i+1}^T v_x(\bar{x}) + o(t)$$

$$= \sum_{i=n_k}^{m(n_k,t)} a_i f^T(x_i) v_x(x_i) + \sum_{i=n_k}^{m(n_k,t)} a_i f^T(x_i)(v_x(\bar{x}) - v_x(x_i))$$

$$+ \sum_{i=n_k}^{m(n_k,t)} a_i v_x^T(\bar{x}) \epsilon_{i+1} + o(t), \tag{2.2.24}$$

where $v_x(\cdot)$ denotes the gradient of $v(\cdot)$ and $o(t) \longrightarrow 0$ as $t \longrightarrow 0$.

For $N \ge M + 1 + \|\overline{x}\| \vee N_0$, condition (2.2.2) implies that

$$\limsup_{k \to \infty} \|v_x^T(\overline{x}) \sum_{i=n_k}^{m(n_k,t)} a_i \epsilon_{i+1}\| = o(t). \tag{2.2.25}$$

By (2.2.15) and (2.2.18) it follows that

$$\| \sum_{i=n_k}^{m(n_k,t)} a_i f^T(x_i) \big(v_x(\overline{x}) - v_x(x_i)\big) \| = o(t). \tag{2.2.26}$$

Then, by (2.2.23) and (2.2.1) from (2.2.24)–(2.2.26) it follows that there are $\alpha > 0$ and $t > 0$ such that

$$v(x_{m(n_k,t)+1}) - v(x_{n_k}) \leq -\alpha t \tag{2.2.27}$$

for all sufficiently large k.

Noticing (2.2.22), from (2.2.27) we derive

$$\limsup_{k\longrightarrow\infty} v(x_{m(n_k,t)+1}) \leq \delta_1 - \alpha t. \tag{2.2.28}$$

However, by (2.2.15) we have

$$\lim_{t\longrightarrow 0} \max_{n_k\leq m\leq m(n_k,t)} |v(x_{m+1}) - v(x_{n_k}))| = 0,$$

which implies that $m(n_k, t) + 1 < m_k$ for small enough t.

This means that $v(x_{m(n_k,t)+1}) \in [\delta_1, \delta_2)$, which contradicts (2.2.28).

Step 4. We now show that the number of truncations is bounded.

By A2.2.2, $v(J)$ is nowhere dense, and hence a nonempty interval $[\delta_1, \delta_2]$ exists such that $[\delta_1, \delta_2] \subset \big(v(x^*), \inf_{\|x\|=c_0} v(x)\big)$ and $d\big([\delta_1, \delta_2], v(J)\big) > 0$. If $\sigma_k \xrightarrow[k\longrightarrow\infty]{} \infty$, then x_k, starting from x^*, will cross the sphere $\{x : \|x\| = c_0\}$ infinitely many times. Consequently, $v(x_k)$ will cross $[\delta_1, \delta_2]$ infinitely often with $\{x_{n_k}\}$ bounded. In Step 3, we have shown this process is impossible. Therefore, starting from some k_0, the algorithm (2.1.1)–(2.1.3) will have no truncations and $\{x_k\}$ is bounded.

This means that the algorithm defined by (2.1.1)–(2.2.3) turns to be the conventional RM algorithm for $k \geq k_0$ and a stronger than (2.2.2) condition is satisfied:

$$\lim_{T\longrightarrow 0} \limsup_{k\longrightarrow\infty} \frac{1}{T} \| \sum_{i=n_k}^{m(n_k,T_k)} a_i \epsilon_{i+1} \| = 0, \quad \forall T_k \in [0, T] \tag{2.2.29}$$

for any $\{n_k\}$ such that x_{n_k} converges.

Step 5. We now show that $v(x_k)$ converges. Let

$$v_1 \triangleq \liminf_{k\longrightarrow\infty} v(x_k) \leq \limsup_{k\longrightarrow\infty} v(x_k) \triangleq v_2.$$

We have to show $v_1 = v_2$.

If $v_1 < v_2$ and one of v_1 and v_2 does not belong to $v(J)$, then $[\delta_1, \delta_2] \subset [v_1, v_2]$ exists such that $d([\delta_1, \delta_2], v(J)) > 0$ and $\delta_2 > \delta_1$. By Step 3 this is impossible. So, both v_1 and v_2 belong to $v(J)$ and

$$\lim_{k\longrightarrow\infty} d(v(x_k), v(J)) = 0. \tag{2.2.30}$$

If we can show that $\{v(x_k)\}$ is dense in $[v_1, v_2]$, then from (2.2.30) it will follow that $v(J)$ is dense in $[v_1, v_2]$, which contradicts to the assumption that $v(J)$ is nowhere dense. This will prove $v_1 = v_2$, i.e., the convergence of $v(x_k)$.

To show that $\{v(x_k)\}$ is dense in $[v_1, v_2]$, it suffices to show that $x_{k+1} - x_k \xrightarrow[k\to\infty]{} 0$. Assume the converse: there is a subsequence

$$\lim_{k\to\infty} \|x_{l_k+1} - x_{l_k}\| \triangleq \beta > 0. \tag{2.2.31}$$

Without loss of generality, we may assume x_{l_k} converges. Otherwise, a convergent subsequence can be extracted, which is possible because $\{x_k\}$ is bounded. However, if we take $t = a_{l_k}$ in (2.2.15), we have

$$\|x_{l_k+1} - x_{l_k}\| \le 2c_2 a_{l_k} \xrightarrow[k\to\infty]{} 0,$$

which contradicts (2.2.31). Thus $v_1 = v_2$ and $v(x_k)$ converges.

Step 6. For proving $d(x_k, J) \xrightarrow[k\to\infty]{} 0$, it suffices to show that all limit points of $\{x_k\}$ belong to J.

Assume the converse: $x_{n_k} \longrightarrow \bar{x} \notin J$, $d(\bar{x}, J) \triangleq \delta > 0$. By (2.2.15) we have

$$d(x_m, J) > \frac{\delta}{2}, \quad \forall m : n_k \le m \le m(n_k, t)$$

for all large k if t is small enough. By (2.2.1) it follows that

$$v_x^T(x_m) f(x_m) < -b < 0, \quad \forall m : n_k \le m \le m(n_k, t),$$

and from (2.2.24)

$$v(x_{m(n_k,t)+1}) - v(x_{n_k}) \le -\frac{bt}{2} \tag{2.2.32}$$

for small enough t. This leads to a contradiction because $v(x_k)$ converges and the left-hand side of (2.2.32) tends to zero as $k \longrightarrow \infty$. Thus, we conclude $d(x_k, J) \xrightarrow[k\to\infty]{} 0$. □

Remark 2.2.3 In (2.1.1)–(2.1.3) the spheres with expanding radiuses $\{M_k\}$ are used for truncations. Obviously, the spheres can be replaced by other expanding sets. At first glance the point x^* in (2.1.1) may be arbitrarily chosen, but actually the restriction is imposed on the existence of c_0 such that $v(x^*) < \inf_{\|x\|=c_0} v(x)$. The condition is obviously satisfied if $v(x) \longrightarrow \infty$ as $\|x\| \longrightarrow \infty$ because the availability of c_0 is not required.

Remark 2.2.4 In the proof of Theorem 2.2.1 it can be seen that the conclusion "$d(x_k, J) \xrightarrow[k\to\infty]{} 0$" remains valid if in A2.2.2 "J is the zero set of $f(\cdot)$" is removed. As a matter of fact, J may be bigger than the zero set of $f(\cdot)$. Of course, it should at least contain the zero set of $v_x(\cdot)$ in order (2.2.1) to be satisfied. It should also be noted that for $d(v(x_k), v(J)) \xrightarrow[k\to\infty]{} 0$ we need not require $v(J)$ to be nowhere dense.

Let us modify A2.2.2 as follows.

A2.2.2' *There is a continuously differentiable function $v(\cdot)$: $\mathbb{R}^l \longrightarrow R$ such that*

$$\sup_{\delta\le d(x,J)\le\Delta} f^T(x)v_x(x) < 0$$

for any $\Delta > \delta > 0$, and $v(J)$ is nowhere dense. Further, x^* used in (2.1.1) is such that $v(x^*) < \inf_{\|x\|=c_0} v(x)$ for some $c_0 > 0$ and $\|x^*\| < c_0$.

A2.2.2" *There is a continuously differentiable function $v(\cdot) : \mathbb{R}^l \longrightarrow R$ such that*

$$\sup_{\delta\le d(x,J)\le\Delta} f^T(x)v_x(x) < 0$$

for any $\Delta > \delta > 0$ and J is closed. Further, x^* used in (2.1.1) is such that $v(x^*) < \inf_{\|x\|=c_0} v(x)$ for some $c_0 > 0$ and $\|x^*\| < c_0$.

Notice that, in A2.2.2' and A2.2.2" the set J is not specified, but it certainly contains the root sets of both $f(\cdot)$ and $v_x(\cdot)$. We may modify Theorem 2.2.1 as follows.

Theorem 2.2.1' *Let $\{x_k\}$ be given by (2.1.1)–(2.1.3) for a given initial value x_0. Assume A2.2.1, A2.2.2',A2.2.3, and A2.2.4 hold. Then $d(x_k, J) \xrightarrow[k\to\infty]{} 0$ for the sample path ω for which A2.2.3 holds.*

Proof. The Proof of Theorem 2.2.1 applies without any change. □

Theorem 2.2.1" *Let $\{x_k\}$ be given by (2.1.1)–(2.1.3) for a given initial value. If A2.2.1, A2.2.2",A2.2.3, and A2.2.4 hold, then $d(v(x_k), v(J)) \xrightarrow[k\to\infty]{} 0$ for the sample path ω for which A2.2.3 holds.*

Proof. We still have Step 1– Step 3 in the proof of Theorem 2.2.1. Let

$$v_1 \triangleq \liminf_{k\to\infty} v(x_k) \le \limsup_{k\to\infty} v(x_k) \triangleq v_2.$$

If v_1 or v_2 or both do not belong to J, then $[\delta_1, \delta_2] \subset (v_1, v_2)$ exists such that $d([\delta_1, \delta_2], v(J)] > 0$ since J is closed. Then $v(x_k)$ would cross $[\delta_1, \delta_2]$ infinitely many times. But, by Step 3 of the Proof for Theorem 2.2.1, this is impossible. Therefore both v_1 and v_2 belong to $v(J)$. □

Theorems 2.2.1 and 2.2.1' only guarantee that the distance between $\{x_k\}$ and the set J tends to zero. As a matter of fact, we have more precise result.

Theorem 2.2.2 *Assume conditions of Theorem 2.2.1 or Theorem 2.2.1' hold. Then for fixed x_0 and ω, for which A2.2.3 holds, a connected subset $J^* \subset \bar{J}$ exists such that*

$$d(x_k, J^*) \xrightarrow[k \to \infty]{} 0,$$

where $\bar{J}$ denotes the closure of J and $\{x_k\}$ is generated by (2.1.1)–(2.1.3).

Proof. Denote by J^* the set of limit points of $\{x_k\}$. Assume the converse: i.e., J^* is disconnected. In other words, closed sets J_1^* and J_2^* exist such that$J^* = J_1^* \cup J_2^*$ and $d(J_1^*, J_2^*) > 0$.

Define

$$\rho \triangleq \frac{1}{3} d(J_1^*, J_2^*).$$

Since $d(x_k, J^*) \xrightarrow[k \to \infty]{} 0$, a k_0 exists such that

$$x_k \in B(J_1^*, \rho) \cup B(J_2^*, \rho), \quad \forall k \geq k_0,$$

where $B(A, \rho)$ denotes the ρ-neighborhood of set A.

Define

$$\begin{aligned} n_0 &= \inf\{k > k_0, d(x_k, J_1^*) < \rho\}, \\ m_l &= \inf\{k > n_l, d(x_k, J_2^*) < \rho\}, \\ n_{l+1} &= \inf\{k > m_l, d(x_k, J_1^*) < \rho\}. \end{aligned}$$

It is clear that $m_l < \infty, n_l < \infty, \forall l$ and

$$x_{n_l} \in B(J_1^*, \rho), \quad x_{n_l - 1} \in B(J_2^*, \rho).$$

Since by $d(J_1^*, J_2^*) = 3\rho$ we have

$$\|x_{n_l} - x_{n_l - 1}\| \geq \rho. \tag{2.2.33}$$

By boundedness of $\{x_{n_l-1}\}$, we may assume that x_{n_l-1} converges. Then, by taking $t = a_{n_l-1}$ in (2.2.15), we derive

$$\|x_{n_l} - x_{n_l-1}\| \leq 2c_2 a_{n_l-1} \xrightarrow[l \to \infty]{} 0,$$

which contradicts (2.2.33) and proves the theorem. □

Corollary 2.2.1 *If J is not dense in any connected set, then under conditions of Theorem 2.2.1, $\{x_k\}$ given by (2.1.1)–(2.1.3) converges to a point in $\bar{J}$. This is because in the present case any connected set in $\bar{J}$ consists of a single point.*

Example 2.2.1 Reconsider the example given in Section 2.1:

$$f(x) = -(x-10)^3, \quad J = \{10\}, \quad x_0 = 0, \quad a_k = \frac{1}{k+1}.$$

It was shown that the RM algorithm rapidly diverges to $\pm\infty$ even in the noise-free case.

We now assume the observations are noise-corrupted:

$$y_{k+1} = f(x_k) + \epsilon_{k+1},$$

where $\{\epsilon_k\}$ is an ARMA process driven by the independent identically distributed normal random variables $w_k \in N(0,\sigma)$,

$$\epsilon_{k+1} + A_1\epsilon_k = w_{k+1} + C_1 w_k,$$

where $A_1 = -0.9, C_1 = 0.5, \sigma = 0.1$.

We use the algorithm (2.1.1)–(2.1.3) with $M_k = 2^{k+1}, x^* = 0.5$. The computation shows

$$x_{10} = 0.5, x_{20} = 43.4, x_{30} = -23.9, x_{40} = 21.9, x_{50} = 7.76, x_{60} = 8.74,$$
$$x_{70} = 8.97, x_{100} = 9.26, x_{200} = 9.46, x_{300} = 9.52, x_{400} = 9.61$$

which tend to the sought-for root 10.

Example 2.2.2 Let $f(x) = \sin x, x^* = 3, M_k = k, a_k = \frac{1}{k}$. Then

$$J = \{k\pi, k = 0, \pm 1, \pm 2, \cdots\}.$$

Clearly, A2.2.1 and A2.2.4 hold. Concerning A2.2.2, we may take $\cos x$ to serve as $v(x)$. Since

$$f(x)v_x(x) = -(\sin x)^2,$$

(2.2.1) is satisfied. The existence of c_0 required in A2.2.2 is obvious, for example, $c_0 = 5\pi/3$:

$$\cos 3 < 0 < \cos(\pm 5\pi/3).$$

Finally, $v(J) = \{-1, 1\}$ is nowhere dense. So A2.2.2 also holds.
Now assume the noise is such that

$$\sum_{k=1}^{\infty} \frac{1}{k} \epsilon_{k+1} < \infty.$$

Then A2.2.3 is satisfied too.

By Corollary 2.2.1, $\{x_k\}$ given by (2.1.1)–(2.1.3) converges to a point $\in J$.

If for the conventional (untruncated) RM algorithm

$$x_{k+1} = x_k + a_k y_{k+1}, \quad y_{k+1} = f(x_k) + \epsilon_{k+1} \tag{2.2.34}$$

it is *a priori* known that $\{x_k\}$ is bounded, then we have the following theorem.

Theorem 2.2.3 *Assume A2.2.1–A2.2.4 hold but in A2.2.2 the requirement: "Further, x^* used in (2.1.1) is such that $v(x^*) < \inf_{\|x\|=c_0} v(x)$ for some $c_0 > 0$ and $\|x^*\| < c_0$." is removed. If $\{x_k\}$ produced by (2.2.34) is bounded, then $d(x_k, J^*) \xrightarrow[k\to\infty]{} 0$ for the sample path ω for which A2.2.3 holds, where J^* is a connected subset of $\bar{J}$.*

Proof. As a matter of fact, by boundedness of $\{c_k\}$, (2.2.3) and (2.2.15) become obvious. Steps 3, 5, and 6 in the proof of Theorem 2.2.1 remain unchanged, while Step 4 is no longer needed. Then the conclusion follows from Theorems 2.2.1 and 2.2.2.

Remark 2.2.5 All theorems concerning SA algorithms with expanding truncations remain valid for $\{x_k\}$ produced by (2.2.34), if $\{x_k\}$ given by (2.2.34) is known to be bounded.

Theorems 2.2.1 and 2.2.2 concern with time-invariant function $f(\cdot)$, but the results can easily be extended to time-varying functions, i.e., to the case where the measurements are carried out for $f_k(x_k)$:

$$y_{k+1} = f_k(x_k) + \epsilon_{k+1},$$

where $f_k(\cdot)$ depends on time k.

Conditions A2.2.2 and A2.2.4 are respectively replaced by the following conditions:

A2.2.2^o *There is a continuously differentiable function $v(\cdot) : \mathbb{R}^l \longrightarrow R$ such that*

$$\sup_k \sup_{\delta \le d(x,J) \le \Delta} f_k^T(x) v_x(x) < 0$$

for any $\Delta > \delta > 0$, *and* $v(J) \triangleq \{v(x) : x \in J\}$ *is nowhere dense, where* $J = \bigcup_{j=1}^{\infty} \bigcap_{k=1}^{\infty} J_{k+j}$ *and* $J_k = \{x : f_k(x) = 0\}$, *and* $v_x(\cdot)$ *denotes the gradient of* $v(\cdot)$; *Further,* x^* *used in (2.1.1) is such that* $v(x^*) < \inf_{\|x_0\|=c_0} v(x)$ *for some* $c_0 > 0$ *and* $\|x^*\| < c_0$;

A2.2.4' $f_k(\cdot), \forall k$ are measurable and uniformly locally bounded, i.e., for any constant $c \geq 0$

$$\sup_k \sup_{\|x\|<c} \|f_k(x)\| < \infty.$$

Theorem 2.2.4 *Let* $\{x_k\}$ *be given by (2.1.1)–(2.1.3) for a given initial value* x_0. *Assume A2.2.1, A2.2.2*^{o}*, and A2.2.4' hold. Then* $d(x_k, J^*) \xrightarrow[k\to\infty]{} 0$ *for the sample path* ω *for which A2.2.3 holds, where* J^* *is a connected subset of* $\bar{J}$.

Proof. It suffices to replace $f(x_k)$ by $f_k(x_k)$ everywhere in the proof for Theorems 2.2.1 and 2.2.2. □

Remark 2.2.6 If it is known that $\{x_k\}$ given by an SA algorithm evolves in a subspace S of $\mathbb{R}^l$, then it suffices to verify A2.2.2, A2.2.2', A2.2.2", and A2.2.2^o in the subspace S in order the corresponding conclusions about convergence of $\{x_k\}$ to hold. For example, in this case A2.2.2 changes to
A2.2.2 (S): There is a continuously differentiable function $v(\cdot) : \mathbb{R}^l \longrightarrow R$ such that $\sup_{\substack{\delta \leq d(x, J\cap S) \\ x\in S}} f^T(x) v_x(x) < 0$ for any $\Delta > \delta > 0$ and $v(S \cap J)$ is nowhere dense.
Further, x^* used in (2.1.1) is such that $v(x^*) < \inf_{\|x\|=c_0} v(x)$ for some $c_0 > 0$ and $\|x^*\| < c_0$. According to Remark 2.2.4, here J is not specified. Then, with A2.2.2 and x_0 replaced by A2.2.2(S) and $x_0 \in S$, respectively, Theorem 2.2.1 incorporating with Theorem 2.2.2 asserts that

$$d(x_k, S \cap \overline{J}) \xrightarrow[k\to\infty]{} 0.$$

2.3. Convergence Under State-Independent Conditions

In the last section we have established convergence theorems under general conditions. These theorems take a sample-path-based form: under A2.2.1, A2.2.2, and A2.2.4 $\{x_k\}$ converges at those sample paths for which A2.2.3 holds. Condition A2.2.3 looks rather complicated, but it is so weak that it is necessary as to be shown later. However, condition A2.2.3 is state-dependent in the sense that the condition itself depends on the behavior of $\{x_k\}$. This makes it not always possible to verify the condition beforehand. We are planning to give convergence theorems under conditions with no state $\{x_k\}$ involved. For this we have to reformulate Theorems 2.2.1 and 2.2.2.

As defined in Section 2.2 $\epsilon_{k+1} = \epsilon_{k+1}(x_k, \omega)$, where $\epsilon_{k+1}(\cdot, \cdot)$ is a measurable function. In lieu of A2.2.3 we introduce the following condition.

A2.3.1 *For any sufficiently large integer $N (\geq N_0)$ there is an $\omega-$set Ω_N with $P\Omega_N = 1$ such that for any $\omega \in \Omega_N$*

$$\lim_{T \longrightarrow 0} \limsup_{k \longrightarrow \infty} \frac{1}{T} \| \sum_{i=n_k}^{m(n_k, T_k)} a_i \epsilon_{i+1}(x_i(\omega), \omega) I_{[\|x_i(\omega)\| \leq N]} \| = 0, \quad \forall T_k \in [0, T] \tag{2.3.1}$$

for any $\{n_k\}$ such that $\{x_{n_k}\}$ converges.

Theorem 2.3.1 *Assume A2.2.1, A2.2.2, A2.2.4, and A2.3.1 hold. Then $d(x_k, J^*) \xrightarrow[k \longrightarrow \infty]{} 0$ a.s. for $\{x_k\}$ generated by (2.1.1)–(2.1.3) with a given initial value x_0, where J^* is a connected subset contained in $\bar{J}$, the closure of J.*

Proof. Let $\Omega' \triangleq \bigcap_{N=N_0}^{\infty} \Omega_N$. It is clear that $P\Omega' = 1 - P(\Omega')^c = 1 - P \bigcup_{N=N_0}^{\infty} \Omega_N^c \geq 1 - \sum_{N=N_0}^{\infty} P\Omega_N^c = 1$, *i.e.*, $P\Omega' = 1$. Then for any $\omega \in \Omega'$, A2.2.3 is fulfilled with N_0 possibly depending on ω, and the conclusion of the theorem follows from Theorems 2.2.1 and 2.2.2. □

We now introduce a state-independent condition on noise.

A2.3.2 *For any $x \in \mathbb{R}^l$, $(\epsilon_k(x, \omega), \mathcal{F}_k)$ is a martingale difference sequence and for some $p \in (1, 2]$*

$$E(\|\epsilon_{k+1}(x, \omega)\|^p | \mathcal{F}_k) \triangleq \sigma_{k+1}(x) < \infty \quad a.s. \quad \forall x, \tag{2.3.2}$$

$$\sup_k \sup_{\|x\| \leq N} \sigma_{k+1}(x) \triangleq \sigma(N) < \infty, \quad \forall N, \tag{2.3.3}$$

where $\{\mathcal{F}_k\}$ is a family of nondecreasing σ-algebras independent of x.

We first give an example satisfying A2.3.2. Let $\{w_k, \mathcal{F}_k\}$ be an m-dimensional $(m \geq 1)$ martingale difference sequence with $\sup_k (E\|w_{k+1}\|^p$

$|\mathcal{F}_k) < \infty$ for some $p \in (1, 2]$ and let $g(x) : (\mathbb{R}^l, \mathcal{B}^l) \longrightarrow (\mathbb{R}^{l\times m}, \mathcal{B}^{l\times m})$ be a measurable and locally bounded function. Then $\epsilon_{k+1}(x, \omega) = g(x)w_{k+1}(\omega)$ satisfies A2.3.2, because

$$E(\|\epsilon_{k+1}(x,\omega)\|^p|\mathcal{F}_k) = \|g(x)\|^p E(\|w_{k+1}\|^p|\mathcal{F}_k) < \infty \quad \forall x$$

and

$$\sup_k \sup_{\|x\|\le N} \sigma_{k+1}(x) = \sup_{\|x\|\le N} \|g(x)\|^p \sup_k E(\|w_{k+1}\|^p|\mathcal{F}_k) < \infty \quad \forall N$$

by assumption.

Theorem 2.3.2 *Let $\{x_k\}$ be given by (2.1.1)–(2.1.3) for a given initial value. Assume A2.2.1, A2.2.2, A2.2.4, and A2.3.2 hold and $\sum_{k=1}^{\infty} a_k^p < \infty$ for p given in A2.3.2. Then $d(x_k, J^*) \xrightarrow[k\to\infty]{} 0$ a.s., where J^* is a connected subset contained in $\bar{J}$.*

Proof. Since $\epsilon_{k+1}(x,\omega)$ is measurable and x_k is $\mathcal{F}_k$-measurable, it follows that $\left(\epsilon_{k+1}(x_k(\omega),\omega), \mathcal{F}_{k+1}\right)$ is adapted. Approximating $\epsilon_{k+1}(x,\omega)$ by simple functions, it is seen that

$$E(\epsilon_{k+1}(x_k,\omega)|\mathcal{F}_k) = E(\epsilon_{k+1}(x,\omega)|\mathcal{F}_k)|_{x=x_k} = 0.$$

Hence, $\left(\epsilon_{k+1}(x_k(\omega),\omega), \mathcal{F}_{k+1}\right)$ is a martingale difference sequence, and

$$E(\|\epsilon_{k+1}(x_k(\omega),\omega)\|^p I_{[\|x_k(\omega)\|\le N]}|\mathcal{F}_k) \le I_{[\|x_k(\omega)\|\le N]}\sigma_{k+1}(x_k) \le \sigma(N) \text{ a.s.}$$

By the convergence theorem for martingale difference sequences, the series

$$\sum_{k=1}^{\infty} a_k \epsilon_{k+1}(x_k(\omega),\omega) I_{[\|x_k(\omega)\|\le N]}$$

converges a.s., which implies that Ω_N with $P\Omega_N = 1$ exists such that for each $\omega \in \Omega_N$

$$\sum_{k=n}^{m} a_k \epsilon_{k+1}(x_k(\omega),\omega) I_{[\|x_k(\omega)\|\le N]}$$

converges to zero as $n \longrightarrow \infty$ uniformly in m.

This means that A2.3.1 holds, and the conclusion of the theorem follows from Theorem 2.3.1. □

In applications it may happen that $f(\cdot)$ is not directly observed. Instead, the time-varying functions $g_k(\cdot)$ are observed, and the observations y_{k+1} may be done not at x_k, but at $x_k + r_k$, i.e., at x_k with bias r_k,

$$y_{k+1} = g_k(x_k + r_k) + \epsilon_{k+1}(x_k + r_k, \omega). \tag{2.3.4}$$

Theorem 2.3.3 *Let $\{x_k\}$ be given by (2.1.1)–(2.1.3) for a given initial value. Assume that A2.2.1, A2.2.2, A2.2.4, and A2.3.2 hold and $\sum_{k=1}^{\infty} a_k^p < \infty$ for p given in A2.3.2. Further, assume $(r_k, \mathcal{F}_k)$ is an adapted sequence, $\{r_k\}$ is bounded by a constant, and for any sufficiently large integer $N(\geq N_0)$ there exists Ω_N with $P\Omega_N = 1$ such that for any $\omega \in \Omega_N$*

$$\lim_{T \longrightarrow 0} \limsup_{k \longrightarrow \infty} \frac{1}{T} \Big\| \sum_{i=n_k}^{m(n_k, T_k)} a_i (g_i(x_i + r_i) - f(x_i)) I_{[\|x_i\| \leq N]} \Big\| = 0 \quad (2.3.5)$$

$$\forall T_k \in [0, T]$$

for any $\{n_k\}$ such that $\{x_{n_k}\}$ converges. Then, $d(x_k, J^*) \xrightarrow[k \longrightarrow \infty]{} 0$ a.s., where J^* is a connected subset contained in $\bar{J}$.

Proof. By assumption $\|r_k\| < c, \forall k$, where c is a constant. Then

$$E(\|\epsilon_{k+1}(x_k(\omega) + r_k(\omega), \omega)\|^p I_{[\|x_k(\omega)\| \leq N]} | \mathcal{F}_k)$$
$$= I_{[\|x_k(\omega)\| \leq N]} \sigma_{k+1}(x_k + r_k) \leq \sigma(N + c),$$

and again by the convergence theorem for martingale difference sequences, the series

$$\sum_{k=1}^{\infty} a_k \epsilon_{k+1}(x_k(\omega) + r_k(\omega), \omega) I_{[\|x_k(\omega)\| \leq N]}$$

convergence a.s. Consequently, there exists Ω' with $P\Omega' = 1$ such that for any $\omega \in \Omega'$ the convergence indicated in (2.3.5) holds and for any integer $N \geq N_0$

$$\sum_{k=n}^{m} a_k \epsilon_{k+1}(x_k(\omega) + r_k(\omega), \omega) I_{[\|x_k(\omega)\| \leq N]}$$

tends to zero as $n \longrightarrow \infty$ uniformly in m.

Therefore, A2.3.1 is fulfilled and the conclusion of the theorem follows from Theorem 2.3.1. □

Remark 2.3.1 The obvious sufficient condition for (2.3.5) is

$$\sup_{\|x\| \leq N} \|g_k(x + r_k) - f(x)\| \xrightarrow[k \longrightarrow \infty]{} 0,$$

which in turn is satisfied, if $g_k(\cdot) \equiv f(\cdot)$, $f(\cdot)$ is continuous and $r_k \xrightarrow[k \longrightarrow \infty]{} 0$.

Remark 2.3.2 Theorems 2.3.2 and 2.3.3 with A2.2.2 and A2.2.4 replaced by A2.2.2^o and A2.2.4', respectively, remain valid, if $f(x_k)$ is replaced by time-varying $f_k(x_k)$.

2.4. Necessity of Noise Condition

Under Conditions A2.2.1–A2.2.4 we have established convergence Theorems for $\{x_k\}$ recursively given by (2.1.1)–(2.1.3). Condition A2.2.1 is a commonly accepted requirement for decreasing step size, while A2.2.2 is a stability condition. This kind of conditions are unavoidable for convergence of SA type algorithms, although it may appear in different forms. Concerning A2.2.4 on $f(\cdot)$, it is the weakest possible: neither continuity nor growth rate of $f(\cdot)$ is required. So, it is natural to ask is it possible to further weaken Condition A2.2.3 on noise? We now answer this question.

Theorem 2.4.1 *Assume $f(\cdot)$ only has one root x^0, i.e., $J = \{x^0\}$ and $f(\cdot)$ is continuous at x_0. Further, assume A2.2.1 and A2.2.2 hold. Then $\{x_k\}$ given by (2.1.1)–(2.1.3) converges to x^0 at those sample paths for which one of the following conditions holds:*

i)

$$\lim_{T\longrightarrow 0}\limsup_{n\longrightarrow\infty}\frac{1}{T}\Big\|\sum_{k=n}^{m(n,t)}a_k\epsilon_{k+1}\Big\| = 0, \quad \forall t\in[0,T]; \tag{2.4.1}$$

ii) ϵ_k can be decomposed into two parts $\epsilon_k = \epsilon_k^{(1)} + \epsilon_k^{(2)}$ such that $\sum_{k=1}^{\infty} a_k\epsilon_{k+1}^{(1)} < \infty$ and $\epsilon_k^{(2)} \xrightarrow[k\longrightarrow\infty]{} 0$.

Conversely, if $x_k \longrightarrow x^0$, then both i) and ii) are satisfied.

Proof. Sufficiency. It is clear that ii) implies i), which in turn implies A2.2.3. Consequently, sufficiency follows from Theorem 2.2.1.

Necessity. Assume $x_k \longrightarrow x^0$. Then $\{x_k\}$ is bounded and (2.1.1)–(2.1.3) turns to be the RM algorithm after a finite number of steps (for $k \ge k^0$). Therefore,

$$\epsilon_{k+1} = \frac{x_{k+1}-x_k}{a_k} - f(x_k) = \epsilon_{k+1}^{(1)} + \epsilon_{k+1}^{(2)},$$

where $\epsilon_{k+1}^{(1)} = \frac{x_{k+1}-x_k}{a_k}$, $\epsilon_{k+1}^{(2)} = -f(x_k)$.

Since $x_k \longrightarrow x^0$ and $f(\cdot)$ is continuous, Condition ii) is satisfied. And, Condition i) being a consequence of ii) also holds. □

Remark 2.4.1 In the case where $J = \{x^0\}$ and $f(\cdot)$ is continuous at x^0, under conditions A2.2.1, A2.2.2, and A2.2.3 by Theorem 2.2.1 we arrive at $x_k \longrightarrow x^0$. Then by Theorem 2.4.1 we derive (2.4.1) which is stronger than A2.2.3. One may ask why a weaker condition A2.2.3 can imply a stronger condition (2.4.1)? Are they equivalent ? The answer is "yes" or "no": Yes, these conditions are equivalent but only under additional conditions A2.2.1, A2.2.2, and continuity of $f(\cdot)$ at x^0 being the unique root of $f(\cdot)$. However, these conditions by themselves are not equivalent because condition A2.2.3 is weaker than (2.4.1) indeed.

We now consider the multi-root case. Instead of the singleton $\{x^0\}$ we now have a root set J. Accordingly, continuity of $f(\cdot)$ at x^0 is replaced by the following condition

$$\lim_{d(x,J)\longrightarrow 0} f(x) = 0 \tag{2.4.2}$$

In order to derive the necessary condition on noise, we consider the linear interpolating function

$$x^0(t) = \frac{t_k - t}{a_k} x_{k-1} + \frac{t - t_{k-1}}{a_k} x_k, \quad t \in [t_{k-1}, t_k],$$

where $t_k = \sum_{i=0}^{k} a_i$. From $x^0(t)$ form a family $\{x_n(t), t \in [0, c]\}$ of functions, where

$$x_n(t) \triangleq x^0(t_n + t), \quad t \in [0, c],$$

where $c > 0$ is a constant.

For any subsequence $\{x_{n_k}(t)\}$ define

$$\limsup_{k\longrightarrow\infty} x_{n_k}(t) \triangleq x(t, \{n_k\}), \tag{2.4.3}$$

where $\{n_k\}$ appearing on the right-hand side of (2.4.3) denotes the dependence of the limit function on the subsequence, and the limsup of a vector sequence is taken component-wise. In general, $x(t, \{n_k\})$ may be discontinuous.

However, if $x_k \longrightarrow x^0$, then

$$\lim_{n\longrightarrow\infty} x_n(t) \triangleq x(t, \{n\}) (\equiv x^0), \tag{2.4.4}$$

which is not only continuous but also differentiable.

Thus, (2.4.2) for the multi-root case corresponds to the continuity of $f(\cdot)$ at x^0 for the single root case, while $d(x_k, J) \xrightarrow[k\longrightarrow\infty]{} 0$ and a certain analytic property of $x(t, \{n_k\})$ correspond to $x_k \longrightarrow x^0$.

Theorem 2.4.2 *Assume (2.4.2), A2.2.1, A2.2.2, and A2.2.4 hold. Then* $\{x_k\}$ *given by (2.1.1)–(2.1.3) is bounded,* $d(x_k, J^*) \xrightarrow[k\longrightarrow\infty]{} 0$, *and the right derivative* $\dot{x}^r(t, \{n_k\})|_{t=0} = 0$ *for any convergent subsequence* $\{x_{n_k}\}$

if and only if condition A2.2.3 is satisfied, where J^* is a connected subset of $\bar{J}$.

Proof. Sufficiency. By Theorem 2.2.1 it follows that $\{x_k\}$ is bounded and $d(x_k, J^*) \xrightarrow[k\to\infty]{} 0$. We only need to show $\dot{x}(t, \{n_k\})|_{t=0} = 0$.

Let $x_{n_k} \xrightarrow[k\to\infty]{} \bar{x}$ be a convergent subsequence. Since $\{x_k\}$ is bounded, the algorithm (2.1.1)–(2.1.3) becomes the one without truncations for large enough k. Therefore,

$$\frac{1}{T}\left(x_{m(n_k,T)+1} - x_{n_k}\right) = \frac{1}{T}\sum_{i=n_k}^{m(n_k,T)} a_i f(x_i) + \frac{1}{T}\sum_{i=n_k}^{m(n_k,T)} a_i \epsilon_{i+1}. \tag{2.4.5}$$

Notice that $x^0(t_k) = x_k$, and hence

$$\begin{aligned}
x_{m(n_k,T)+1} &= x^0\left(t_{m(n_k,T)+1}\right) = x^0\Big(\sum_{i=1}^{m(n_k,T)+1} a_i\Big) \\
&= x^0\Big(\sum_{i=1}^{n_k} a_i + \sum_{i=n_k+1}^{m(n_k,T)+1} a_i\Big) = x^0\Big(t_{n_k} + \sum_{i=n_k+1}^{m(n_k,T)+1} a_i\Big) \\
&= x_{n_k}\Big(\sum_{i=n_k+1}^{m(n_k,T)+1} a_i\Big) = x_{n_k}(T + o(1))
\end{aligned} \tag{2.4.6}$$

where $o(1) \longrightarrow 0$ as $k \longrightarrow \infty$.

Then from (2.4.5) we have

$$\begin{aligned}
&\Big\|\lim_{T\to 0}\limsup_{k\to\infty}\frac{1}{T}\left(x_{n_k}(T+o(1)) - x_{n_k}\right)\Big\| \\
&\le \lim_{T\to 0}\limsup_{k\to\infty}\frac{1}{T}\Big\|\sum_{i=n_k}^{m(n_k,T)} a_i f(x_i)\Big\| + \lim_{T\to 0}\limsup_{k\to\infty}\frac{1}{T}\Big\|\sum_{i=n_k}^{m(n_k,T)} a_i\epsilon_{i+1}\Big\|.
\end{aligned} \tag{2.4.7}$$

In (2.4.5) the last term tends to zero by A2.2.3 because $\{x_k\}$ is bounded and hence the indicator in (2.2.2) can be removed for sufficiently large k. By (2.4.2) the first term on the right-hand side of (2.4.7) also tends to zero as $k \longrightarrow \infty$. The left-hand side of (2.4.7) is $\Big\|\lim_{T\to 0}\frac{1}{T}\left(x(T, \{n_k\}) - \bar{x}\right)\Big\|$. Consequently,

$$\dot{x}^r(t, \{n_k\})|_{t=0} = 0.$$

Necessity. We now assume $\{x_k\}$ is bounded, $d(x_k, J^*) \xrightarrow[k\to\infty]{} 0$, and $\dot{x}^r(t, \{n_k\})|_{t=0} = 0$ for any convergent subsequence $\{x_{n_k}\}$, and want to show A2.2.3. Let $x_{n_k} \longrightarrow \bar{x}$. For any $T_k \in [0, T]$ from (2.4.5) we have

$$\frac{1}{T}\sum_{i=n_k}^{m(n_k,T_k)} a_i\epsilon_{i+1} = -\frac{1}{T}\sum_{i=n_k}^{m(n_k,T_k)} a_i f(x_i) + \frac{1}{T}(x_{m(n_k,T_k)+1} - x_{n_k}) \quad (2.4.8)$$

for sufficiently large k.

The first term on the right-hand side of (2.4.8) tends to zero as $k \longrightarrow \infty$ by (2.4.2) and $d(x_k, J^*) \xrightarrow[k\to\infty]{} 0$. So, to verify A2.2.3 it suffices to show that

$$\lim_{T\to 0}\limsup_{k\to\infty}\frac{1}{T}\big(x_{m(n_k,T_k)+1} - x_{n_k}\big). \quad (2.4.9)$$

From (2.4.6) it is seen that

$$x_{m(n_k,T_k)+1} = x_{n_k}\big(T_k + o(1)\big), \quad T_k \in (0, T], \quad (2.4.10)$$

where $o(1) \longrightarrow 0$ as $k \longrightarrow \infty$.

The assumption $\dot{x}^r(t, \{n_k\})|_{t=0} = 0$ means that

$$\limsup_{k\to\infty} x_{n_k}(t) - \overline{x} = o(t), \quad (2.4.11)$$

where $t > 0$, and $o(t) \xrightarrow[t\to 0]{} 0$.

Noticing the continuity of $x_{n_k}(\cdot)$, from (2.4.10) and (2.4.11) it follows that

$$\limsup_{k\to\infty} x_{n_k}\big(T_k + o(1)\big) = \overline{x} + o(T),$$

which incorporating with $x_{n_k} \xrightarrow[k\to\infty]{} \overline{x}$ yields (2.4.9). Thus, we have

$$\lim_{T\to 0}\limsup_{k\to\infty}\frac{1}{T}\left\|\sum_{i=n_k}^{m(n_k,T_k)} a_i\epsilon_{i+1}\right\| = 0 \quad \forall T_k \in [0, T] \quad (2.4.12)$$

for any $\{n_k\}$ such that $\{x_{n_k}\}$ converges.

By the boundedness of $\{x_k\}$, (2.4.12) is equivalent to (2.2.2), and the proof is completed. □

Corollary 2.4.1 *Assume (2.4.2), A2.2.1, A2.2.2, and A2.2.4 hold, and assume J is not dense in any connected set. Then $\{x_k\}$ given by (2.1.1)–(2.1.3) converges to some point in J if and only if A2.2.3 holds.*

This corollary is a direct generalization of Theorem 2.4.1. The sufficiency part follows from Corollary 2.2.1, while the necessity part follows from Theorem 2.4.2 if notice that convergence of $\{x_k\}$ implies $\dot{x}^r(t, \{n\})|_{t=0} = 0$.

2.5. Non-Additive Noise

In the algorithm (2.1.1)–(2.1.3) the noise ϵ_{k+1} in observation y_{k+1} is additive. In this section we continue considering (2.1.1)–(2.1.2) but in lieu of (2.1.3) we now have the non-additive noise

$$y_{k+1} = f(x_k, \xi_{k+1}), \tag{2.5.1}$$

where ξ_{k+1} is the observation noise at time $k+1$.

The problem is that under which conditions does the algorithm defined by (2.1.1), (2.1.2), and (2.5.1) converge to J, the root set of $f(\cdot)$, which is the average of $f(\cdot,\cdot)$ with respect to its second argument? To be precise, let $f(\cdot,\cdot)$ be an $I\!R^l \times I\!R^l \longrightarrow I\!R^l$ measurable function and let $F(\cdot)$ be a distribution function in $I\!R^m$. The function $f(\cdot)$ is defined by

$$f(x) = \int_{-\infty}^{\infty} f(x,z)dF(z). \tag{2.5.2}$$

It is clear that the observation given by (2.5.1) can formally be expressed by the one with additive noise:

$$y_{k+1} = f(x_k) + \epsilon_{k+1}, \quad \epsilon_{k+1} = f(x_k, \xi_{k+1}) - f(x_k), \tag{2.5.3}$$

and Theorems 2.2.1 and 2.2.2 can still be applied. The basic problem is how to verify A2.2.3. In other words, under which conditions on $f(\cdot,\cdot)$ and $\{\xi_k\}$, does $\{\epsilon_{k+1}\}$ given by (2.5.3) satisfy A2.2.3?

Before describing conditions to be used we first introduce some notations. We always take the regular version of conditional probability. This makes conditional distributions introduced later are well-defined.

Let $F_k(z)$ be the distribution function of ξ_k, and $F_{k+1}(z;\mathcal{F}_1^j)$ be the conditional distribution of ξ_{k+1} given $\mathcal{F}_1^j$, $k \geq j$, where $\mathcal{F}_k^j = \sigma\{\xi_i, k \leq i \leq j\}$.

Further, let us introduce the following coefficients,

$$\phi_k(i) = \sup_{B\in\mathcal{F}_{k+i}^{\infty}} \operatorname*{esssup}_{\omega} |P(B|\mathcal{F}_1^k) - P(B)|, \tag{2.5.4}$$

$$\psi_k = \sup_{A\in\mathcal{B}^m} \left|P(\xi_k \in A) - \int_A dF(x)\right|, \tag{2.5.5}$$

where $\mathcal{B}^m$ denotes the Borel σ-algebra in $I\!R^m$ and for a random variable $\eta(\omega)$, $\operatorname{esssup}_\omega \eta(\omega) = \inf_{\mathcal{N}} \sup_{\omega\in\Omega\cap\mathcal{N}^c} \eta(\omega)$ where $\mathcal{N}$ runs over all sets with probability zero.

$\phi_k(i)$ is known as the mixing coefficient of $\{\xi_k\}$ and it measures the dependence between $\{\xi_1,\ldots,\xi_k\}$ and $\{\xi_{k+i},\xi_{k+i+1},\ldots\}$. It is clear that ψ_k measures the closeness of the distribution of ξ_k to $F(x)$.

The following conditions will be needed.

A2.5.1 $a_k > 0,\ \sum\limits_{k=1}^{\infty} a_k = \infty,\ \sum\limits_{k=1}^{\infty} a_k^2 < \infty;$

A2.5.2 *(=A2.2.2);*

A2.5.3 *$f(\cdot,\cdot)$ is a measurable function and is locally Lipschitz-continuous in the first argument, i.e., for any fixed $L > 0$;*

$$\|(f(x,z) - f(y,z))I_{[\|x\|\le L, \|y\|\le L]}\| \le c_L\|x-y\|g(z), \tag{2.5.6}$$

where c_L is a constant depending on L;

A2.5.4 *(Noise Condition)*

i) $\{\xi_k\}$ is a ϕ-mixing process with mixing coefficient $\phi_n(k) \longrightarrow 0$ as $k \longrightarrow \infty$ uniformly in n;

ii)

$$\sup_k E[(g^2(\xi_{k+1}) + \|f(0,\xi_{k+1})\|^2)|\mathcal{F}_1^k] \triangleq \mu^2 < \infty, \quad E\mu^2 < \infty \tag{2.5.7}$$

$$\int_{-\infty}^{\infty} (g^2(z) + \|f(0,z)\|^2)dF(z) \triangleq \lambda^2 < \infty, \tag{2.5.8}$$

where $g(\cdot)$ is defined in (2.5.6);

iii) $\psi_k \longrightarrow 0$ as $k \longrightarrow \infty$.

Theorem 2.5.1 *Assume A2.5.1–A2.5.4. Then for $\{x_k\}$ generated by (2.1.1), (2.1.2), and (2.5.1)*

$$d(x_k, J^*) \xrightarrow[k\longrightarrow\infty]{} 0 \ a.s.,$$

where J^* is a connected subset of $\overline{J}$.

The proof consists in verifying Condition A2.2.3 satisfied a.s. by $\{\epsilon_k\}$ given in (2.5.3). Then the theorem follows from Theorems 2.2.1 and 2.2.2.

We first prove lemmas.

Lemma 2.5.1 *Assume A2.5.1, A2.5.3, and A2.5.4 hold. Then there is an Ω_0 with $P\Omega_0 = 1$ such that for any $\omega \in \Omega_0$ and any bounded subsequence $\{x_{n_k}\}$ of $\{x_k\}$, say, $\|x_{n_k}\| \le L/2$, $n_k \longrightarrow \infty$ as $k \longrightarrow \infty$,* *(without loss of generality assume $c_L L \geq 1$) there exists an integer k_0 such that for all $k \geq k_0$,*

$$\|x_i - x_{n_k}\| \leq ct, \quad \forall i : n_k \leq i \leq m(n_k, t), \quad \forall t \in [0, T], \tag{2.5.9}$$

if T is small enough, where $\{x_k\}$ is given by (2.1.1), (2.1.2), and (2.5.1), $c = (1 + \sqrt{2} c_L L \mu)$, and $m(k,t)$ is given by (1.3.2).

Proof. For any $L > 0$, set

$$g_L(z) \triangleq \sup_{\|x\| \leq L} \|f(x,z)\|. \tag{2.5.10}$$

By setting $y = 0$ in (2.5.6), it is clear that

$$g_L(z) \leq c_L L g(z) + \|f(0,z)\|. \tag{2.5.11}$$

From (2.5.7), it follows that

$$\begin{aligned} \sup_k E g_L^2(\xi_{k+1}) &\leq \sup_k \left\{ 2c_L^2 L^2 E g^2(\xi_{k+1}) + 2E\|f(0, \xi_{k+1})\|^2 \right\} \\ &\leq 2(1 \vee c_L^2 L^2) E\mu^2 \leq 2c_L^2 L^2 E\mu^2, \end{aligned} \tag{2.5.12}$$

and

$$\sup_k E(g_L^2(\xi_{k+1}) | \mathcal{F}_1^k) \leq 2c_L^2 L^2 \mu^2, \tag{2.5.13}$$

where (and hereafter) L is taken large enough so that $c_L L \geq 1$.

Since $E(\mu^2 | \mathcal{F}_1^k)$ is a convergent martingale, there is a $\nu^2 < \infty$ a.s. such that

$$E(\mu^2 | \mathcal{F}_1^k) \leq \nu^2 < \infty, \quad \forall k \geq 1. \tag{2.5.14}$$

From (2.5.13) and $\sum_{k=1}^{\infty} a_k^2 < \infty$, it is clear that for any integer L the series of martingale differences

$$\sum_{k=1}^{\infty} a_k \{ g_L(\xi_{k+1}) - E(g_L(\xi_{k+1}) | \mathcal{F}_1^k) \} < \infty \tag{2.5.15}$$

converges a.s.

Denote by Ω_L the ω-set where the above series converges, and set

$$\Omega_0 = \bigcap_{L=1}^{\infty} \Omega_L \cap [\mu < \infty, \nu < \infty].$$

It is clear that $P\Omega_0 = 1$.

Let ω be fixed and $\omega \in \Omega_L \cap [\mu < \infty, \nu < \infty]$ with $c_L L \geq 1$ and $\|x_{n_k}\| \leq L/2$. Then for any integer $p \geq n_k$ by (2.5.13) we have

$$\begin{aligned}
&\|\sum_{i=n_k}^{p} a_i f(x_i, \xi_{i+1}) I_{[\|x_i\| \leq L]}\| \leq \|\sum_{i=n_k}^{p} a_i g_L(\xi_{i+1})\| \\
&\leq \|\sum_{i=n_k}^{p} a_i [g_L(\xi_{i+1}) - E(g_L(\xi_{i+1})|\mathcal{F}_1^i)]\| + \|\sum_{i=n_k}^{p} a_i E(g_L(\xi_{i+1})|\mathcal{F}_1^i)\| \\
&\leq \|\sum_{i=n_k}^{p} a_i [g_L(\xi_{i+1}) - E(g_L(\xi_{i+1})|\mathcal{F}_1^i)]\| + \sqrt{2} c_L L \mu \sum_{i=n_k}^{p} a_i, \qquad (2.5.16)
\end{aligned}$$

where the first term on the right-hand side tends to zero as $k \longrightarrow \infty$ by (2.5.15).

Assume i_0 is sufficiently large such that

i) $M_{\sigma_{k_i}} > L$ for $\forall i \geq i_0$ if $\sigma_k \longrightarrow \infty$ as $k \longrightarrow \infty$, or

ii) $\lim_{k \longrightarrow \infty} \sigma_k = \sigma_{k_{i_0}}$ if $\lim_{k \longrightarrow \infty} \sigma_k < \infty$.

We note that in case ii) there will be no truncation in (2.1.1) for $k > k_{i_0}$.

Assume $k > k_{i_0}$ and fix a small enough T such that $cT < L/2$. Let $t \in [0, T]$ be arbitrarily fixed.

We prove (2.5.9) by induction. It is clear (2.5.9) is true for $i = n_k$.

Assume (2.5.9) is true for $i = n_k, n_k + 1, \ldots, s < m(n_k, t)$ and there is no truncation for $i = n_k + 1, \ldots, s$ if $n_k < s < m(n_k, t)$. Noticing $\|x_i\| \leq \|x_i - x_{n_k}\| + \|x_{n_k}\| < L$, $i = n_k, \ldots, s$, we have, by (2.5.16)

$$\begin{aligned}
\|\sum_{i=n_k}^{s} a_i f(x_i, \xi_{i+1})\| &= \|\sum_{i=n_k}^{s} a_i f(x_i, \xi_{i+1}) I_{[\|x_i\| \leq L]}\| \\
&\leq \|\sum_{i=n_k}^{s} a_i (g_L(\xi_{i+1}) - E(g_L(\xi_{i+1})|\mathcal{F}_1^i))\| + \sqrt{2} c_L L \mu t \\
&< (1 + \sqrt{2} c_L L \mu) t = ct < L/2,
\end{aligned}$$

if k is large enough.

This means that at time $s+1$ there is no truncation in (2.1.1), and

$$\|x_{s+1} - x_{n_k}\| = \|\sum_{i=n_k}^{s} a_i f(x_i, \xi_{i+1})\| < ct,$$

$$\|x_{s+1}\| < L/2 + ct < L. \qquad \square$$

Lemma 2.5.2 *Assume A2.5.1, A2.5.3, and A2.5.4 hold. There is an $\Omega' \subset \Omega_0$ with $P\Omega' = 1$, such that if $\omega \in \Omega'$ and if $\{x_{n_k}, n_k \longrightarrow \infty$ as*

$k \longrightarrow \infty\}$ is a bounded subsequence of $\{x_k\}$ produced by (2.1.1), (2.1.2), and (2.5.1), then

$$\lim_{T\longrightarrow 0}\sup_{t\in[0,T]}\limsup_{k\longrightarrow\infty}\frac{1}{T}\Big\|\sum_{i=n_k}^{m(n_k,t)} a_i\big(f(x_i,\xi_{i+1})-Ef(x,\xi_{i+1})|_{x=x_i}\big)\Big\|=0 \tag{2.5.17}$$

Proof. Write

$$\sum_{i=n_k}^{m(n_k,t)} a_i[f(x_i,\xi_{i+1})-Ef(x,\xi_{i+1})|_{x=x_i}]=(I)+(II)+(III)+(IV), \tag{2.5.18}$$

where

$$(I)=\sum_{i=n_k}^{m(n_k,t)} a_i[f(x_i,\xi_{i+1})-E(f(x_i,\xi_{i+1})|\mathcal{F}_1^{i-j})], \tag{2.5.19}$$

$$(II)=\sum_{i=n_k}^{m(n_k,t)} a_i[E(f(x_i,\xi_{i+1})|\mathcal{F}_1^{i-j})-E(f(x_{n_k},\xi_{i+1})|\mathcal{F}_1^{i-j})], \tag{2.5.20}$$

$$(III)=\sum_{i=n_k}^{m(n_k,t)} a_i[E(f(x_{n_k},\xi_{i+1})|\mathcal{F}_1^{i-j})-Ef(x,\xi_{i+1})|_{x=x_{n_k}}], \tag{2.5.21}$$

$$(IV)=\sum_{i=n_k}^{m(n_k,t)} a_i[Ef(x,\xi_{i+1})-f(y,\xi_{i+1})]|_{x=x_{n_k},y=x_i}]. \tag{2.5.22}$$

By (2.5.13), for $j>0$ we have

$$\begin{aligned}\sup_i E(g_L^2(\xi_{i+1})|\mathcal{F}_1^{i-j})=&\sup_i E(E(g_L^2(\xi_{i+1})|\mathcal{F}_1^{i})|\mathcal{F}_1^{i-j})\\ \le&2c_L^2L^2\sup_i E(\mu^2|\mathcal{F}_1^{i-j}),\end{aligned}$$

which converges to a finite limit as $i\longrightarrow\infty$ by the martingale convergence theorem.

Therefore, for any integers L and j

$$\sum_{i=1}^{\infty} a_i[f(x_i,\xi_{i+1})I_{[\|x_i\|\le L]}-E(f(x_i,\xi_{i+1})I_{[\|x_i\|\le L]}|\mathcal{F}_1^{i-j})]<\infty \tag{2.5.23}$$

converges a.s.

Therefore, there is $\Omega' \subset \Omega_0$ with $P\Omega' = 1$ such that (2.5.23) holds for any integers L and j.

Let ω be fixed, $\omega \in \Omega'$. By Lemma 2.5.1, $\|x_i\| \leq L$, $\forall i : n_k \leq i \leq m(n_k, t)$ for small t.

Then

$$(I) = \sum_{i=n_k}^{m(n_k,t)} a_i \big[f(x_i, \xi_{i+1}) I_{[\|x_i\| \leq L]} - E(f(x_i, \xi_{i+1}) I_{[\|x_i\| \leq L]} | \mathcal{F}_1^{i-j}] \xrightarrow[k \longrightarrow \infty]{} 0 \tag{2.5.24}$$

for any j by (2.5.23).

We now estimate (II). By Lemma 2.5.1 we have the following,

$$\begin{aligned}
\|(II)\| = & \| \sum_{i=n_k}^{m(n_k,t)} a_i [E\big(f(x_i, \xi_{i+1}) I_{[\|x_i\| \leq L} | \mathcal{F}_1^{i-j}\big) - E(f(x_{n_k}, \xi_{i+1}) | \mathcal{F}_1^{i-j})] \| \\
\leq & \| \sum_{i=n_k}^{m(n_k,t)} a_i c_L E[\|x_i - x_{n_k}\| g(\xi_{i+1}) | \mathcal{F}_1^{i-j}] \| \\
\leq & c_L t \sum_{i=n_k}^{m(n_k,t)} a_i [E((1 + \sqrt{2} c_L L \mu)^2 | \mathcal{F}_1^{i-j}]^{\frac{1}{2}} \cdot [E(g^2(\xi_{i+1}) | \mathcal{F}_1^{i-j})]^{\frac{1}{2}}.
\end{aligned}$$

Noticing (2.5.7) and (2.5.14), we then have

$$\|(II)\| \leq c_L (2 + 4 c_L^2 L^2 \nu^2) \nu t^2, \quad \forall t \in [0, T]. \tag{2.5.25}$$

Similarly, by Lemma 2.5.1 and (2.5.7)

$$\|(IV)\| \leq \sum_{i=n_k}^{m(n_k,t)} a_i c_L \|x_{n_k} - x_i\| E g(\xi_{i+1}) \leq c c_L \mu t^2, \quad \forall t \in [0, T]. \tag{2.5.26}$$

Combining (2.5.18), (2.5.24), and (2.5.26) leads to

$$\begin{aligned}
\lim_{T \longrightarrow 0} \sup_{t \in [0,T]} \limsup_{k \longrightarrow \infty} \frac{1}{T} \| \sum_{i=n_k}^{m(n_k,t)} a_i (f(x_i, \xi_{i+1}) - E(f(x, \xi_{i+1}) |_{x = x_i} \| \\
\leq \lim_{T \longrightarrow 0} \sup_{t \in [0,T]} \limsup_{k \longrightarrow \infty} \frac{1}{T} \|(III)\|.
\end{aligned} \tag{2.5.27}$$

Therefore, to prove the lemma it suffices to show that the right-hand side of (2.5.27) is zero.

Applying the Jordan-Hahn decomposition to the signed measure,

$$dG_{i+1}(z,\omega) \triangleq dF_{i+1}(z;\mathcal{F}_1^{i-j}) - dF_{i+1}(z), \quad i > j,$$

and noticing that $\{\xi_k\}$ is a ϕ-mixing process with mixing coefficient $\phi_k(i)$, we know that there is a Borel set D in $\mathbb{R}^m$, such that for any Borel set A in $\mathbb{R}^m$

$$\int_A dG^+_{i+1,j}(z,\omega) = \int_{A\cap D^c} dG_{i+1,j}(z,\omega) \le \phi_{i-j}(j+1), \tag{2.5.28}$$

$$\int_A dG^-_{i+1,j}(z,\omega) = \int_{A\cap D} dG_{i+1,j}(z,\omega) \le \phi_{i-j}(j+1), \tag{2.5.29}$$

and

$$dG_{i+1,j}(z,\omega) = dG^+_{i+1,j}(z,\omega) - dG^-_{i+1,j}(z,\omega), \tag{2.5.30}$$

$$dG^+_{i+1,j}(z,\omega) + dG^-_{i+1,j}(z,\omega) = dF_{i+1}(z;\mathcal{F}_1^{i-j}) + dF_{i+1}(z).$$

Then, we have the following,

$$\begin{aligned}\|(III)\| =& \Big\| \sum_{i=n_k+j}^{m(n_k,t)} a_i \int_{-\infty}^{\infty} f(x_{n_k},z)(dF_{i+1}(z;\mathcal{F}_1^{i-j}) - dF_{i+1}(z)) \\ &+ \sum_{i=n_k}^{n_k+j-1} a_i [E(f(x_{n_k},\xi_{i+1})|\mathcal{F}_1^{i-j}) - E(f(x,\xi_{i+1}))|_{x=x_{n_k}}]\Big\|,\end{aligned}$$

where

$$\begin{aligned}&\Big\| \sum_{i=n_k+j}^{m(n_k,t)} a_i \int_{-\infty}^{\infty} f(x_{n_k},z)\Big(dF_{i+1}(z;\mathcal{F}_1^{i-j}) - dF_{i+1}(z)\Big)\Big\| \\ &\le \sum_{i=n_k+j}^{m(n_k,t)} a_i \int_{-\infty}^{\infty} g_L(z)\Big(dG^+_{i+1,j}(z,\omega) + dG^-_{i+1,j}(z,\omega)\Big) \\ &\le \sum_{i=n_k+j}^{m(n_k,t)} a_i \phi^{\frac{1}{2}}_{i-j}(j+1)\Big[\Big(\int_{-\infty}^{\infty} g_L^2(z) dG^+_{i+1,j}(z,\omega)\Big)^{\frac{1}{2}}\end{aligned}$$

$$\Big(\int_{-\infty}^{\infty} g_L^2(z)dG_{i+1,j}^-(z,\omega)\Big)^{\frac{1}{2}}\Big]$$

$$\le \sqrt{2}\sum_{i=n_k+j}^{m(n_k,t)} a_i\phi_{i-j}^{\frac{1}{2}}(j+1)\big[E(g_L^2(\xi_{i+1})|\mathcal{F}_1^{i-j}) + Eg^2(\xi_{i+1})\big]^{\frac{1}{2}}$$

$$\le 2c_L L\sum_{i=n_k+j}^{m(n_k,t)} a_i\phi_{i-j}^{\frac{1}{2}}(j+1)(\nu^2+E\mu^2)^{\frac{1}{2}}.$$

For any given $\epsilon > 0$, there is a j such that

$$\phi_{i-j}^{\frac{1}{2}}(j+1)2c_L L(\nu^2+E\mu^2)^{\frac{1}{2}} < \epsilon, \quad \forall i \ge n_k.$$

For any fixed j, by (2.5.13), (2.5.14), and $a_k \xrightarrow[k\to\infty]{} 0$, it follows that

$$\lim_{k\to\infty}\|\sum_{i=n_k}^{n_k+j-1} a_i[E(f(x_{n_k},\xi_{i+1})|\mathcal{F}_1^{i-j}) - E(f(x,\xi_{i+1}))|_{x=x_{n_k}}]\|$$

$$\le \lim_{k\to\infty}\|\sum_{i=n_k}^{n_k+j-1} a_i[E(g_L(\xi_{i+1})|\mathcal{F}_1^{i-j}) + Eg_L(\xi_{i+1})] = 0.$$

Therefore,

$$\lim_{T\to 0}\sup_{t\in[0,T]}\limsup_{k\to\infty}\frac{1}{T}\|(III)\| < \epsilon.$$

Since $\epsilon > 0$ may be arbitrarily small, this combining with (2.5.27) proves the lemma. □

Proof of Theorem 2.5.1.

For proving the theorem it suffices to show that A2.2.3 is satisfied by $\epsilon_{i+1} = f(x_i,\xi_{i+1}) - f(x_i)$ a.s. By Lemma 2.5.2, we need only to prove that

$$\lim_{T\to 0}\sup_{t\in[0,T]}\limsup_{k\to\infty}\frac{1}{T}\|\sum_{i=n_k}^{m(n_k,t)} a_i(Ef(x,\xi_{i+1})|_{x=x_i} - f(x_i))\| = 0 \tag{2.5.31}$$

for $\omega \in \Omega'$, if $\{x_{n_k}\}$ is a bounded subsequence, and $n_k \longrightarrow \infty$ as $k \longrightarrow \infty$.

Assume $\|x_{n_k}\| \le L/2$, $k = 1, 2, \ldots$.

Applying the Jordan-Hahn decomposition to the signed measure,

$$dG_{i+1}(z) \triangleq dF_{i+1}(z) - dF(z),$$

we conclude that

$$
\begin{aligned}
&\| \sum_{i=n_k}^{m(n_k,t)} a_i (Ef(x,\xi_{i+1})|_{x=x_i} - f(x_i)\| \\
&= \| \sum_{i=n_k}^{m(n_k,t)} a_i [\int_{-\infty}^{\infty} f(x_i,z)(dF_{i+1}(z) - dF(z)]\| \\
&\le \| \sum_{i=n_k}^{m(n_k,t)} \sqrt{2} a_i \psi_{i+1}^{\frac{1}{2}} [Eg_L^2(\xi_{i+1}) + \int_{-\infty}^{\infty} g_L^2(z) dF(z)]^{\frac{1}{2}} \\
&\le 2c_L L \sum_{i=n_k}^{m(n_k,t)} a_i \psi_{i+1}^{\frac{1}{2}} (E\mu^2 + \lambda^2)^{\frac{1}{2}},
\end{aligned} \tag{2.5.32}
$$

where for the last inequality (2.5.8) and (2.5.12) are invoked. Since $\psi_i \longrightarrow 0$ as $i \longrightarrow \infty$, the right-hand side of (2.5.32) tends to zero as $k \longrightarrow \infty$ for any $t \in [0,T]$. This proves (2.5.31) and completes the proof of Theorem 2.5.1. □

Remark 2.5.1 From the expression (2.5.3) for observation it is seen that the observation with non-additive noise can be reduced to the additive but state-dependent noise which was considered in Section 2.3. However, Theorem 2.5.1 is not covered by Theorems in Section 2.3 and *vice versa*.

2.6. Connection Between Trajectory Convergence and Property of Limit Points

In the multi-root case, what we have established so far is that the distance between $\{x_k\}$ given by (2.1.1)–(2.1.3) and J^*, a connected subset of $\overline{J}$, converges to zero under various sets of conditions.

As pointed out in Corollary 2.2.1, if J is not dense in any connected set, then x_k converges to a point belonging to $\overline{J}$. However, it is still not clear how does $\{x_k\}$ behave when J is dense in some connected set? The following example shows that $\{x_k\}$ still may not converge, although $d(x_k, J) \xrightarrow[k\to\infty]{} 0$.

Example 2.6.1 Let

$$
f(x) = \begin{cases} -x, & x \le 0, \\ 0, & x \in [0,1], \\ -(x-1), & x \ge 1, \end{cases} \quad J = [0,1],
$$

and let

$$\epsilon_1 = 1, \quad \epsilon_2 = -1, \quad \epsilon_{2^k+1} = \cdots = \epsilon_{2^{k+1}} = (-1)^{k-1}\sqrt{\frac{1}{2^k}}.$$

Take step sizes as follows

$$a_0 = 1, \quad a_1 = 1, \quad a_{2^k} = \cdots = a_{2^{k+1}-1} = \sqrt{\frac{1}{2^k}}.$$

We apply the RM algorithm (2.2.34) with $x_0 = 0$.

As $v(\cdot)$, we may take

$$v(x) = \begin{cases} x^2, & x \leq 0, \\ 0, & x \in [0,1], \\ (x-1)^2, & x \geq 1. \end{cases}$$

Then, all conditions A2.2.1–A2.2.4 are satisfied.

Notice that

$$\sum_{i=2^s}^{2^{s+1}-1} a_i \epsilon_{i+1} = \frac{1}{2^s} \sum_{i=2^s}^{2^{s+1}-1} (-1)^{s-1} = \begin{cases} -1, & \text{if } s \text{ is even}, \\ 1, & \text{if } s \text{ is odd}, \end{cases} \tag{2.6.1}$$

and

$$\begin{aligned} x_{n+1} =& \sum_{i=0}^{n} a_i \epsilon_{i+1} = 1 + \sum_{i=1}^{n} a_i \epsilon_{i+1} \\ =& 1 + \sum_{s=0}^{k-1} \sum_{s=2^s}^{2^{s+1}-1} a_i \epsilon_{i+1} + \sum_{i=2^k}^{n} a_i \epsilon_{i+1}, \end{aligned} \tag{2.6.2}$$

where k is such that $2^k \leq n < 2^{k+1}$.

By (2.6.1), it is clear that in (2.6.2)

$$1 + \sum_{s=0}^{k-1} \sum_{i=2^s}^{2^{s+1}-1} a_i \epsilon_{i+1} = 0, \quad \sum_{i=2^k}^{n} a_i \epsilon_{i+1} > 0 \text{ for odd } k,$$

and

$$1 + \sum_{s=0}^{k-1} \sum_{i=2^s}^{2^{s+1}-1} a_i \epsilon_{i+1} = 1, \quad \sum_{i=2^k}^{n} a_i \epsilon_{i+1} \leq 0 \text{ for even } k.$$

Therefore, $\{x_k\}$ is bounded and $d(x_k, J) \xrightarrow[k \to \infty]{} 0$ by Theorem 2.2.4. As a matter of fact, $\{x_k\}$ changes from one to zero and then from zero to one, and this process repeats forever with decreasing step sizes.

Thus, $\{x_k\}$ is dense in [0,1]. This phenomenon hints that for trajectory convergence of $\{x_k\}$ the stability-like condition A2.2.2 is not enough; a stronger stability is needed.

Definition 2.6.1 *A point $\overline{x} \in J$, i.e., a root of $f(\cdot)$, is called dominantly stable for $f(\cdot)$, if there exist a $\delta_0 > 0$ and a positive measurable function $c(\cdot) : (0, +\infty) \longrightarrow (0, +\infty)$, which is bounded in the interval $[\delta, \delta_0]$ and satisfies the following condition*

$$f^T(x)(x - \overline{x}) \leq -c(\|x - \overline{x}\|)\|f(x)\| \tag{2.6.3}$$

for all $x \in B(\overline{x}, \delta_0)$, the ball centered at $\overline{x}$ with radius δ_0.

Remark 2.6.1 The dominant stability implies stability. To see this, it suffices to take $v(x) = \|x - \overline{x}\|^2$ as the Lyapunov function. Then

$$\frac{1}{2}v_x^T(x)f(x) = f^T(x)(x - \overline{x}) \leq -c(\|x - \overline{x}\|)\|f(x)\| < 0, \quad \forall x \notin J.$$

The dominant stability of $\overline{x}$, however, is not necessary for asymptotic stability.

Remark 2.6.2 Equality (2.6.3) holds for any $x \in J$, whatever $\overline{x}$ is. Therefore, all interior points of J are dominantly stable for $f(\cdot)$. Further, for a boundary point $\overline{x}$ of J to be dominantly stable for $f(\cdot)$, it suffices to verify (2.6.3) for $x \in B(\overline{x}, \delta_0) \cap J^c$ with small δ_0, i.e., all x that are close to $\overline{x}$ and outside J.

Example 2.6.2 Let

$$f(x) = \begin{cases} -(\|x\|^2 - 1)x, & \text{if } \|x\| > 1, \\ 0, & \text{otherwise.} \end{cases}$$

In fact, $f(x)$ is the gradient of

$$L(x) = \begin{cases} -\frac{1}{4}(\|x\|^2 - 1)^2, & \text{if } \|x\| > 1, \\ 0, & \text{otherwise.} \end{cases}$$

In this example $J = \{x : \|x\| \leq 1]$. We now show that all points of J are dominantly stable for $f(\cdot)$. For this, by Remark 2.6.2, it suffices to show that all $\overline{x}$ with $\|\overline{x}\| = 1$ are dominantly stable for $f(\cdot)$, and for this, it in turn suffices to show (2.6.3) for any x with $\|x\| > 1$ and $\|x - \overline{x}\| = \delta$ for small enough $\delta > 0$. Denoting by $L(x, y)$ the angle between vectors x and y, we have for $\|x\| > 1$

$$\begin{aligned} f^T(x)(x - \overline{x}) &= -(\|x\|^2 - 1)x^T(x - \overline{x}) = -\|f(x)\|\frac{x^T(x - \overline{x})}{\|x\|} \\ &= -\|f(x)\|\|x - \overline{x}\| \cos \angle(x, x - \overline{x}). \end{aligned}$$

It is clear that

$$\inf_{\|x-\overline{x}\|=\delta, \|x\|>1} \cos(x, x-\overline{x}) > 0$$

for all small enough $\delta > 0$. Therefore, all points in J are dominantly stable for $f(\cdot)$.

Theorem 2.6.1 *Assume A2.2.1, A2.2.2, and A2.2.4 hold. If for a given* ω, $\sum_{i=1}^{\infty} a_i \epsilon_{i+1}$ *is convergent and a limit point* $\overline{x}$ *of* $\{x_k\}$ *generated by (2.1.1)–(2.1.3) is deminantly stable for* $f(\cdot)$, *then for this trajectory* $x_k \xrightarrow[k\to\infty]{} \overline{x}$.

Proof. For any $\delta \in (0, \delta_0/3)$, define

$$\begin{aligned} n_0 &= \inf\{k > 0, x_k \in B(\overline{x}, \delta)\}, \\ m_i &= \inf\{k > n_i, x_k \in \overline{B(\overline{x}, 2\delta)} \backslash B(\overline{x}, \delta)\}, \\ l_i &= \inf\{k > m_i, x_k \notin \overline{B(\overline{x}, 2\delta)} \backslash B(\overline{x}, \delta)\}, \\ n_{i+1} &= \inf\{k \geq l_i, x_k \in B(\overline{x}, \delta)\}, \end{aligned}$$

where δ_0 is the one indicated in Definition 2.6.1.

It is clear that n_0 is well-defined, because there is a convergent subsequence: $x_{n_k} \longrightarrow \overline{x}$ and $x_{n_k} \in \mathcal{B}(\overline{x}, \delta)$ for any k greater than some k_0. If for any $\delta > 0$, $m_i = \infty$ for some i, then $x_k \xrightarrow[k\to\infty]{} \overline{x}$ by arbitrariness of δ. Therefore, for proving the theorem, it suffices to show that, for any small $\delta > 0$, an N_0 exists such that "$m_i < \infty$" implies "$l_i = n_{i+1}$" if $m_i > N_0$.

Since $\sum_{i=1}^{\infty} a_i \epsilon_{i+1} < \infty$ implies A2.2.3, all conditions of Theorem 2.2.1 are satisfied. By the boundedness of $\{x_k\}$ we may assume that l_i is large enough so that the truncations no longer exist in (2.1.1)–(2.1.3) for $k \geq m_i - 1$. It then follows that

$$\begin{aligned} \|x_{l_i} - \overline{x}\|^2 =& \|x_{l_i} - x_{l_i-1} + x_{l_i-1} - \overline{x}\|^2 \\ =& a_{l_i-1}^2 \|f(x_{l_i-1})\|^2 + 2a_{l_i-1}^2 f^T(x_{l_i-1})\epsilon_{l_i} \\ &+ 2a_{l_i-1} f^T(x_{l_i-1})(x_{l_i-1} - \overline{x}) + \|x_{l_i-1} - \overline{x} + a_{l_i-1}\epsilon_{l_i}\|^2. \end{aligned} \tag{2.6.4}$$

Notice that for any $x \in \overline{\mathcal{B}(\overline{x}, 2\delta)} \backslash \mathcal{B}(\overline{x}, \delta)$, $c(\|x - \overline{x}\|) \geq c_1 > 0$ and $\|f(x)\|$ is bounded, and hence by (2.6.3)

$$\begin{aligned} & a_k^2 \|f(x)\|^2 + 2a_k^2 f^T(x)\epsilon_{k+1} + 2a_k f^T(x)(x - \overline{x}) \\ & \leq a_k \|f(x)\| (a_k \|f(x)\| + \|2a_k \epsilon_{k+1}\| - 2c(\|x - \overline{x}\|) \leq 0 \end{aligned} \tag{2.6.5}$$

$\forall k \geq N_1$, for some N_1, because $\sum_{k=1}^{\infty} a_k \epsilon_{k+1}$ is convergent and $a_k \xrightarrow[k \to \infty]{} 0$.

Further,

$$\begin{aligned} &\|x_{l_i-1} - \overline{x} + a_{l_i-1}\epsilon_{l_i}\|^2 = \|x_{l_i-1} - x_{l_i-2} + x_{l_i-2} - \overline{x} + a_{l_i-1}\epsilon_{l_i}\|^2 \\ &\quad = \|a_{l_i-2} f(x_{l_i-2}) + x_{l_i-2} - \overline{x} + a_{l_i-2}\epsilon_{l_i-1} + a_{l_i-1}\epsilon_{l_i}\|^2. \qquad (2.6.6) \end{aligned}$$

An argument similar to that used for (2.6.5) leads to

$$a_{l_i-2}^2 \|f(x_{l_i-2})\|^2 + 2a_{l_i-2} f^T(x_{l_i-2})[(x_{l_i-2} - \overline{x}) + a_{l_i-2}\epsilon_{l_i-1} + a_{l_i-1}\epsilon_{l_i}] \leq 0,$$

if N_1 is large enough.

Then from (2.6.6) we have

$$\|x_{l_i-1} - \overline{x} + a_{l_i-1}\epsilon_{l_i}\|^2 \leq \|x_{l_i-2} - \overline{x} + a_{l_i-2}\epsilon_{l_i-1} + a_{l_i-1}\epsilon_{l_i}\|^2. \qquad (2.6.7)$$

From (2.6.4) and (2.6.7) we see that we can inductively obtain

$$\|x_{l_i} - \overline{x}\|^2 \leq \|x_{m_i} - \overline{x} + \sum_{k=m_i}^{l_i-1} \epsilon_k \epsilon_{k+1}\|^2.$$

Then, noticing $\|x_{m_i} - \overline{x}\| \leq \delta$ by definitions of m_i, we have

$$\begin{aligned} &\|x_{l_i} - \overline{x}\|^2 \\ &\quad \leq 3\|\sum_{k=m_i}^{l_i-1} a_k \epsilon_{k+1}\|^2 + \frac{3}{2}\|x_{m_i} - \overline{x}\|^2 \\ &\quad \leq 3\|\sum_{k=m_i}^{l_i-1} a_k \epsilon_{k+1}\|^2 + \frac{3}{2}\left[\|a_{m_i-1}(f(x_{m_i-1}) + \epsilon_{m_i})\| + \|x_{m_i-1} - \overline{x}\|\right]^2 \\ &\quad \leq 3\|\sum_{k=m_i}^{l_i-1} a_k \epsilon_{k+1}\|^2 + \frac{3}{2}[5a_{m_i-1}^2(\|f(x_{m_i-1})\| + \|\epsilon_{m_i}\|)^2 \\ &\qquad + \frac{5}{4}\|x_{m_i-1} - \overline{x}\|^2] \\ &\quad \leq 3\|\sum_{k=m_i}^{l_i-1} a_k \epsilon_{k+1}\|^2 + 15a_{m_i-1}^2\left(\|f(x_{m_i-1})\|^2 + \|\epsilon_{m_i})\|^2\right) + \frac{15}{8}\delta^2, \end{aligned} \qquad (2.6.8)$$

where the elementary inequality

$$\|x + y\|^2 \leq (1 + a)\|x\|^2 + (1 + \frac{1}{a})\|y\|^2, \quad \forall a > 0,$$

is used with $a = 2$ for the first inequality in (2.6.8), and with $a = 5$ for the third inequality in (2.6.8). Because $\|f(x_{m_i-1})\|$ is bounded, $\|\sum_{k=m_i}^{l_i-1} a_k\epsilon_{k+1}\| \xrightarrow[i\to\infty]{} 0$ and $\|a_{m_i-1}\epsilon_{m_i}\| \xrightarrow[i\to\infty]{} 0$, an N_2 exists such that

$$\|x_{l_i} - \overline{x}\|^2 \le \frac{31}{16}\delta^2, \quad \forall m_i > N_2.$$

This means that $l_i = n_{i+1}$ and completes the proof. □

For convergence of SA algorithms we have imposed the stability-like condition A2.2.2 for $d(x_k, J^*) \xrightarrow[k\to\infty]{} 0$ and the dominant stability condition (2.6.3) for trajectory convergence. It is natural to ask does a limit point of trajectory possess a certain stability property? The following example gives the negative answer.

Example 2.6.3 Let

$$f(x) = \begin{cases} -x, & x \le 0, \\ x, & 0 < x \le \frac{1}{2}, \\ -x+1, & x > \frac{1}{2}. \end{cases}$$

It is straightforward to check that

$$v(x) = \begin{cases} x^2 + \frac{1}{2}, & x \le 0, \\ -x^2 + \frac{1}{2}, & 0 < x \le \frac{1}{2}, \\ (x-1)^2, & x > \frac{1}{2} \end{cases}$$

satisfies A2.2.2. Take $a_k = 1/(k+1)$, $k \ge 0$, $\epsilon_1 = -1$, $\epsilon_k = \rho_k < 0$, $\forall k \ge 1$, where $\{\rho_k\}$ is a sequence of mutually independent random variables such that $\sum_{k=0}^{\infty} a_k\rho_{k+1} < \infty$ a.s. Then $J = \{0, 1\}$ with 1 being a stable attractor for $\dot{x} = f(x)$ and all A2.2.1–A2.2.4 are satisfied. Take $M_0 = \|x_0\| + 1$. Then by Theorem 2.2.1 it follows that $d(x_k, J) \xrightarrow[k\to\infty]{} 0$ a.s. Since $\epsilon_k < 0$, $x_k \le 0$, x_k must converge to 0 a.s. Zero, however, is unstable for $f(\cdot)$.

In this example $\{x_k\}$ converges to a limit, which is independent of initial values and unstable, although conditions A2.2.1–A2.2.4 hold. This strange phenomenon happens because

$$F_k(x_k) \triangleq x_k + a_k f(x_k) + a_k\epsilon_{k+1}$$

as a function of x_k is singular for some k, $k = 0, 1, \ldots$ in the sense that it restricts the algorithm to evolve only in a certain set of $\mathbb{R}^l$. Therefore, in order the limit of $\{x_k\}$ to be stable, imposing a certain regularity condition on $F_k(x_k)$ and some restrictions on noises is unavoidable.

As in Section 2.3, assume that observation noise is $\epsilon_{k+1} = \epsilon_{k+1}(x_k, \omega)$ with $\epsilon_{k+1}(x, \omega)$ being a measurable function defined on $\mathbb{R}^l \times \Omega$. Set

$$F_k(x) = x + a_k f(x) + a_k \epsilon_{k+1}(x, \omega).$$

Let us introduce the following conditions:

A2.6.1 *For a given ω, $F_k(x)$ is a surjection for any $k = 0, 1, \ldots$;*

A2.6.2 *For any ω and k, $\epsilon_k(x, \omega)$ is continuous in x, and for any ω, x, and $\delta > 0$,*

$$\lim_{T \longrightarrow 0} \limsup_{k \longrightarrow \infty} \frac{1}{T} \sup_{\substack{t \in [0,T] \\ x_k \in \mathcal{B}(x,\delta)}} \| \sum_{i=k}^{m(k,t)} a_i \epsilon_{i+1}(x_i, \omega) \| = 0, \tag{2.6.9}$$

where $\mathcal{B}(x, \delta)$ denotes the ball centered at x with radius δ.

It is clear, that A2.6.2 is equivalent to A2.6.2':

A2.6.2' *For any ω and any compact set $K \subset \mathbb{R}^l$*

$$\lim_{T \longrightarrow 0} \limsup_{k \longrightarrow \infty} \frac{1}{T} \sup_{\substack{t \in [0,T] \\ x_k \in K}} \Big\| \sum_{i=k}^{m(k,t)} a_i \epsilon_{i+1}(x_i, \omega) \Big\| = 0.$$

Before formulating Theorem 2.6.2 we first give some remarks on Conditions A2.6.1 and A2.6.2.

Remark 2.6.3 If ϵ_i does not depend on x_{i-1}, then in (2.6.9) $x_k \in \mathcal{B}(x, \delta)$ can be removed when taking supremum. In Condition A2.2.3 $\{x_{n_k}\}$ is a convergent subsequence, and hence $\{x_{n_k}\}$ is automatically located in a compact set. In Theorems in Sections 2.2, 2.3, 2.4, and 2.5, the initial value x_0 is fixed, and hence for fixed ω, $\{x_{n_k}\}$ is a fixed sequence. In contrast to this, in Theorem 2.6.2 we will consider the case where the initial value arbitrarily varies, and hence x_k for any fixed k may be any point in $\mathbb{R}^l$. If x_k in (2.6.9) were not restricted to a compact set (i.e., with "$x_k \in \mathcal{B}(x, \delta)$" removed in (2.6.9)), then the resulting condition would be too strong. Therefore, to put "$x_k \in \mathcal{B}(x, \delta)$" in (2.6.9) is to make the condition reasonable.

Remark 2.6.4 If $F_k(\cdot)$ is continuous and if $\frac{x^T F(x)}{\|x\|} \longrightarrow \infty$ as $\|x\| \longrightarrow \infty$, then $F_k(\cdot)$ is a surjection.

By this property, $F_k(\cdot)$ is a surjection for a large class of $f(\cdot)$. For example, let ϵ_k be free of x, and let the growth rate of $\|f(x)\|$ be not faster than linear as $\|x\| \longrightarrow \infty$. Then with $\{a_k\}$ satisfying A2.2.1 we have $x^T F_k(x)/\|x\| \longrightarrow \infty$ as $\|x\| \longrightarrow \infty$ for all $k \geq 0$. Hence, A2.6.1 holds. In the case where the growth rate of $\|f(x)\|$ is faster than linear as $\|x\| \longrightarrow \infty$ and $x^T f(x) \geq 0$, $\forall x : \|x\| \geq M$ for some $M > 0$, we also have that $x^T F_k(x)/\|x\| \longrightarrow \infty$ as $\|x\| \longrightarrow \infty$ for all $k \geq 0$ and A2.6.1 holds.

In what follows by stability of a set J_0 for $\dot{x} = f(x)$ we mean it in the Lyapunov sense, i.e., a nonnegative continuously differentiable function $L(\cdot)$ exists such that $L(x) = \text{const}\ (\triangleq L(J_0))\ \forall x \in J_0$, $L(x) > L(J_0)$, and $L_x^T(x)f(x) < 0$, $\forall x \in \mathcal{B}(J_0, \delta) \cap J_0^c$ for some $\delta > 0$, where

$$\mathcal{B}(J_0, \delta) = \{x : d(x, J_0) \leq \delta\}.$$

Theorem 2.6.2 *Assume A2.2.1, A2.2.2, and A2.6.2 hold, and that $f(\cdot)$ is continuous and for a given ω A2.6.1 holds. If $\{x_k\}$ defined by (2.1.1)–(2.1.3) with any initial value x_0 converges to a limit $\overline{x}$ independent of x_0, then $\overline{x}$ belongs to the unique stable set of $\dot{x} = f(x)$.*

Proof. Since by A2.2.2 $v(x^*) < \inf_{\|x\|=c_0} v(x)$ and $\|x^*\| < c_0$, by continuity of $v(x)$, $\hat{x}$ exists with $\|\hat{x}\| < c_0$ such that $v(\hat{x}) = \min_{\|x\| \leq c_0} v(x)$. Hence, $v_x(\hat{x}) = 0$. By continuity of $f(\cdot)$, J is closed, and hence by A2.2.2,

$$f^T(x)v_x(x) < 0, \quad \forall x \notin J.$$

Since $f^T(\hat{x})v_x(\hat{x}) = 0$, we must have $\hat{x} \in J$. Denote by J_0 the connected subset of J containing $\hat{x}$. The minimizer set of $v(\cdot)$ that contains $\hat{x}$ is closed and is contained in J_0. Since $\{v(x) : x \in J_0\}$ is a connected set and by A2.2.2 $v(J)$ is nowhere dense, $v(J_0)$ is a constant.

By continuity of $f(\cdot)$, all connected root-sets are closed and they are separated. Thus, there exists a $\delta > 0$ such that $\mathcal{B}(J_0, \delta) \cap (J \backslash J_0) = \emptyset$, i.e., $\mathcal{B}(J_0, \delta)$ contains no root of $f(\cdot)$ other than those located in J_0.

Set

$$L(x) = v(x) - v(\hat{x}).$$

Then $L(x) > 0$ and $f^T(x)L_x(x) = f^T(x)v_x(x) < 0$, $\forall x \in \mathcal{B}(J_0, \delta) \cap J_0^c$.

Therefore, by definition, J_0 is stable for $\dot{x} = f(x)$.

We have to show that $\overline{x} \in J_0$ and J_0 is the unique stable root-set.

Let A_h be the connected set of $\{x : v(x) < v(\hat{x}) + h, h > 0\}$ such that A_h contains J_0. By continuity of $v(\cdot)$, for an arbitrary small $\delta > 0$, $h_2 > h_1 > 0$ exist such that $h_2 < \delta$, $A_{h_2} \subset \mathcal{B}(J_0, \delta)$ and the distance between the interval $[v(\hat{x}) + h_1, v(\hat{x}) + h_2]$ and the set $v(J)$ is positive; i.e., $d([v(\hat{x}) + h_1, v(\hat{x}) + h_2], v(J)) > 0$.

We first show that, for any $\omega, x \in I\!R^l$, and $\hat{\delta} > 0$, there exist $c_1 > 0$, $T_1 > 0$, and k_T such that, for any $k > k_T$ if $x_k(x_0) \in \mathcal{B}(x, \hat{\delta})$, then

$$\|x_m(x_0) - x_k(x_0)\| \le c_1 T, \quad \forall T \in [0, T_1], \qquad \forall m : k \le m \le m(k,T) + 1. \tag{2.6.10}$$

By Theorem 2.2.1, for $k \ge k_T$ with k_T sufficiently large there will be no truncation for (2.1.1)–(2.1.3), and

$$x_m(x_0) - x_k(x_0) = \sum_{i=k}^{m-1} a_i f(x_i(x_0)) + \sum_{i=k}^{m-1} a_i \epsilon_{i+1}(x_i(x_0), \omega). \tag{2.6.11}$$

For any $M > 0$, let $c \triangleq \underline{\max_{y \in \mathcal{B}(x, M+\hat{\delta})}} |f(y)|$. By A2.6.2, sufficiently small $T_1 < M/3c$ and large enough k_T exist such that for any $k \ge k_T$

$$\|\sum_{i=k}^{m-1} a_i \epsilon_{i+1}(x_i, \omega)\| < cT, \quad \forall T \in [0, T_1], \quad \forall m : k < m \le m(k,T) + 1.$$

If $x_i \in \mathcal{B}(x, M+\hat{\delta})$ for $i : k \le i \le m-1$, $\forall k$, then (2.6.10) immediately follows by setting $c_1 = 2c$. Assume $x_i \notin \mathcal{B}(x, M+\hat{\delta})$ for some $i : k < i \le m$. Let $i_0 > k$ be the first such one. Then

$$\|x_{i_0}(x_0) - x_k(x_0)\| > M. \tag{2.6.12}$$

By (2.6.11), however,

$$\begin{aligned}\|x_{i_0}(x_0) - x_k(x_0)\| &\le \sum_{i=k}^{i_0-1} a_i \|f(x_i(x_0))\| + \|\sum_{i=k}^{i_0-1} a_i \epsilon_{i+1}(x_i(x_0), \omega)\| \\ &\le cT + cT \le M,\end{aligned}$$

which contradicts (2.6.12). Thus $x_i \in \mathcal{B}(x, \hat{\delta} + M)$, $\forall i \le m$, and (2.6.10) is verified.

For a given ω, we now prove the existence of k_0 such that $x_k(\omega, x_0) \in A_{h_2}$ for any $k \ge k_0$ if $x_{k_0}(\omega, x_0) \in A_{h_1}$, where the dependence of $x_k(\omega, x_0)$ on ω and on the initial value x_0 is emphasized. For simplicity of writing, $x_k(\omega, x_0)$ is written as $x_k(x_0)$ in the sequel.

Assume the assertion is not true; i.e., for any N, x_0^N exists such that $x_N(x_0^N) \in A_{h_1}$ and $x_{N'}(x_0^N) \notin A_{h_2}$ for some $N' > N$.

Suppose $N \le k_N < k'_N \le N'$ and $x_{k_N}(x_0^N) \in \overline{A}_{h_1}$, $x_{k'_N}(x_0^N) \in \overline{A}^c_{h_2}$, $x_i(x_0^N) \in A_{h_2} \backslash \overline{A}_{h_1}$, i.e., $v(\hat{x}) + h_1 \le v(x_i(x_0^N)) \le v(\hat{x}) + h_2$, $\forall i : k_N < i < k'_N$.

If there exists an $x' \in J_0$ with $\|x'\| > c_0$, then $x'' \in J_0$ with $\|x''\| = c_0$ exists because J_0 is connected and $\hat{x} \in J_0$ with $\|\hat{x}\| < c_0$.

This yields a contradictory inequality:

$$v(x'') = v(J_0) > v(x^*) \geq v(\hat{x}) = v(J_0),$$

where the first inequality follows from A2.2.2 while the second inequality is because $\hat{x}$ is the minimizer of $v(\cdot)$.

Consequently, $\|x\| \leq c_0$ for any $x \in J_0$ and

$$x_{k_N}(x_0^N) \in \mathcal{B}(J_0, \delta) \subset \{x : \|x\| \leq c_0 + \delta\},$$

and a subsequence of $\{x_{k_N}(x_0^N)\}$ exists, also denoted by $\{x_{k_N}(x_0^N)\}$ for notational simplicity, such that $\{x_{k_N}(x_0^N)\} \xrightarrow[N\to\infty]{} \widetilde{x}$. By the continuity of $v(x)$

$$v(\widetilde{x}) = v(\hat{x}) + h_1.$$

Hence, $d(\widetilde{x}, J) \stackrel{\Delta}{=} \widetilde{\delta} > 0$ by the fact $d([v(\hat{x}) + h_1, v(\hat{x}) + h_2], v(J)) > 0$.

By (2.6.10) and the fact $x_{k_N}(x_0^N) \longrightarrow \widetilde{x}$ we can choose sufficiently small T and large enough N such that

$$d(x_m(x_0^N), J) \geq \frac{\widetilde{\delta}}{2} \tag{2.6.13}$$

and $m(k_N, T) < k'_N$, i.e.,

$$v(\hat{x}) + h_1 \leq v(x_m(x_0^N)) \leq v(\hat{x}) + h_2 \tag{2.6.14}$$

for any $m : k_N \leq m \leq m(k_N, T) + 1$. By (2.6.10), ξ exists with the property $\|\xi - x_{k_N}(x_0^N)\| \leq c_1 T$ such that

$$\begin{aligned}
&v(x_{m(k_N,T)+1}(x_0^N)) - v(x_{k_N}(x_0^N)) \\
&= (x_{m(k_N,T)+1}(x_0^N)) - x_{k_N}(x_0^N))^T v_x(\widetilde{x}) \\
&\quad + (x_{m(k_N,T)+1}(x_0^N)) - x_{k_N}(x_0^N))^T (v_x(\xi) - v_x(\widetilde{x})).
\end{aligned} \tag{2.6.15}$$

Because $x_{k_N}(x_0^N) \longrightarrow \widetilde{x}$ as $N \longrightarrow \infty$, for sufficiently large N, $\|\xi - \widetilde{x}\| \leq 2c_1 T$, by (2.6.10) the last term of (2.6.15) is $o(T)$. Then

$$\begin{aligned}
&v(x_{m(k_N,T)+1}(x_0^N)) - v(x_{k_N}(x_0^N)) \\
&= \sum_{i=k_N}^{m(k_N,T)} a_i y_{i+1}^T v_k(\widetilde{x}) + o(T)
\end{aligned}$$

$$
\begin{aligned}
&= \sum_{i=k_N}^{m(k_N,T)} a_i f^T(x_i(x_0^N))v_x(x_i(x_0^N)) + \sum_{i=k_N}^{m(k_N,T)} a_i v_x^T(\widetilde{x})\epsilon_{i+1}(x_i,\omega) \\
&+ \sum_{i=k_N}^{m(k_N,T)} a_i f^T(x_i(x_0^N))(v_x(\widetilde{x}) - v_x(x_i(x_0^N))) + o(T). \qquad (2.6.16)
\end{aligned}
$$

By (2.6.10) and the continuity of $v_x(x)$, the third term on the right hand side of (2.6.16) is $o(T)$, and by A2.6.2 (Since $x_{k_N}(x_0^N) \longrightarrow \widetilde{x}$, $x_{k_N}(x_0^N) \in \mathcal{B}(\widetilde{x},\delta)$ with $\delta > 0$ for all sufficiently large N.), the norm of the second term on the right-hand side of (2.6.16) is also $o(T)$ as $N \longrightarrow \infty$. Hence by A2.2.2 and (2.6.13), some $\alpha > 0$ exists such that the right-hand side of (2.6.16) is less than $-\alpha T$ for all sufficiently large N if T is small enough. By noticing $x_{k_N}(x_0^N) \xrightarrow[N\longrightarrow\infty]{} \widetilde{x}$ and $v(\widetilde{x}) = v(\hat{x}) + h_1$ mentioned above, from (2.6.14) it follows that the left-hand side of (2.6.16) tends to a nonnegative limit as $N \longrightarrow \infty$. The obtained contradiction shows that k_0 exists such that $x_k(x_0) \in A_{h_2}$ for any $k \geq k_0$ if $x_{k_0}(x_0) \in A_{h_1}$. With fixed ω, for any k_0 by A2.6.1 x_0 exists such that $x_{k_0}(\omega, x_0) \in A_{h_1}$. By $A_{h_2} \subset \mathcal{B}(J_0,\delta)$ and the arbitrary smallness of $\delta > 0$, from this it follows that $d(x_k, J_0) \xrightarrow[k\longrightarrow\infty]{} 0$. Since $x_k \longrightarrow \overline{x}$ by assumption, we have $\overline{x} \in J_0$, which means that $\overline{x}$ is stable. If another stable set J_0' existed such that $J_0' \cap J_0 = \emptyset$, then by the same argument $\overline{x}$ would belong to J_0'. The contradiction shows that the uniqueness of the stable set. □

2.7. Robustness of Stochastic Approximation Algorithms

In this section for the single root case, i.e, the case $J = \{x^0\}$, we consider the behavior of SA algorithms when conditions for convergence of algorithms to x^0 are not exactly satisfied. It will be shown that a "small" violation of conditions will cause no big effect on the behavior of the algorithm.

The following result known as Kronecker lemma will be used several times in the sequel. We state it separately for convenience of reference.

Kronecker Lemma. *If* $\sum_{i=1}^{\infty} \frac{1}{b_i} M_i < \infty$, *where* $\{b_i\}$ *is a sequence of positive numbers nondecreasingly diverging to infinity and* $\{M_i\}$ *is a sequence of matrices, then* $\sum_{i=1}^{n} M_i = o(b_n)$.

Proof. Set $N_0 = 0$, $N_n = \sum_{i=1}^{n} \frac{1}{b_i} M_i$, $b_0 = 0$. Since $N_n \xrightarrow[n\longrightarrow\infty]{} N < \infty$, for any $\epsilon > 0$ there is n_ϵ such that $\|N_i - N\| < \epsilon$ if $i \geq n_\epsilon$. Then it follows that

$$\begin{aligned}\|\frac{1}{b_n}\sum_{i=1}^{n} M_i\| &= \|\frac{1}{b_n}\sum_{i=1}^{n} b_i(N_i - N_{i-1})\| \\ &= \|N_n + \frac{1}{b_n}\sum_{i=2}^{n}(b_{i-1} - b_i)N_{i-1}\| \\ &= \|N_n - \frac{b_n - b_1}{b_n}N + \frac{1}{b_n}\sum_{i=2}^{n}(b_{i-1} - b_i)(N_{i-1} - N)\| \\ &\le \|N_n - N\| + \frac{b_1}{b_n}\|N\| \\ &\quad + \frac{1}{b_n}\sum_{i=2}^{n_\epsilon}(b_{i-1} - b_i)\|N_{i-1} - N\| + \epsilon \longrightarrow 0\end{aligned}$$

as $n \longrightarrow \infty$ and then $\epsilon \longrightarrow 0$ □

We still consider the algorithm given by (2.1.1)–(2.1.3), where x_k denotes the estimate for x^0 at time k, but x^0 may not be the exact root of $f(\cdot)$. As a matter of fact, the following set of conditions will be used to replace A2.2.1–A2.2.4:

A2.7.1 $a_k > 0$, $\{a_k\}$ *nonincreasingly tends to zero,* $\sum_{k=1}^{\infty} a_k = \infty$, *and* k_0 *exists such that*

$$\frac{1}{a_{k+1}} - \frac{1}{a_k} \le 1, \text{ i.e., } a_k - a_{k+1} \le a_k a_{k+1} \text{ for } k \ge k_0; \tag{2.7.1}$$

A2.7.2 *There exists a nonnegative twice continuously differentiable function* $v(\cdot)$ *such that* $v(x^0) = 0$, $\lim_{\|x\|\longrightarrow\infty} v(x) = \infty$, *and*

$$v_x^T(x)f(x) < 0, \quad \forall x : \|x - x^0\| > e \ge 0; \tag{2.7.2}$$

A2.7.3 *For sample path* ω *the observation noise* $\{\epsilon_k\}$ *satisfies the following condition*

$$\limsup_{k\longrightarrow\infty} \|a_k \sum_{i=1}^{k} \epsilon_{i+1}\| \le \epsilon < \infty; \tag{2.7.3}$$

A2.7.4 $f(\cdot) : \mathbb{R}^l \longrightarrow \mathbb{R}^l$ *is continuous, but* $x^0 \in \mathbb{R}^l$ *is not necessary to be the root of* $f(\cdot)$.

Comparing A2.7.1–A2.7.4 with A2.2.1–A2.2.4, we see the following conditions required here are not assumed in Section 2.2: nonincreasing property of $\{a_k\}$, condition (2.7.1), nonnegativity of $v(\cdot)$, divergence of $v(\cdot)$ to infinity and continuity of $f(\cdot)$, but e in (2.7.2), ϵ in (2.7.3), and $\|f(x^0)\|$ are allowed to be greater than zero.

Concerning $a_k \sum_{i=1}^{k} \epsilon_{i+1}$, we note that from the convergence of $\sum_{i=1}^{\infty} a_i \epsilon_{i+1} < \infty$ it follows that i) A2.2.3 holds and ii) by the Kronecker lemma $a_k \sum_{i=1}^{k} \epsilon_{i+1} \xrightarrow[k\to\infty]{} 0$ because $\{a_k\}$ is nonincreasing. We will demonstrate how does the deviation from x^0 of the estimate given by (2.1.1)–(2.1.3) depend on ϵ and e?

For x^* used in (2.1.1) define $d_0 \triangleq v(x^*)$. Since $v(x) \longrightarrow \infty$ as $\|x\| \longrightarrow \infty$, an M can be taken sufficiently large such that

$$\|x^*\| < M, \quad \inf\{v(x) : \|x\| > M\} \triangleq d > d_0. \tag{2.7.4}$$

Let the initial truncation bound M_0 used in (2.1.1) and (2.1.2) be large enough such that

$$M_0 > M + 8. \tag{2.7.5}$$

Take real numbers $\delta_2 > \delta_1 > 0$ such that

$$[\delta_1, \delta_2] \subset (d_0, d) \text{ and } \frac{d - d_0}{2} < \delta_1. \tag{2.7.6}$$

Since $v(\cdot)$ is continuous, an $e^* \in (0, 8)$ exists such that

$$\{x : v(x) \geq \delta_1\} \subset \{x : \|x - x^0\| > e^*\}. \tag{2.7.7}$$

Denote

$$D \triangleq \{x : \delta_1 \leq v(x) \leq \delta_2\}, \tag{2.7.8}$$

$$f_1 \triangleq \max_{\|x\| \leq M+8} \|f(x)\|, \tag{2.7.9}$$

$$U = \{x : \|x\| \leq M + 6\}, \tag{2.7.10}$$

and

$$r_1 = \max_{x \in U} \|v_x(x)\|, \quad r_2 = \max_{x \in U} \|v_{xx}(x)\|, \tag{2.7.11}$$

where $v_{xx}(x)$ denotes the matrix consisting of the second partial derivatives of $v(\cdot)$.

Since $\delta_2 < d$, we have $\|x\| \leq M$ for any $x \in D$, and hence $D \subset U$.

Set

$$a \triangleq -\max_{x \in D} f^T(x) v_x(x). \tag{2.7.12}$$

We will only consider those e in (2.7.2) for which $e \leq e^*$ where e^* is given in (2.7.7). From (2.7.7) and (2.7.8) it is seen that $\|x - x^0\| > e^* \geq e$, $\forall x \in D$. Consequently, by (2.7.2), a given by (2.7.12) is positive.

By continuity of $f(\cdot)$ and $v(\cdot)$, $\alpha > 0$, $\beta > 0$, $\epsilon^* \in (0, T)$, and T^* exist such that the following inequalities hold:

$$T^* \leq \frac{1}{1+f_1}, \tag{2.7.13}$$

$$-\alpha > -a + \frac{1}{2} r_2 (2 + f_1)^2 T^*, \tag{2.7.14}$$

$$-\beta > -\alpha + r_1 \max_{\substack{x,y \in U \\ \|x-y\| \leq 3\epsilon^* + T^*(2+2f_1)}} \|f(x) - f(y)\|, \tag{2.7.15}$$

$$\max_{\substack{\|x-y\| \leq 4\epsilon^* \\ x,y \in U}} |v(x) - v(y)| < \delta_2 - \delta_1, \tag{2.7.16}$$

$$\epsilon^* < \beta T^* / (8r_1 + 14r_2), \quad \delta_1 + r_1[7\epsilon^* + (f_1 + 2\epsilon^*)T^*] < \delta_2. \tag{2.7.17}$$

By A2.7.3 for $\epsilon \leq \epsilon^*$, $K (\geq k_0)$ can be taken sufficiently large such that

$$a_k \leq \epsilon^*/2f_1. \quad a_k \|\sum_{i=1}^{k} \epsilon_{i+1}\| \leq \frac{11}{8}\epsilon^* \text{ for } k \geq K. \tag{2.7.18}$$

Lemma 2.7.1 *Assume A2.7.1, A2.7.2, A2.7.4 hold with ϵ given in (2.7.3) being less than or equal to ϵ^* : $\epsilon \leq \epsilon^* \leq 1$. If for $\{x_k\}$ given by (2.1.1)–(2.1.3) with (2.7.5) fulfilled, $\|x_k\| \leq M$ for some $k \geq K$, where K is given in (2.7.18), then for any $T \in [0, \frac{1}{1+f_1}]$,*

$$\|\sum_{i=k}^{m+1} a_i y_{i+1}\| \leq 6\epsilon^* + 2, \quad \forall m : k \leq m \leq m(k, T). \tag{2.7.19}$$

Proof. Because $m(k, T)$ is nondecreasing as T increases, it suffices to prove the lemma for $T = \frac{1}{1+f_1}$.

Assume the converse: there exists an m_1 such that

$$m_1 = \min\{m : k \leq m \leq m(k, T), \|\sum_{i=k}^{m+1} a_i y_{i+1}\| > 6\epsilon^* + 2\}.$$

Then for any $m : k \le m \le m_1$ we have

$$\|x_k + \sum_{i=k}^{m} a_i y_{i+1}\| \le \|x_k\| + \|\sum_{i=k}^{m} a_i y_{i+1}\| \le M + 6\epsilon^* + 2 < M_0 < M_{\sigma_k}, \tag{2.7.20}$$

and hence

$$x_{m+1} = x_m + a_m y_{m+1}, \quad \forall m : k \le m \le m_1, \tag{2.7.21}$$

which incorporating with the definition of m_1 leads to

$$\|x_{m_1+1} - x_k + a_{m_1+1} y_{m_1+2}\| = \|\sum_{i=k}^{m_1+1} a_i y_{i+1}\| > 6\epsilon^* + 2. \tag{2.7.22}$$

On the other hand, from (2.7.20) and (2.7.21) it follows that

$$\|x_m\| \le M + 6\epsilon^* + 2, \quad \forall m : k \le m \le m_1 + 1. \tag{2.7.23}$$

From (2.7.9) we have

$$\|f(x_m)\| \le f_1, \quad \forall m : k \le m \le m_1 + 1. \tag{2.7.24}$$

By a partial summation we have

$$\left\|\sum_{i=k}^{m_1} a_i \epsilon_{i+1}\right\| = \left\|a_{m_1} \sum_{i=1}^{m_1} \epsilon_{i+1} - a_k \sum_{i=1}^{k-1} \epsilon_{i+1} + \sum_{i=k}^{m_1-1} (a_i - a_{i+1}) \sum_{j=1}^{i} \epsilon_{j+1}\right\|. \tag{2.7.25}$$

Applying (2.7.3) to the first two terms on the right-hand side of (2.7.25), and (2.7.1) and (2.7.3) to the last term we find

$$\|\sum_{i=k}^{m_1} a_i \epsilon_{i+1}\| \le \frac{11}{4}\epsilon^* + \sum_{i=k}^{m_1-1} a_{i+1} \|a_i \sum_{j=1}^{i} \epsilon_{j+1}\| \le \frac{11}{4}\epsilon^* + \frac{11}{8} T\epsilon^*. \tag{2.7.26}$$

From (2.7.24) and (2.7.26) it then follows that

$$\begin{aligned}
&\|x_{m_1+1} - x_k + a_{m_1+1} y_{m_1+2}\| \\
&\quad \le \|x_{m_1+1} - x_k\| + \|a_{m_1+1}(f(x_{m_1+1}) + \epsilon_{m_1+2}\| \\
&\quad \le \|\sum_{i=k}^{m_1} a_i f(x_i)\| + \|\sum_{i=k}^{m_1} a_i \epsilon_{i+1}\| + a_{m_1+1} f_1 \\
&\qquad + a_{m_1+1} \|\sum_{i=1}^{m_1+1} \epsilon_{i+1} - \sum_{i=1}^{m_1} \epsilon_{i+1}\| \\
&\quad \le T f_1 + \frac{11}{4}\epsilon^* + \frac{11}{8} T\epsilon^* + \frac{\epsilon^*}{2} + \frac{11}{4}\epsilon^* = (f_1 + \frac{11}{8}\epsilon^*)T + 6\epsilon^* < 2 + 6\epsilon^*,
\end{aligned}$$

which contradicts (2.7.22). This proves the lemma. □

Lemma 2.7.2 *Under the conditions of Lemma 2.7.1, for any $T \in [0, \frac{1}{1+f_1})$ the following estimate holds:*

$$\|x_m - x_k\| < 3\epsilon^* + T(f_1 + 2\epsilon^*), \quad \forall m : k \leq m \leq m(k,T) + 1. \quad (2.7.27)$$

Proof. Since $\|x_k\| \leq M$, by Lemma 2.7.1 we have

$$\|x_k + \sum_{i=k}^{m} a_i y_{i+1}\| \leq M + 6\epsilon^* + 2 \leq M_0 \leq m_{\sigma_k},$$
$$\forall m : k \leq m \leq m(k,T) + 1,$$

and hence

$$x_{m+1} = x_m + a_m y_{m+1}, \quad \forall m : k \leq m \leq m(k,T) + 1,$$
$$\|x_m\| \leq M + 6\epsilon^* + 2, \quad \|f(x_m)\| \leq f_1.$$

Consequently, we have

$$\|x_m - x_k\| \leq \| \sum_{i=k}^{m-1} a_i (f(x_i) + \epsilon_{i+1})\| \leq f_1 T$$
$$+ \left\| a_{m-1} \sum_{i=1}^{m-1} \epsilon_{i+1} - a_k \sum_{i=1}^{k-1} \epsilon_{i+1} + \sum_{i=k}^{m-2} (a_i - a_{i+1}) \sum_{j=1}^{i} \epsilon_{j+1} \right\|$$
$$\leq f_1 T + \frac{11}{4}\epsilon^* + \frac{11}{8} T\epsilon^* < 3\epsilon^* + T(f_1 + 2\epsilon^*),$$
$$\forall m : k \leq m \leq m(k,T) + 1.$$

□

Lemma 2.7.3 *Assume A2.7.1–A2.7.4 hold and e^* satisfies (2.7.7). Then for the sample path for which A2.7.3 holds, a σ that is independent of $\epsilon \in [0, \epsilon^*]$ and $e \in [0, e^*]$ exists such that*

$$\sup_k \sigma_k \leq \sigma < \infty,$$

in other words, $\{x_k\}$ given by (2.1.1)–(2.1.3) is bounded.

Proof. Let r_0 be a sufficiently large integer such that

$$a_{r_0} < T^*, \quad r_0 > K, \quad (2.7.28)$$

where K is given by (2.7.18).

Assume the lemma is not true. Then there exist $\epsilon \in [0, \epsilon^*]$ and k such that $\sigma_k > r_1 \triangleq r_0 + 2$. Let k_0 be the maximal integer satisfying the following equality:

$$r_0 = \sum_{i=1}^{k_0-1} I_{[\|x_i + a_i y_{i+1}\| > M_{\sigma_i}]}. \tag{2.7.29}$$

Then by definition we have

$$\sigma_{k_0} = r_0, \quad \sigma_{k_0+1} = r_0 + 1, \quad x_{k_0+1} = x^*, \tag{2.7.30}$$

and by (2.7.28) and (2.7.29),

$$k_0 > r_0 > K.$$

We first show that under the converse assumption there must be an $m_0 \geq k_0 + 2$ such that

$$\|x_{m_0}\| > M. \tag{2.7.31}$$

Otherwise, $\|x_k\| \leq M$ for any $k \geq k_0 + 2$, and from (2.7.24) it follows that

$$\begin{aligned}
\|x_k + a_k y_{k+1}\| \leq & M + \|a_k f(x_k)\| + \|a_k \epsilon_{k+1}\| \\
\leq & M + a_k f_1 + \|a_k \sum_{i=1}^{k} \epsilon_{i+1} - a_k \sum_{i=1}^{k-1} \epsilon_{i+1}\| \\
\leq & M + \frac{\epsilon^*}{2} + \frac{11}{4}\epsilon^* \\
\leq & M + 4 < M_0 < M_{\sigma_k}, \quad \forall k \geq k_0 + 2.
\end{aligned} \tag{2.7.32}$$

This together with (2.7.30) implies

$$\sigma_k \equiv \sigma_{k_0+2} \leq 1 + \sigma_{k_0+1} = r_0 + 2 = r_1, \quad \forall k \geq k_0 + 2,$$

which contradicts with the converse assumption.

Hence (2.7.31) must be held.

By the definition of $d_0 = v(x^*)$, (2.7.6), and (2.7.30) we have

$$v(x_{k_0+1}) = v(x^*) < \delta_1. \tag{2.7.33}$$

Since $\|x_{m_0}\| > M$ by (2.7.31), from (2.7.4) and (2.7.6) it follows that

$$v(x_{m_0}) \geq d > \delta_2, \quad m_0 \geq k_0 + 2. \tag{2.7.34}$$

We now show $m_0 > k_0 + 2$. For this it suffices to prove $v(x_{k_0+2}) < \delta_2$ by noticing (2.7.34).

Since $\|x_{k_0+1}\| = \|x^*\| < M$, similar to (2.7.32) we have

$$\|x_{k_0+1} + a_{k_0+1}y_{k_0+2}\| \leq M + 4 < M_0 \leq M_{\sigma_k}, \quad \forall k \geq k_0 + 2, \tag{2.7.35}$$

and hence

$$x_{k_0+2} = x_{k_0+1} + a_{k_0+1}y_{k_0+2}, \quad \|x_{k_0+2}\| \leq M + 4. \tag{2.7.36}$$

From (2.7.32) and (2.7.36) it is seen that

$$\begin{aligned}\|x_{k_0+2} - x_{k_0+1}\| \leq & a_{k_0+1}\|f(x_{k_0+1})\| + a_{k_0+1}\|\epsilon_{k_0+2}\| \\ \leq & a_{k_0+1}f_1 + \|a_{k_0+1}\sum_{i=1}^{k_0+1}\epsilon_{i+1} - a_{k_0+1}\sum_{i=1}^{k_0}\epsilon_{i+1}\| \\ \leq & \frac{\epsilon^*}{2} + \frac{11}{4}\epsilon^* = \frac{13}{4}\epsilon^* < 4\epsilon^*,\end{aligned} \tag{2.7.37}$$

where for the second inequality, (2.7.9) and $\|x_{k_0+1}\| < M$ are used, while for the last inequality (2.7.18) is invoked.

Paying attention to (2.7.10), we have $x_{k_0+1} \in U$ and $x_{k_0+2} \in U$, and by (2.7.16)

$$v(x_{k_0+2}) < v(x_{k_0+1}) + \delta_2 - \delta_1.$$

Then by (2.7.32) we see $v(x_{k_0+2}) < \delta_2$, and (2.7.34) becomes

$$v(x_{m_0}) > \delta_2, \quad \forall m_0 > k_0 + 2. \tag{2.7.38}$$

Thus, we can define

$$\begin{aligned}k_1 &= \max\{i : v(x_{i-1}) < \delta_1, \quad k_0 + 2 \leq i < m_0\}, \\ k_2 &= \max\{i : v(x_{i-1}) > \delta_2, \quad k_0 + 2 < i \leq m_0\},\end{aligned}$$

and have

$$v(x_{k_1-1}) < \delta_1, \quad v(x_{k_2}) > \delta_2, \tag{2.7.39}$$

$$\delta_1 \leq v(x_i) \leq \delta_2, \quad \forall i : k_1 \leq i \leq k_2 - 1. \tag{2.7.40}$$

Taking $k = k_1$, $T = T^*$ in Lemmas 2.7.1 and 2.7.2, and paying attention to (2.7.4) and $v(x_{k_1}) \leq \delta_2 < d$, we know $\|x_{k_1}\| \leq M$. By Lemmas 2.7.1 and 2.7.2, from (2.7.28) we see $a_{r_0} < T^*$. From (2.7.28)–(2.7.30) we have obtained $k_0 > r_0 > K$, which together with the definition of k_1 implies $k_1 \geq k_0 + 2 > r_0$ and hence $a_{k_1} < T^*$. Therefore, $m(k, T^*)$ is well defined, and by the Taylor's expansion we have

$$\begin{aligned}v(x_{m(k_1,T^*)+1}) - v(x_{k_1}) = & (x_{m(k_1,T^*)+1} - x_{k_1})^T v_x(x_{k_1}) \\ & + \frac{1}{2}(x^T_{m(k_1,T^*)+1} - x^T_{k_1})v_{xx}(\eta)(x_{m(k_1,T^*)+1} - x_{k_1}),\end{aligned} \tag{2.7.41}$$

where $\eta \in \mathbb{R}^l$ with components located in-between x_{k_1} and $x_{m(k_1,T^*)+1}$.

We now show that $v(x_{m(k_1,T^*)+1}) < \delta_1$, which, as to be shown, implies a contradiction.

By Lemma 2.7.2 we have

$$\begin{aligned} \|x_{m(k_1,T^*)+1} - x_{k_1}\| &< 3\epsilon^* + T^*(f_1 + 2\epsilon^*), \\ \|\eta - x_{k_1}\| &\le 3\epsilon^* + T^*(f_1 + 2\epsilon^*), \end{aligned} \tag{2.7.42}$$

and hence

$$\begin{aligned} \|\eta\| &\le M + 3\epsilon^* + T^*(f_1 + 2\epsilon^*) \le M + 3\epsilon^* + \frac{f_1 + 2\epsilon^*}{f_1 + 1} \\ &\le M + 5\epsilon^* + 1 \le M + 6. \end{aligned}$$

By (2.7.10) it follows that $\eta \in U$, and $\|v_{xx}(\eta)\| \le r_2$ by (2.7.11). Using Lemma 2.7.1, we continue (2.7.41) as follows:

$$\begin{aligned} &v(x_{m(k_1,T^*)+1}) - v(x_{k_1}) \\ &\le \sum_{i=k_1}^{m(k_1,T^*)} a_i\big(f^T(x_i) + \epsilon_{i+1}^T\big)v_x(x_{k_1}) + \frac{r_2}{2}\big(3\epsilon^* + T^*(f_1 + 2\epsilon^*)\big)^2 \\ &\le \sum_{i=k_1}^{m(k_1,T^*)} a_i f^T(x_{k_1})v_x(x_{k_1}) + \sum_{i=k_1}^{m(k_1,T^*)} a_i\Big(f^T(x_i) - f^T(x_{k_1})\Big)v_x(x_{k_1}) \\ &\quad + \sum_{i=k_1}^{m(k_1,T^*)} a_i \epsilon_{i+1}^T v_x(x_{k_1}) + \frac{r_2}{2}\big(3\epsilon^* + T^*(f_1 + 2\epsilon^*)\big)^2. \end{aligned} \tag{2.7.43}$$

Noticing $v(x_{k_1-1}) < \delta_1 < \delta_2$, we see $\|x_{k_1-1}\| < M$, $x_{k_1-1} \in U$.

It is clear that (2.7.35) and (2.7.37) remain valid with $k_0 + 1$ replaced by $k_1 - 1$. Hence, similar to (2.7.37) we have

$$x_{k_1} \in U \text{ and } \|x_{k_1} - x_{k_1-1}\| \le \frac{13}{4}\epsilon^*.$$

By (2.7.11) and the Taylor's expansion we have

$$|v(x_{k_1}) - v(x_{k_1-1})| \le \|v_x(\theta)\| \cdot \|x_{k_1} - x_{k_1-1}\| \le \frac{13}{4} r_1 \epsilon^*,$$

and consequently,

$$v(x_{k_1}) \le v(x_{k_1-1}) + \frac{13}{4} r_1 \epsilon^* < \delta_1 + \frac{13}{4} r_1 \epsilon^*. \tag{2.7.44}$$

By (2.7.40), $x_{k_1} \in D$. Substituting (2.7.44) into (2.7.43) and using (2.7.12) lead to

$$v(x_{m(k_1,T^*)+1}) \leq \delta_1 + \frac{13}{4} r_1 \epsilon^* - aT^* + a_{m(k_1,T^*)+1} |f^T(x_{k_1}) v_x(x_{k_1})|$$
$$+ r_1 T^* \max_{k_1 \leq i \leq m(k_1,T^*)} \|h(x_{k_1}) - h(x_i)\| + r_1 \Big\| \sum_{i=k_1}^{m(k_1,T^*)} a_i \epsilon_{i+1} \Big\|$$
$$+ \frac{r_2}{2} [3\epsilon^* + T^*(f_1 + 2\epsilon^*)]^2.$$

Estimating $\sum_{i=k_1}^{m(k_1,T^*)} a_i \epsilon_{i+1}$ by the treatment similar to that used for (2.7.26) yields

$$v(x_{m(k_1,T^*)+1}) \leq \delta_1 + \frac{13}{4} r_1 \epsilon^* - aT^* + a_{k_1} f_1 r_1$$
$$+ r_1 T^* \max_{k_1 \leq i \leq m(k_1,T^*)} \|f(x_{k_1}) - f(x_i)\|$$
$$+ (\frac{11}{4}\epsilon^* + \frac{11}{8} T^* \epsilon^*) r_1 + \frac{r_2}{2} [3\epsilon^* + T^*(f_1 + 2\epsilon^*)]^2. \tag{2.7.45}$$

Noticing $\|x_{k_1}\| \leq M$, by Lemma 2.7.2 we find that

$$\|x_m - x_{k_1}\| < 3\epsilon^* + T^*(f_1 + 2\epsilon^*) \leq 3\epsilon^* + T^*(f_1 + 2) \tag{2.7.46}$$

and

$$\|x_m\| < M + 3\epsilon^* + T^*(f_1 + 2\epsilon^*) < M + 5\epsilon^* + 1 \leq M + 6,$$
$$\forall m : k_1 \leq m \leq m(k_1, T^*) + 1.$$

Hence, $x_m \in U$, and by (2.7.15) from (2.7.45) it follows that

$$v(x_{m(k_1,T^*)+1}) \leq \delta_1 + \frac{13}{4} r_1 \epsilon^* - aT^* + a_{k_1} f_1 r_1 + T^*(\alpha - \beta)$$
$$+ r_1 \epsilon^* (\frac{11}{4} + \frac{11}{8} T^*) + \frac{r_2}{2} [3\epsilon^* + T^*(f_1 + 2\epsilon^*)]^2.$$

Using (2.7.14), from the above estimate we have

$$v(x_{m(k_1,T^*)+1}) \leq \delta_1 + \frac{13}{4} r_1 \epsilon^* - T^*[\alpha + \frac{1}{2} r_2 (f_1 + 2)^2 T^*] + a_{k_1} f_1 r_1$$
$$+ T^*(\alpha - \beta) + r_1 \epsilon^* (\frac{11}{4} + \frac{11}{8} T^*) + \frac{9}{2} r_2 \epsilon^{*2} + 3 r_2 \epsilon^* T^* (f_1 + 2\epsilon^*)$$
$$+ \frac{r_2}{2} T^{*2} (f_1 + 2\epsilon^*)^2 \leq \delta_1 - \beta T^* + r_1 \epsilon^* (6 + \frac{11}{8} T^*) + a_{k_1} f_1 r_1$$
$$+ r_2 \epsilon^* (\frac{9}{2} \epsilon^* + 3T^*(f_1 + 2\epsilon^*)).$$

From (2.7.18) it follows that $a_{k_1} f_1 \leq \frac{\epsilon^*}{2}$. Taking notice of (2.7.13) by (2.7.17) we derive

$$v(x_{m(k_1,T^*)+1}) \leq \delta_1 - \beta T^* + 8r_1\epsilon^* + 14r_2\epsilon^* < \delta_1. \tag{2.7.47}$$

On the other hand, by Lemma 2.7.2 and (2.7.11), (2.7.17), and (2.7.44) it follows that

$$\begin{aligned} v(x_m) =& v(x_{k_1}) + v_x^T(\phi)(x_m - x_{k_1}) \\ \leq& \delta + \frac{13}{4} r_1\epsilon^* + r_1[3\epsilon^* + T^*(f_1 + 2\epsilon^*)] < \delta_2 \end{aligned} \tag{2.7.48}$$

where $\phi \in U$.

From (2.7.39), (2.7.40), and (2.7.48) we see that

$$m(k_1, T^*) + 1 < k_2,$$

and hence $v(x_{m(k_1,T^*)+1}) \geq \delta_1$, which contradicts with (2.7.47). This means that the converse assumption of the lemma cannot be held. □

Corollary 2.7.1 *From Lemma 2.7.3 it follows that there exist $\epsilon^* > 0$, $e^* > 0$, and σ, which is independent of e and ϵ arbitrarily varying in intervals $e \in [0, e^*]$ and $\epsilon \in [0, \epsilon^*]$ such that*

$$\|x_k\| \leq M_\sigma, \tag{2.7.49}$$

and for $k \geq K_1$ with K_1 sufficiently large the algorithm (2.1.1)–(2.1.3) turns to an ordinary RM algorithm:

$$x_{k+1} = x_k + a_k(f(x_k) + \epsilon_{k+1}). \tag{2.7.50}$$

Set

$$\max_{\|x\| \leq M_0 + 8} v(x) = L_0, \qquad \max_{\|x\| \leq M_\sigma} \|f(x)\| = L_3, \tag{2.7.51}$$

$$\max_{\|x\| \leq M_\sigma + 8} \|v_x(x)\| = L_1, \qquad \max_{\|x\| \leq M_\sigma + 8} \|v_{xx}(x)\| = L_2. \tag{2.7.52}$$

Take $\rho \in [e, e^*]$ and denote

$$\alpha(\rho) = \frac{1}{L_0} \min_{\substack{\|x - x^0\| \geq \rho \\ \|x\| \leq M_\sigma}} -f^T(x)v_x(x). \tag{2.7.53}$$

By A2.7.2, $\alpha(\rho) > 0$. Set

$$\beta(\rho) = \max_{\|x - x^0\| \leq \rho} f^T(x)v_x(x). \tag{2.7.54}$$

If $e = 0$ in (2.7.2), then $\beta(\rho) \le 0$. In the general case ($e \ge 0$), $\beta(\rho)$ may be positive.

Theorem 2.7.1 *Assume A2.7.1–A2.7.4 hold and $\{x_k\}$ is given by (2.1.1)–(2.1.3) with (2.7.5) held. Then there exist $\epsilon^* \in (0,1]$, $e^* \in (0,8)$ and a nondecreasing, left-continuous function $h(\cdot)$ defined on $[0,\infty)$ such that for the sample path for which A2.7.3 holds,*

$$\limsup_{k\longrightarrow\infty} \|x_k - x^0\| \le h\left(\frac{3}{\alpha(\rho)}(\beta(\rho)\wedge 0) + \frac{3L_4\epsilon}{\alpha(\rho)} + 3L_1\rho + 24L_1\epsilon\right) + 8\epsilon, \quad \forall \rho \in [e, e^*] \tag{2.7.55}$$

whenever $e \le e^*$ and $\epsilon \le \epsilon^*$, where e and ϵ are the ones appearing in (2.7.2) and (2.7.3), respectively. As a matter of fact, $h(t)$ can be taken as the inverse function of $m(r) \triangleq \min_{\|x-x^0\|\ge r} v(x)$: $h(t) = \min\{r : m(r) = t\}$.

Proof. Given u_0, recursively define

$$u_{k+1} = u_k + a_k(-u_k + \epsilon_{k+1}).$$

We now show that $K_2(\ge K_1)$ exists such that

$$\sup_{k\ge K_2} \|u_k\| \le 8\epsilon. \tag{2.7.56}$$

Set $s_i = \sum_{j=1}^{i} \epsilon_{j+1}$ and assume $\prod_{j=k+i}^{k} (1-a_j) = 1$, $\forall i \ge 1$.
From the recursion of $\{u_k\}$, we have

$$\begin{aligned}\|u_{k+1}\| &= \|\prod_{i=K_1}^{k}(1-a_i)u_{K_1} + \sum_{i=K_1}^{k}\prod_{j=i+1}^{k}(1-a_j)a_i\epsilon_{i+1} \\ &\le \|u_{K_1}\|\exp[-\sum_{i=K_1}^{k} a_i] + \|\sum_{i=K_1}^{k}\prod_{j=i+1}^{k}(1-a_j)a_i(s_i - s_{i-1})\|.\end{aligned} \tag{2.7.57}$$

Assume K_1 is large enough such that by A2.7.3

$$\|a_k s_k\| \le 2\epsilon, \quad \forall k \ge K_1 - 1. \tag{2.7.58}$$

By a partial summation, from (2.7.57) we find that

$$\begin{aligned}
\|u_{k+1}\| \leq & o(1) + \|a_k s_k - \prod_{j=K_1+1}^{k} (1-a_j) a_{K_1} s_{K_1-1} \\
& + \sum_{i=K_1-1}^{k-1} \prod_{j=i+1}^{k} (1-a_j) a_i s_i - \sum_{i=K_1-1}^{k-1} \prod_{j=i+2}^{k} (1-a_j) a_{i+1} s_i\| \\
\leq & o(1) + 4\epsilon + \sum_{i=K_1-1}^{k-1} \prod_{j=i+2}^{k} (1-a_j) \|[(1-a_{i+1})a_i - a_{i+1}] s_i\| \\
\leq & o(1) + 4\epsilon + 2\epsilon \sum_{i=K_1-1}^{k-1} |1 - a_{i+1} - \frac{a_{i+1}}{a_i}| \exp(-\sum_{j=i+2}^{k} a_j),
\end{aligned} \tag{2.7.59}$$

where (2.7.58) is invoked.

By (2.7.1) we see

$$-a_{k+1} \leq 1 - \frac{a_{k+1}}{a_k} - a_{k+1} \leq 0. \tag{2.7.60}$$

Without loss of generality, we may assume $a_k < \frac{1}{2}$, $\forall k \geq K_1$. Then by (2.7.1) we have

$$\frac{a_k}{a_{k+1}} \leq 1 + a_k < \frac{3}{2}. \tag{2.7.61}$$

Applying (2.7.60) and (2.7.61) to (2.7.59) leads to

$$\begin{aligned}
\|u_{k+1}\| \leq & o(1) + 4\epsilon + 2\epsilon \exp(-\sum_{j=1}^{k} a_j) \sum_{i=K_1-1}^{k-1} a_{i+1} \exp(\sum_{j=1}^{i+1} a_j) \\
\leq & o(1) + 4\epsilon + 3\epsilon \exp(-\sum_{j=1}^{k} a_j) \sum_{i=K_1-1}^{k-1} a_{i+2} \exp(\sum_{j=1}^{i+1} a_j) \\
\leq & o(1) + 4\epsilon + 3\epsilon \exp(-\sum_{j=1}^{k} a_j) \sum_{i=K_1-1}^{k-1} (e^{a_{i+2}} - 1) \exp(\sum_{j=1}^{i+1} a_j) \\
= & o(1) + 4\epsilon + 3\epsilon \exp(-\sum_{j=1}^{k} a_j)(\exp(\sum_{j=1}^{k+1} a_j) - \exp(\sum_{j=1}^{K_1} a_j)) \\
\leq & o(1) + 4\epsilon + 3\epsilon e^{a_{k+1}},
\end{aligned} \tag{2.7.62}$$

and hence

$$\limsup_{k \longrightarrow \infty} \|u_k\| \leq 7\epsilon,$$

which implies (2.7.56).

For $k \geq K_2$, $\|x_k - u_k\| \leq M_\sigma + 8$, and by (2.7.53)

$$\begin{aligned} f^T(x_k)v_x(x_k)I_{[\|x_k - x^0\| \geq \rho]} &\leq -\alpha(\rho)L_0 I_{[\|x_k - x^0\| \geq \rho]} \\ &\leq -\alpha(\rho)v(x_k - u_k)I_{[\|x_k - x^0\| \geq \rho]}. \end{aligned}$$

Taking this into account for $k \geq K_2$, by (2.7.51)–(2.7.54) and the Taylor's expansion we have

$$\begin{aligned} & v(x_{k+1} - u_{k+1}) \\ & = v(x_k - u_k) + a_k(f^T(x_k) + u_k^T)v_x(x_k - u_k) \\ & \quad + \frac{1}{2}a_k^2(f^T(x_k) + u_k^T)v_{xx}(\zeta_k)(f(x_k) + u_k) \\ & \leq v(x_k - u_k) + a_k f^T(x_k)(v_x(x_k) - v_{xx}(\eta_k)u_k) + a_k 8L_1\epsilon \\ & \quad + L_2(L_3^2 + 64\epsilon^2)a_k^2 \\ & \leq (1 - \alpha(\rho)a_k)v(x_k - u_k) + a_k(0 \vee f^T(x_k)v_x(x_k))I_{[\|x_k - x^0\| < \rho]} \\ & \quad + \alpha(\rho)a_k v(x_k - u_k)I_{[\|x_k - x^0\| < \rho]} + 8\epsilon a_k L_2 L_3 \\ & \quad + a_k 8L_1\epsilon + L_2(L_3^2 + 64\epsilon^2)a_k^2. \end{aligned} \tag{2.7.63}$$

For $\|x_k - x^0\| < \rho$ we have

$$\|x^0\| \leq \|x^0 - x_k + x_k\| \leq M_\sigma + \rho.$$

Therefore, in the following Taylor's expansion

$$v(x_k - u_k)I_{[\|x_k - x^0\| < \rho]} = [v(x^0) + v_x(\eta)(x_k - u_k - x^0)]I_{[\|x_k - x^0\| < \rho]}$$

we have $\|\eta\| \leq \min(\|x^0\|, \|x_k - u_k\|) \leq M_\sigma + 8$ and hence $\|v_x(\eta)\| \leq L_1$, and

$$v(x_k - u_k)I_{[\|x_k - x^0\| < \rho]} \leq L_1(\rho + 8\epsilon). \tag{2.7.64}$$

Denote

$$L_4 = 8(L_2 L_3 + L_1), \quad L_5 = L_2(L_3^2 + 64\epsilon^2).$$

From (2.7.63) and (2.7.64) it then follows that

$$\begin{aligned}
&v(x_{k+1}-u_{k+1})\\
&\le (1-\alpha(\rho)a_k)v(x_k-u_k)+a_k(0\vee\beta(\rho))+L_1(\rho+8\epsilon)\alpha(\rho)a_k\\
&\quad +L_4\epsilon a_k+L_5a_k^2\\
&\le v(x_{K_2}-u_{K_2})\exp\Big[-\sum_{i=K_2}^{k}\alpha(\rho)a_i\Big]+\sum_{i=K_2}^{k}\prod_{j=i+1}^{k}(1-\alpha(\rho)a_j)\\
&\quad\cdot\Big[a_i((0\vee\beta(\rho))+L_1(\rho+8\epsilon)\alpha(\rho)+L_4\epsilon+L_5a_i\Big].
\end{aligned}$$

Similar to (2.7.62), we see that

$$\begin{aligned}
&\sum_{i=K_2}^{k}\prod_{j=i+1}^{k}(1-\alpha(\rho)a_j)a_i\\
&\le\frac{3}{2}\sum_{i=K_2}^{k}a_{i+1}\exp\Big[-\sum_{j=i+1}^{k}\alpha(\rho)a_j\Big]\\
&\le\frac{3}{2\alpha(\rho)}\exp\Big[-\sum_{j=1}^{k}\alpha(\rho)a_j\Big]\sum_{i=K_2}^{k}(e^{\alpha(\rho)a_{i+1}}-1)\exp\Big[\sum_{j=1}^{i}\alpha(\rho)a_j\Big]\\
&\le\frac{3}{2\alpha(\rho)}e^{\alpha(\rho)a_{k+1}}.
\end{aligned}$$

Consequently, we arrive at

$$\begin{aligned}
v(x_{k+1}-u_{k+1})\le& v(x_{K_2}-u_{K_2})\exp[-\sum_{i=K_2}^{k}\alpha(\rho)a_i]\\
&+\frac{3}{2\alpha(\rho)}e^{\alpha(\rho)a_{k+1}}((0\vee\beta(\rho))+L_4\epsilon+L_5a_{K_2})\\
&+\frac{3}{2}e^{\alpha(\rho)a_{k+1}}L_1(\rho+8\epsilon)\triangleq t_k.
\end{aligned}\tag{2.7.65}$$

Define

$$m(r)=\min_{\|x-x^0\|\ge r}v(x),\quad r\ge 0.\tag{2.7.66}$$

It is clear that $m(r)$ is nondecreasing as r increases and $m(r)\xrightarrow[r\to 0]{}m(0)=0$.

Take r_k such that $m(r_k)=2t_k$. Then we have

$$v(x_{k+1}-u_{k+1})\le\frac{m(r_k)}{2}<m(r_k).\tag{2.7.67}$$

Define function $h(\cdot)$

$$h(t) = \min\{r : m(r) = t\}.$$

It is clear that $f(\cdot)$ is left-continuous, nondecreasing and $f(t) \xrightarrow[t \longrightarrow 0^+]{} 0$.

From (2.7.66) and (2.7.67) it follows that

$$\|x_{k+1} - u_{k+1} - x^0\| < r_k = h(2t_k),$$

which implies, by (2.7.57) and the definition of t_k,

$$\limsup_{k \longrightarrow \infty} \|x_k - x^0\| < h\Big(\frac{3}{\alpha(\rho)}((0 \vee \beta(\rho)) + L_4\epsilon) + 3L_1(\rho + 8\epsilon)\Big) + 8\epsilon.$$

□

Corollary 2.7.2 *If $e = 0$ in (2.7.2) ($f(x^0)$ may not be zero), then $\beta(\rho) \leq 0$ and the right-hand side of (2.7.55) will be*

$$h(\frac{3L_4\epsilon}{\alpha(\rho)} + 3L_1\rho + 24L_1\epsilon) + 8\epsilon \xrightarrow[e \longrightarrow 0]{} h(3L_1\rho).$$

Since $e = 0$, ρ may be arbitrarily small and hence the estimation error $\|x_k - x^0\|$ may be arbitrarily small. If, in addition, $\epsilon = 0$ in A2.7.3, then tending $\epsilon \longrightarrow 0$ and then $\rho \longrightarrow 0$ in both sides of (2.7.55) we derive

$$\limsup_{k \longrightarrow \infty} \|x_k - x^0\| = 0.$$

In the case where $e > 0$, by tending $\epsilon \longrightarrow 0$, the right-hand side of (2.7.55) converges to

$$h(\frac{3}{\alpha(\rho)}(\beta(\rho) \vee 0) + 3L_1\rho).$$

Consequently, as $\epsilon \longrightarrow 0$ the estimation error depends on how big is $\frac{3}{\alpha(\rho)}(\beta(\rho) \vee 0) + 3L_1\rho$. If $e \longrightarrow 0$ in (2.7.2), then ρ can also be taken arbitrarily small and the estimation error depends on the magnitude of $\frac{3}{\alpha(\rho)}(\beta(\rho) \vee 0) + 3L_1\rho)$.

2.8. Dynamic Stochastic Approximation

So far we have discussed the root-searching problem for an unknown function, which is unchanged during the process of estimation. We now consider the case where the unknown functions together with their roots change with time. To be precise, Let $\{f_k(\cdot)\}$ be a sequence of unknown functions $f_k(\cdot) : \mathbb{R}^l \longrightarrow \mathbb{R}^l$ with roots θ_k, i.e., $f_k(\theta_k) = 0$, $k = 1, 2, \ldots$. Let x_k be the estimate for θ_k at time k based on the observations $\{y_j, j \le k\}$.

Assume the evolution of the roots θ_k satisfies the following equation

$$\theta_{k+1} = g_k(\theta_k) + w_{k+1}, \tag{2.8.1}$$

where $g_k(\cdot) : \mathbb{R}^l \longrightarrow \mathbb{R}^l$ are known functions, while $\{w_k\}$ is a sequence of dynamic noises.

The observations are given by

$$y_{k+1} = f_{k+1}(g_k(x_k)) + \epsilon_{k+1}, \tag{2.8.2}$$

where $\{\epsilon_k\}$ is the observation noise and ϵ_{k+1} is allowed to depend on $(x_k - \theta_k)$.

In what follows the discussion is for a fixed sample, and the analysis is purely deterministic. Let us arbitrarily take x_1 as the estimate for θ_1 and define

$$h_1(x) = g_1(x), \quad h_k(x) = g_k(h_{k-1}(x)), \qquad \text{for } k = 2, 3, \ldots. \tag{2.8.3}$$

From equation (2.8.1), we see that $h_k(x_1)$ may serve as a rough estimate for θ_{k+1}. In the sequel, we will impose some conditions on $\{g_k(\cdot)\}$ and $\{\epsilon_k\}$ so that

$$\|h_k(x_1) - \theta_{k+1}\| < \eta < \infty, \quad \forall k = 1, 2, \ldots, \tag{2.8.4}$$

where η is an unknown constant. Therefore, $x_{k+1} - h_k(x_1)$ should not diverge to infinity. But η is unknown, so we will use the expanding truncation technique.

Take a sequence of increasing numbers $\{M_i\}$ satisfying

$$M_i > 0, \quad M_{i+1} > M_i, \text{ and } \quad \lim_{i \longrightarrow \infty} M_i = \infty.$$

Let $\{x_k\}$ be recursively defined by the following algorithm:

$$\begin{aligned} x_{k+1} =& (g_k(x_k) + a_k y_{k+1}) I_{[\|g_k(x_k) + a_k y_{k+1} - h_k(x_1)\| \le M_{\sigma_k}]} \\ &+ h_k(x_1) I_{[\|g_k(x_k) + a_k y_{k+1} - h_k(x_1)\| > M_{\sigma_k}]}, \end{aligned} \tag{2.8.5}$$

$$\sigma_k = \sum_{i=1}^{k-1} I_{[\|g_i(x_i) + a_i y_{i+1} - h_i(x_1)\| > M_{\sigma_i}]}, \tag{2.8.6}$$

where σ_k denotes the number of truncations in (2.8.5) occurred until time k.

We list conditions to be used.

A2.8.1 $a_k > 0$, $a_k \xrightarrow[k\to\infty]{} 0$, *and* $\sum_{k=1}^{\infty} a_k = \infty$.

A2.8.2 $f_k(\cdot) : \mathbb{R}^l \longrightarrow \mathbb{R}^l$, $f_k(\theta_k) = 0$, *$f_k(\cdot)$ is measurable and for any $c > 0$ a constant $\alpha(c)$, possibly depending on c, exists so that* $\|f_k(\theta_k + \xi)\| < \alpha(c)$ *for* $\forall \xi$ *with* $\|\xi\| \le c$, $\forall k = 1, 2, \ldots$.

A2.8.3 $g_k(\cdot) : \mathbb{R}^l \longrightarrow \mathbb{R}^l$ *is known such that* $\|d_k(x)\| \le \gamma_k \|x - \theta_k\|$, $\forall x$ *for* $k = 1, 2, \ldots$, *where* $d_k(x) \triangleq g_k(x) - g_k(\theta_k) - (x - \theta_k)$, $\gamma_k = o(a_k)$, *and* $\sum_{k=1}^{\infty} \gamma_k < \infty$.

A2.8.4 $\|w_k\| = o(a_k)$, *and* $\sum_{k=1}^{\infty} \|w_k\| < \infty$.

A2.8.5 *There is a continuously differentiable function* $v(\cdot) : \mathbb{R}^l \longrightarrow R$ *such that* $v(x) \ne 0$ *for* $\forall x \ne 0$, $v(0) = 0$ *and for any* $0 < r_1 < r_2 < \infty$

$$\sup_k \sup_{r_1 \le \|x - \theta_k\| \le r_2} f_k^T(x) v_x(x - \theta_k) < -a,$$

where a is a positive constant possibly depending on r_1 and r_2. A constant $r > \eta$ exists such that

$$\sup_{\|y\| \le \eta} v(y) < \sup_{\|x\| = r} v(x), \tag{2.8.7}$$

where η is an unknown constant that is an upper bound for $\|h_k(x_1) - \theta_{k+1}\|$, $k = 1, 2, \ldots$.

A2.8.6 *For any convergent subsequence* $\{x_{k_i} - \theta_{k_i}\}$ *the observation noise satisfies*

$$\lim_{T\longrightarrow 0} \limsup_{i\longrightarrow\infty} \frac{1}{T} \|\sum_{j=k_i}^{m(k_i,t)} a_j \epsilon_{j+1}\| = 0, \quad \forall t \in [0, T]$$

where $m(k, T) = \max\{m : \sum_{i=k}^{m} a_i \le T\}$.

Remark 2.8.1 Condition A2.8.2 implies the local boundedness, but the upper bound should be uniform with respect to k. In A2.8.3, $d_k(x)$ measures the difference between the estimation error $(x - \theta_k)$ and the prediction error $g_k(x) - g_k(\theta_k)$. In general, $\|g_k(x) - g_k(\theta_k)\|$ is greater than $\|x - \theta_k\|$. For example, $g_k(x) = c + x$, $\epsilon_k \equiv 0$, then A2.8.3 holds with $\gamma_k \equiv 0$, $\theta_{k+1} = c + \theta_k$. A2.8.4 means that in the root dynamics, the noise should be vanishing.

As A2.2.3, Condition A2.8.6 is about existence of a Lyapunov function. To impose such kind a condition is unavoidable in convergence analysis of SA algorithms. Inequality (2.8.7) is an easy condition. For example, if $v(x) \longrightarrow \infty$ as $\|x\| \longrightarrow \infty$, then this condition is automatically satisfied. The noise condition A2.8.6 is similar to A2.2.3.

Before analyzing convergence property of the algorithm (2.8.5), (2.8.6), and (2.8.2) we give an example of application of dynamic stochastic approximation.

Example 2.8.1 Assume that the chemical product is produced in a batch mode, and the product quality or quantity of the kth batch depends on the temperature in batch. When the temperature equals the ideal one, then the product is optimized. Let $f_k(\cdot)$ denote the deviation of the temperature from its optimal value for the kth batch, where x denotes the control parameter, which may be, for example, the pressure in batch, the quantity of catalytic promoter, the raw material proportion and others. The deviation reduces to zero if the control x equals its optimal value θ_k, i.e., $f_k(\theta_k) = 0$. Because of the environment change, the optimal parameter θ_k may change from batch to batch. Assume

$$\theta_{k+1} = g_k(\theta_k) + w_k,$$

where $g_k(\cdot)$ is known and w_k is the noise.

Let x_k be the estimate for θ_k. Then $g_k(x_k)$ may serve as a prediction for θ_{k+1}. Apply $g_k(x_k)$ as the control parameter for the $(k+1)$th batch. Assume that the temperature deviation of $f_{k+1}(g_k(x_k))$ for the $(k+1)$th batch can be observed, but the observation y_{k+1} may be corrupted by noise, i.e., $y_{k+1} = f_{k+1}(g_k(x_k)) + \epsilon_{k+1}$, where ϵ_{k+1} is the observation noise.

Then we can apply algorithm (2.8.5), (2.8.6), and (2.8.2) to estimate θ_k. Under conditions A2.8.1–A2.8.6, by Theorem 2.8.1 to be proved in this section, the estimate x_k is consistent, i.e., $\|x_k - \theta_k\| \xrightarrow[k\longrightarrow\infty]{} 0$.

Theorem 2.8.1 *Under Conditions A2.8.1–A2.8.6 the estimation error $\Delta_k = x_k - \theta_k$ tends to zero as $k \longrightarrow \infty$, where $\{x_k\}$ is given by (2.8.5), (2.8.6), and (2.8.2).*

To prove the theorem we start with lemmas.

Lemma 2.8.1 *Under A2.8.3 and 2.8.4, the sequence* $\{h_k(x_1) - \theta_{k+1}\}$ *is bounded for any* x_1.

Proof. By A2.8.3 and A2.8.4 from (2.8.1) it follows that

$$\begin{aligned}\|h_k(x_1) - \theta_{k+1}\| &= \|h_k(x_1) - g_k(\theta_k) - w_k\| \\ &= \|g_k(h_{k-1}(x_1)) - g_k(\theta_k) - w_k\| \\ &= \|d_k(h_{k-1}(x_1)) - h_{k-1}(x_1) - \theta_k - w_k\| \\ &\le (1+\gamma_k)\|h_{k-1}(x_1) - \theta_k\| + \|w_k\| \\ &\le \prod_{i=1}^{k}(1+\gamma_i)\|g_1(x_1) - \theta_1\| + \sum_{i=1}^{k}\prod_{j=i+1}^{k}(1+\gamma_j)\|\epsilon_i\| \\ &\le \prod_{i=1}^{\infty}(1+\gamma_i)\|g_1(x_1) - \theta_1\| + \sum_{i=1}^{\infty}\prod_{j=1}^{\infty}(1+\gamma_j)\|w_i\| \\ &\triangleq \eta < \infty, \quad \forall k, \end{aligned} \tag{2.8.8}$$

where $\prod_{j=1}^{\infty}(1+\gamma_j) < \infty$ is implied by $\sum_{j=1}^{\infty}\gamma_j < \infty$. □

Lemma 2.8.2 *Assume A2.8.1–A2.8.4 and A2.8.6 hold. Let* $\{\Delta_{k_i}\}$ *be a convergent subsequence such that* $\Delta_{k_i} \longrightarrow \overline{\Delta}$ *as* $i \longrightarrow \infty$. *Then, there are a sufficiently small* $T > 0$ *and a sufficiently large integer* i_0 *such that for* $i \ge i_0$

$$\begin{aligned} x_{m+1} &= g_m(x_m) + a_m y_{m+1} \\ \|\Delta_m - \Delta_{k_i}\| &\le ct \end{aligned} \tag{2.8.9}$$

for $\forall m : k_i \le m \le m(k_i, t)$, $\forall t \in [0, T]$, *where* c *is a constant independent of* i.

Proof. In the case $\sigma_k \longrightarrow \sigma < \infty$ as $k \longrightarrow \infty$, $\{g_k(x_k) + a_k y_{k+1} - h_k(x_1)\}$ is bounded, and hence $\{x_{k+1} - h_k(x_1)\}$ is bounded. By Lemma 2.8.1, $\{h_k(x_1) - \theta_{k+1}\}$ is bounded. Therefore, $\{\Delta_k\}$ is bounded. For large i, $k_i > \sigma$, and

$$x_{k+1} = g_k(x_k) + a_k y_{k+1}, \quad \forall k \ge k_i. \tag{2.8.10}$$

The following expression (2.8.11) and estimate (2.8.12) will frequently be used. By (2.8.1) and A2.8.3 we have

$$\begin{aligned} &g_k(x_k) + a_k f_{k+1}(g_k(x_k)) \\ &\quad = g_k(x_k) - g_k(\theta_k) - w_k + \theta_{k+1} \end{aligned}$$

$$
\begin{aligned}
&+ a_k f_{k+1}(g_k(x_k) - g_k(\theta_k) - w_k + \theta_{k+1}) \\
=&d_k(x_k) + \Delta_k - w_k + \theta_{k+1} \\
&+ a_k f_{k+1}(\theta_{k+1} + d_k(x_k) + \Delta_k - w_k)
\end{aligned} \tag{2.8.11}
$$

and

$$
\begin{aligned}
\|g_k(x_k) + a_k y_{k+1} - \theta_{k+1}\| \le & (a + \gamma_k)\|\Delta_k\| + \|w_k\| \\
& + a_k \alpha((a+\gamma_k)\|\Delta_k\| + \|w_k\|) + \|a_k \epsilon_{k+1}\|.
\end{aligned} \tag{2.8.12}
$$

Substitution of (2.8.12) into (2.8.10) leads to

$$
\begin{aligned}
\|\Delta_{m+1} - \Delta_{k_i}\| \le & \sum_{j=k_i}^{m} \gamma_j \|\Delta_j\| + \sum_{j=k_i}^{m} \|w_j\| \\
& + \sum_{j=k_i}^{m} a_j \alpha\big((a+\gamma_j)\|\Delta_j\| + \|w_j\|\big) + \|\sum_{j=k_i}^{m} a_j \epsilon_{j+1}\|.
\end{aligned}
$$

By boundedness of $\{\Delta_j\}$ and A2.8.3, $\sum_{j=k_i}^{m} \gamma_j \|\Delta_j\| = \sum_{j=k_i}^{m} o(a_j) = \frac{ct}{4}$ for some $c > 0$. By A2.8.4, $\sum_{j=k_i}^{m} \|w_j\| = \frac{ct}{4}$, while the last term is also less than $\frac{ct}{4}$ by A2.8.6.

Without loss of generality, we may assume $\alpha((1+r_j)\|\Delta_j\|+\|w_j\|) < \frac{c}{4}$. Therefore, $\|\Delta_{m+1} - \Delta_{k_i}\| \le ct$, and the lemma is true for the case $\sigma_i \longrightarrow \sigma < \infty$ as $i \longrightarrow \infty$.

We now consider the case $\sigma_i \longrightarrow \infty$ as $i \longrightarrow \infty$. Let i_0 be so large that for $i \ge i_0$

$$
\begin{aligned}
&\|\Delta_{k_i}\| < \|\overline{\Delta}\| + 1, \\
&\|\sum_{j=k_i}^{m(k_i,t)} a_j \epsilon_{j+1}\| \le c_1 T, \quad \forall t \in [0, T]
\end{aligned} \tag{2.8.13}
$$

with c_1 being a constant, and

$$
\gamma_j (1 + \|\overline{\Delta}\|) < a_j, \quad \|w_j\| < a_j, \quad \forall \ge k_{i_0}, \tag{2.8.14}
$$

$$
M_{\sigma_{k_{i_0}}} > 2\|\overline{\Delta}\| + 5 + 2c_1 T + \eta + \alpha(2\|\overline{\Delta}\| + 5), \tag{2.8.15}
$$

where η is given by (2.8.8).

Without loss of generality we may assume

$$a_j < 1, \quad \forall j > k_{i_0}. \tag{2.8.16}$$

Define $c = 2 + c_1 + \alpha(2\|\overline{\Delta}\| + 5)$, and take T so small that $Tc < 1$. We prove the lemma by induction.

By (2.8.8) and (2.8.12), we have

$$\begin{aligned} &\|g_{k_{i_0}}(x_{k_{i_0}} + a_{k_{i_0}}(f_{k_{i_0}+1}(g_{k_{i_0}}(x_{k_{i_0}})) + \epsilon_{k_{i_0}+1}) - h_{k_{i_0}}(x_1)\| \\ &\quad \le 2\|\Delta_{k_{i_0}}\| + 1 + \eta + \alpha(2\|\Delta_{k_{i_0}}\| + 1) + c_1 T \le M_{\sigma_{k_{i_0}}}. \end{aligned} \tag{2.8.17}$$

Therefore, at time $k_{i_0} + 1$, there is no truncation. Then by (2.8.11) and (2.8.12) we have

$$\begin{aligned} \|\Delta_{k_{i_0}+1} - \Delta_{k_{i_0}}\| &\le \gamma_{k_{i_0}}\|\Delta_{k_{i_0}}\| + a_{k_{i_0}} + a_{k_{i_0}}\alpha(2\|\Delta_{k_{i_0}}\| + 1) + c_1 t \\ &\le \gamma_{k_{i_0}}(\|\overline{\Delta}\| + 2) + a_{k_{i_0}} + a_{k_{i_0}}\alpha(2\|\overline{\Delta}\| + 2) + 1) + c_1 t \\ &\le 2a_{k_{i_0}} + a_{k_{i_0}}\alpha(2\|\overline{\Delta}\| + 5) + c_1 t < ct, \end{aligned} \tag{2.8.18}$$

where (2.8.14) and (2.8.15) have been used.

Let the conclusions of the lemma hold for $m = k_i, \dots, k < m(k_i, t)$. We prove that it also holds for $m = k + 1$. Again by (2.8.12), we have

$$\begin{aligned} \|g_k(x_k) + a_k y_{k+1} - h_k(x_1)\| \le& 2(\|\overline{\Delta}\| + 2) + 1 + \eta + \alpha(2\|\overline{\Delta}\| + 5) + 2c_1 t \\ <& M_{\sigma_{k_{i_0}}} < M_{\sigma_k} \text{ for } k > k_{i_0}. \end{aligned} \tag{2.8.19}$$

Hence there is no truncation at time $k + 1$. By the inductive assumption, (2.8.11) and (2.8.12), it follows that

$$\begin{aligned} &\|\Delta_{k+1} - \Delta_{k_i}\| \\ &\le \sum_{j=k_i}^{k} (\gamma_j\|\Delta_j\| + \|w_j\| + a_j\alpha(1 + \gamma_j)\|\Delta_j\| + \|w_j\|) + \|\sum_{j=k_i}^{k} a_j\epsilon_{j+1}\| \\ &\le 2\sum_{j=k_i}^{k} a_j + \sum_{j=k_i}^{k} a_j\alpha(2\|\overline{\Delta} + 5) + c_1 t < ct, \end{aligned}$$

where (2.8.13) and (2.8.14) are invoked.

Therefore, the conclusions of the lemma are also true for $k + 1$. This completes the proof. □

Lemma 2.8.3 *Assume A2.8.1–A2.8.6 hold. Then the number of truncations in (2.8.5) is finite and $\{\Delta_i\}$ is bounded.*

Proof. Using the argument in the proof of Lemma 2.8.2, the boundedness of $\{\Delta_k\}$ follows from the boundedness of the number of truncations. Hence, it suffices to show that $\sigma_k \longrightarrow \sigma < \infty$ as $k \longrightarrow \infty$.

Assume the converse: $\sigma_k \longrightarrow \infty$ as $k \longrightarrow \infty$. This means that the sequence $\{g_k(x_k) + a_k y_{k+1} - h_k(x_1)\}$ is unbounded. Let $\{k_i + 1\}$ be the sequence of truncation times. We prove that $\{\Delta_k\}$ is also unbounded if $\sigma_k \longrightarrow \infty$.

Assume $\{\Delta_k\}$ is bounded. Then $\{\Delta_{k_i}\}$ is also bounded. From $\{\Delta_{k_i}\}$ we select a convergent subsequence, denoted by the same $\{\Delta_{k_i}\}$ for notational simplicity, such that $\Delta_{k_i} \longrightarrow \overline{\Delta}$. By assumption, truncation happens at the next time $k_i + 1$. The obtained contradiction shows the unboundedness of $\{\Delta_k\}$ in the case $\sigma_k \longrightarrow \infty$.

Since $\sigma_k \longrightarrow \infty$, algorithm (2.8.5) returns back to $h_k(x_1)$ for infinitely many times. Let $x_{k_i} = h_{k_i-1}(x_1)$, $i = 1, 2, \ldots$. Then $\Delta_{k_i} = h_{k_i-1}(x_1) - \theta_{k_i}$. By Lemma 2.8.1, $\{\Delta_{k_i}\}$ is bounded and by (2.8.8), $\|\Delta_{k_i}\| \leq \eta$.

Because $\{\Delta_k\}$ is unbounded, starting from k_i, $\{\Delta_k\}$ will exit the ball with radius r, where r is given by (2.8.7). Therefore, there is an interval $[\delta_1, \delta_2] \subset (\sup_{\|y\| \leq \eta} v(y), \inf_{\|x\|=r} v(x))$, and for any i, there is a sequence, Δ_{m_i}, Δ_{m_i+1}, $\ldots$, Δ_{l_i} such that $k_i \leq m_i$, $v(\Delta_{m_i}) \leq \delta_1$, $\delta_1 < v(\Delta_j) < \delta_2$ for $\forall j : m_i < j < l_i$, and $v(\Delta_{l_i}) \geq \delta_2$. In other words, the values of $v(\cdot)$ at the sequence $\{\Delta_{m_i}, \Delta_{m_i+1}, \ldots, \Delta_{l_i}\}$ cross the interval $[\delta_1, \delta_2]$ from the left. It is clear that $\|\Delta_{m_i}\| < r$, $\forall i = 1, 2, \ldots$. Select from $\{\Delta_{m_i}\}$ a convergent subsequence denoted still by $\{\Delta_{m_i}\}$ such that $\Delta_{m_i} \longrightarrow \overline{\Delta}$ as $i \longrightarrow \infty$. It is clear that $\eta \leq \|\overline{\Delta}\| \leq r$.

From now on, assume i is large enough and T is small enough so that Lemma 2.8.2 is applicable and it is valid with k_i replaced by m_i.

Since Δ_{m_i} converges, by A2.8.5 and (2.8.12) it follows that $\Delta_{m_i+1} - \Delta_{m_i} \longrightarrow 0$ as $i \longrightarrow \infty$. Hence we have

$$\lim_{i \longrightarrow \infty} v(\Delta_{m_i}) = v(\overline{\Delta}) = \delta_1. \tag{2.8.20}$$

By Lemma 2.8.2, $\|\Delta_j - \Delta_{m_i}\| \leq cT$ for $\forall j : m_i \leq j \leq m(m_i, T)$. Noticing $\eta \leq \|\overline{\Delta}\|$, for small T we then have

$$\|\Delta_j\| > \frac{\eta}{2}, \quad \forall j : m_i \leq j \leq m(m_i, T). \tag{2.8.21}$$

In the following Taylor's expansion $\widetilde{\Delta} \in \mathbb{R}^l$ is located in-between Δ_{m_i} and $\Delta_{m(m_i,T)}$, and by Lemma 2.8.2, $\|\widetilde{\Delta}\| \leq cT + \|\overline{\Delta}\| + 1$. By (2.8.9) and (2.8.11) we have

$$\begin{aligned}
&v(\Delta_{m(m_i,T)}) - v(\Delta_{m_i}) \\
&= v_x^T(\widetilde{\Delta}) \sum_{j=m_i}^{m(m_i,T)-1} [d_j(x_j) - w_j + a_j(f_{j+1}(g_j(x_j)) + \epsilon_{j+1})] \\
&= v_x^T(\widetilde{\Delta})\Big[\sum_{j=m_i}^{m(m_i,T)-1} (d_j(x_j) - w_j) + \sum_{j=m_i}^{m(m_i,T)-1} a_j \epsilon_{j+1}\Big] \\
&\quad + \sum_{j=m_i}^{m(m_i,T)-1} a_j (v_x^T(\widetilde{\Delta}) - v_x^T(g_j(x_j) - \theta_{j+1})) f_{j+1}(g_j(x_j)) \\
&\quad + \sum_{j=m_i}^{m(m_i,T)-1} a_j v_x^T(g_j(x_j) - \theta_{j+1}) f_{j+1}(g_j(x_j)). \qquad (2.8.22)
\end{aligned}$$

Notice that by Lemma 2.8.2 and (2.8.13)

$$\begin{aligned}
\|a_j y_{j+1}\| &\le a_j \alpha(2\|\overline{\Delta}\| + 5) + \|\sum_{k=m_i}^{j} a_k \epsilon_{k+1} - \sum_{k=m_i}^{j-1} a_k \epsilon_{k+1}\| \\
&\le a_j \alpha(2\|\overline{\Delta}\| + 5) + 2c_1 T < \frac{\eta}{4} \qquad (2.8.23)
\end{aligned}$$

for sufficiently large m_i. From (2.8.21) and (2.8.23), it follows that

$$\|g_j(x_j) - \theta_{j+1}\| = \|g_j(x_j) + \Delta_{j+1} - x_{j+1}\| > \frac{\eta}{4}, \quad \text{for } j = m_i, \ldots, m(m_i,T) - 1. \qquad (2.8.24)$$

On the other hand, by Lemma 2.8.2

$$\begin{aligned}
\|g_j(x_j) - \theta_{j+1}\| &< \frac{\eta}{4} + \|\Delta_{j+1}\| \le \frac{\eta}{4} + \|\Delta_{j+1} - \Delta_{m_i}\| + \|\Delta_{m_i}\| \\
&\le \frac{\eta}{4} + cT + r. \qquad (2.8.25)
\end{aligned}$$

Identifying r_1 and r_2 in A2.8.5 to $\frac{\eta}{4}$ and $\frac{\eta}{4} + cT + r$, respectively, we can find $a > 0$ such that

$$v_x^T(g_j(x_j) - \theta_{j+1}) f_{j+1}(g_j(x_j)) < -a, \quad \forall j : m_i \le j \le m(m_i,T) - 1 \qquad (2.8.26)$$

by A2.8.5.

Let us consider the right-hand side of (2.8.22). Noticing $\|d_j(x_j)\| \leq \gamma_j\|\Delta_j\| \leq \gamma_j(cT + \|\overline{\Delta}\| + 1)$, by A2.8.3 and A2.8.4 we have

$$\lim_{i \longrightarrow \infty} \sum_{j=m_i}^{m(m_i,T)-1} (d_j(x_j) - w_j) = 0. \tag{2.8.27}$$

By A2.8.6,

$$\lim_{T \longrightarrow 0} \limsup_{i \longrightarrow \infty} \| \sum_{j=m_i}^{m(m_i,T)-1} a_j \epsilon_{j+1} \| = o(T). \tag{2.8.28}$$

Noticing that

$$\begin{aligned}
&\|\widetilde{\Delta} - (g_j(x_j) - \theta_{j+1})\| \\
\leq &\|\widetilde{\Delta} - \Delta_{m_i}\| + \|\Delta_j - \Delta_{m_i}\| + \|g_j(x_j) - \theta_{j+1} - \Delta_j\| \\
\leq &2cT + \|d_j(x_j) - w_j\| \leq 2cT + \gamma_j(cT + \|\overline{\Delta}\| + 1) + \|w_j\| \longrightarrow 0.
\end{aligned} \tag{2.8.29}$$

$$\forall j : m_i \leq j \leq m(m_i, T) - 1.$$

as $i \longrightarrow \infty$ and $T \longrightarrow 0$, by continuity of $v_x(\cdot)$ we find that $v_x^T(\widetilde{\Delta}) - v_x^T(g_j(\theta_j) - \theta_{j+1})$ tends to zero as $i \longrightarrow \infty$ and $T \longrightarrow 0$.

Since $\|f_{j+1}(g_j(x_j))\| \leq \alpha(2\|\overline{\Delta}\| + 5)$, the sum of the first and second terms on the right-hand side of (2.8.22) is $o(T)$ as $i \longrightarrow \infty$ and $T \longrightarrow 0$. This combining with (2.8.26) yields the following conclusion that for $i \geq i_0$ with sufficiently large i_0 and for small enough T from (2.8.22) it follows that

$$v(\Delta_{m(m_i,T)}) - v(\Delta_{m_i}) \leq -\frac{a}{2}T. \tag{2.8.30}$$

By (2.8.20), tending i to infinity, from (2.8.30) we derive

$$\limsup_{i \longrightarrow \infty} v(\Delta_{m(m_i,T)}) \leq \delta_1 - \frac{a}{2}T. \tag{2.8.31}$$

By Lemma 2.8.2 we have

$$\lim_{T \longrightarrow 0} \max_{m_i \leq m \leq m(m_i,T)} \|v(\Delta_m) - v(\Delta_{m_i})\| = 0. \tag{2.8.32}$$

However, by definition, $v(\Delta_{m_i}) \leq \delta_1$, $\delta_1 < v(\Delta_j) < \delta_2$ for $j : m_i < j < l_i$ and $v(\Delta_{l_i}) \geq \delta_2$. Hence from (2.8.32), we must have $m(m_i, T) < l_i$ if T is small enough. Therefore, $v(\Delta_{m(m_i,T)}) \in [\delta_1.\delta_2]$. This contradicts (2.8.31). The obtained contradiction shows that $\lim_{k \longrightarrow \infty} \sigma_k < \infty$ □

Theorem 2.8.2 *Assume A2.8.1–A2.8.6 hold. Then the estimation error* $\Delta_k = x_k - \theta_k$ *tends to zero as* $k \longrightarrow \infty$.

Proof. We first show that $v(\Delta_k)$ converges. Assume the converse:

$$v_1 = \liminf_{k \longrightarrow \infty} v(\Delta_k) < \limsup_{k \longrightarrow \infty} v(\Delta_k) = v_2, \tag{2.8.33}$$

where $-\infty < v_1 < v_2 < \infty$ because $\{\Delta_k\}$ is bounded by Lemma 2.8.3. It is clear that there exists an interval $[\delta_1, \delta_2]$ that does not contain zero such that $[\delta_1, \delta_2] \subset (v_1, v_2)$. Without loss of generality, assume $0 < \delta_1 < \delta_2$. From A2.8.6, it follows that there are infinitely many sequences such that $v(\Delta_{m_i}) \leq \delta_1$, $v(\Delta_{l_i}) \geq \delta_2$ and that $v(\Delta_j) \in (\delta_1, \delta_2)$ for $j : m_i < j < l_i$, $i = 1, 2, \ldots$.

Without loss of generality we may assume $\{\Delta_{m_i}\}$ converges: $\Delta_{m_i} \longrightarrow \Delta'$. Since $v(\Delta') = \delta_1 > 0$, a $\xi > 0$ exists such that $\|\Delta'\| \geq \xi$, and by Lemma 2.8.2, $\|\Delta_j\| > \xi/2$, $\forall j : m_i \leq j \leq m(m_i, T)$. Completely the same argument as that used for (2.8.22)–(2.8.32) leads to a contradiction. Hence $v(\Delta_k)$ is convergent.

We now show that $\Delta_k \longrightarrow 0$ as $k \longrightarrow \infty$. Assume the converse: there is a subsequence $\Delta_{m_i} \longrightarrow \Delta' \neq 0$. By the same argument we again arrive at (2.8.30). Tending $i \longrightarrow \infty$, by convergence of $\{v(\Delta_k)\}$, we obtain a contradictory inequality $0 \leq -aT/2$. This implies that $\Delta_k \longrightarrow 0$ as $k \longrightarrow \infty$. □

The following theorem is similar to Theorem 2.4.1.

Theorem 2.8.3 *Assume A2.8.1–A2.8.5 hold and* $f_k(\cdot)$ *is continuous at* θ_k *uniformly in* k. *Then* $x_k - \theta_k \longrightarrow 0$ *as* $k \longrightarrow \infty$ *if and only if A2.8.6 holds. Furthermore, under conditions A2.8.1–A2.8.5, the following three conditions are equivalent.*

1) Condition A2.8.6;

2) $\lim_{T \longrightarrow 0} \limsup_{k \longrightarrow \infty} \frac{1}{T} \| \sum_{i=k}^{m(k,T)} a_i \epsilon_{i+1} \| = 0, \ \forall t \in [0, T]$;

3) ϵ_{k+1} *can be decomposed into two parts:* $\epsilon_{k+1} = \epsilon'_{k+1} + \epsilon''_{k+1}$ *so that* $\sum_{k=1}^{\infty} a_k \epsilon'_{k+1} < \infty$ *and* $\epsilon''_k \longrightarrow 0$ *as* $k \longrightarrow \infty$.

Proof. Assume $x_k - \theta_k \longrightarrow 0$ as $k \longrightarrow \infty$. Then $\{\Delta_k\}$ is bounded. We have shown in the proof of Lemma 2.8.3 that the number of truncations must be finite if $\{\Delta_k\}$ is bounded. Therefore, starting from some k_0, the algorithm (2.8.5) becomes

$$x_{k+1} = g_k(x_k) + a_k y_{k+1}, \quad k \geq k_0.$$

From (2.8.11) we have

$$\epsilon_{k+1} = \frac{\Delta_{k+1} - \Delta_k}{a_k} + \frac{d_k(x_k) - w_k}{a_k} + f_{k+1}(\theta_{k+1} + d_k(x_k) + \Delta_k - w_k).$$

Set

$$\epsilon'_{k+1} = \frac{\Delta_{k+1} - \Delta_k}{a_k}$$

$$\epsilon''_{k+1} = \frac{d_k(x_k) - \epsilon_k}{a_k} + f_{k+1}(\theta_{k+1} + d_k(x_k) + \Delta_k - w_k).$$

By A2.8.3 and A2.8.4 and $\Delta_k \longrightarrow 0$, $(d_k(x_k) - w_k)/a_k \longrightarrow 0$ as $k \longrightarrow \infty$, while $f_{k+1}(\theta_{k+1} + d_k(x_k) + \Delta_k - w_k)$ tends to zero because $f_k(\cdot)$ is uniformly continuous at θ_k and $d_k(x_k) + \Delta_k - w_k \longrightarrow 0$. Consequently, 3) holds.

On the other hand, it is clear that 3) implies 2), which in turn implies A2.8.6. By Theorem 2.8.1, under A2.8.1–A2.8.5, Condition A2.8.6 implies $x_k - \theta_k \longrightarrow 0$ as $k \longrightarrow \infty$.

Thus, the equivalence of 1)–3) has been justified under A2.8.1–A2.8.5.

□

2.9. Notes and References

The initial version of SA algorithms with expanding truncations and its associated analysis method were introduced in [27], where the algorithm was called SA with randomly varying truncations. Convergence results of this kind of algorithms can also be found in [14, 28]. Theorems given in Section 2.2 are the improved versions of those given in [14, 27, 28]. Theorems in Section 2.3 can be found in [18]. Necessity of the noise condition is proved in [24, 94] for the single-root case, and in [17] for the multi-root case.

Convergence results of SA algorithms with additive noise can be found in [16]. Concerning the measure theory, we refer to [31, 76, 84]. Results given in Section 2.6 can be found in [48], and some related problems are discussed in [3]. For the proof of Remark 2.6.4 we refer to Theorem 3.3 in [34]. Example 2.6.1 can be found in [93]. Robustness of SA algorithms is presented in [24]. The dynamic SA was considered in [38, 39, 91], but the results presented in Section 2.8 are given in [25].

Chapter 3

ASYMPTOTIC PROPERTIES OF STOCHASTIC APPROXIMATION ALGORITHMS

In Chapter 2 we were mainly concerned with the path-wise convergence analysis for SA algorithms with expanding truncations. Conditions were given to guarantee $d(x_k, J) \xrightarrow[k\longrightarrow\infty]{} 0$, where J denotes the root set of the unknown function, and x_k the estimate for unknown root given by the algorithm.

In this chapter, for the case where J consists of a singleton x^0 we consider the convergence rate of x_k to x^0, asymptotic normality of $x_k - x^0$ and asymptotic efficiency of the estimate.

Assume $f(\cdot) : \mathbb{R}^l \longrightarrow \mathbb{R}^l$ is differentiable at x^0. Then as $x \longrightarrow x^0$

$$f(x) = F(x - x^0) + \delta(x), \tag{3.0.1}$$

where $\delta(x^0) = 0$ and $\delta(x) = o(\|x - x^0\|)$.

It turns out that the convergence rate heavily depends on whether or not F is degenerate. Roughly speaking, in the case where the step size in (2.1.1) $a_k = \frac{1}{k}$, the convergence rate of $\|x_n - x^0\|$ is $o(n^{-\delta})$ for some positive $\delta > 0$ when F is nondegenerate, and $O((\log n)^{-\alpha})$ for some $\alpha > 0$ when F vanishes.

It will be shown that $\frac{x_k - x^0}{\sqrt{a_k}}$ is asymptotically normal and the covariance matrix of the limit distribution depends on the matrix D if in (2.1.1) the step size a_k is replaced by $a_k D$. If F in (3.0.1) is available, then D can be defined to make the limiting covariance matrix minimal, i.e., to make the estimate efficient. However, this is not the case in SA. To overcome the difficulty one way is to derive the approximate value of F by estimating it, but for this one has to impose rather heavy conditions on $f(\cdot)$. Efficiency here is derived by using a sequence of slowly decreasing step sizes, and the averaged estimate appears asymptotically efficient.

3.1. Convergence Rate: Nondegenerate Case

In this section, we give the rate of convergence of $\|x_k - x^0\|$ to zero in the case F in (3.0.1) is nondegenerate, where x_k is given by (2.1.1)–(2.1.3). It is worth noting that F is the coefficient for the first order in the Taylor's expansion for $f(\cdot)$.

The following conditions are to be used.

A3.1.1 $a_k > 0$, $a_k \xrightarrow[k\to\infty]{} 0$, $\sum_{k=1}^{\infty} a_k = \infty$, *and*

$$\frac{a_k - a_{k+1}}{a_k a_{k+1}} \xrightarrow[k\to\infty]{} \alpha \geq 0. \tag{3.1.1}$$

A3.1.2 *A continuously differentiable function* $v(\cdot) : \mathbb{R}^l \longrightarrow R$ *exists such that*

$$\sup_{\delta \leq \|x - x^0\| \leq \Delta} f^T(x) v_x(x) < 0 \tag{3.1.2}$$

for any $\Delta > \delta > 0$, *and* $v(x^*) < \inf_{\|x\|=c_0} v(x)$ *for some* $c_0 > 0$ *with* $c_0 > \|x^*\|$, *where* x^* *is used in (2.1.1).*

A3.1.3 *For the sample path* ω *under consideration the observation noise* $\{\epsilon_k\}$ *in (2.1.3) can be decomposed into two parts* $\epsilon_k = \epsilon'_k + \epsilon''_k$ *such that*

$$\sum_{k=1}^{\infty} a_k^{1-\delta} \epsilon'_{k+1} < \infty, \quad \epsilon''_{k+1} = O(a_k^{\delta}) \tag{3.1.3}$$

for some $\delta \in (0, 1]$.

A3.1.4 $f(\cdot)$ *is measurable and locally bounded, and is differentiable at* x^0 *such that as* $x \longrightarrow x^0$

$$f(x) = F(x - x^0) + \delta(x), \quad \delta(x^0) = 0, \quad \delta(x) = o(\|x - x^0\|). \tag{3.1.4}$$

The matrix F *is stable (This implies nondegeneracy of* F*.), in addition,* $F + \alpha\delta I$ *is also stable, where* α *and* δ *are given by (3.1.1) and (3.1.3), respectively.*

By stability of a matrix we mean that all its eigenvalues are with negative real parts.

Remark 3.1.1 We now compare A3.1.1–A3.1.4 with A2.2.1–A2.2.4. Because of additional requirement (3.1.1), A3.1.1 is stronger than A2.2.1, but it is automatically satisfied if $a_k = \frac{a}{k}$ with $a > 0$. In this case α in (3.1.1) equals $\frac{1}{a}$. Also, (3.1.1) is satisfied if $a_k = \frac{a}{k^{\frac{1}{2}+\beta}}$ with $a > 0$, $\beta > 0$. In this case $\alpha = 0$. Take $\delta > 0$ sufficiently small such that $\beta > \frac{\delta}{2(1-\delta)}$. Then $(\frac{1}{2}+\beta)2(1-\delta) > 1$ and $\sum\limits_{k=1}^{\infty} a_k^{2(1-\delta)} = \sum\limits_{k=1}^{\infty} \frac{a}{k^{2(1-\delta)(\frac{1}{2}+\beta)}} < \infty$. Assume $\{\epsilon_k, \mathcal{F}_k\}$ is a martingale difference sequence with $\sup\limits_k E(\|\epsilon_{k+1}\|^2|\mathcal{F}_k) < \infty$. Then by the convergence theorem for martingale difference sequences, $\sum\limits_{k=1}^{\infty} a_k^{(1-\delta)}\epsilon_{k+1} < \infty$ a.s. Therefore (3.1.3) is satisfied a.s. with $\epsilon''_k \equiv 0$. Condition A3.1.4 assumes differentiability of $f(\cdot)$ at x^0, which is not required in A2.2.4.

Lemma 3.1.1 *Let $\{H_k\}$ and H be $l \times l$-matrices. Assume H is stable and $H_k \xrightarrow[k\to\infty]{} H$. If $\{a_k\}$ satisfies A3.1.1 and l-dimensional vectors $\{e_k\}$, $\{\nu_k\}$ satisfy the following conditions*

$$\sum_{k=1}^{\infty} a_k e_{k+1} < \infty \quad \text{and} \quad \nu_k \xrightarrow[k\to\infty]{} 0, \tag{3.1.5}$$

then $\{x_k\}$ defined by the following recursion with arbitrary initial value x_0 tends to zero:

$$x_{k+1} = x_k + a_k H_k x_k + a_k(e_{k+1} + \nu_{k+1}). \tag{3.1.6}$$

Proof. Set

$$\Phi_{k,j} \triangleq (I + a_k H_k) \cdots (I + a_j H_j), \quad \Phi_{j,j+1} \triangleq I. \tag{3.1.7}$$

We now show that there exist constants $c_0 > 0$ and $c > 0$ such that

$$\|\Phi_{k,j}\| \le c_0 \exp[-c\sum_{i=j}^{k} a_i], \quad \forall k \ge j, \forall j \ge 0. \tag{3.1.8}$$

Let S be any $l \times l$ negative definite matrix. Consider

$$P = -\int_0^{\infty} e^{H^T t} S e^{Ht} dt.$$

Since H is stable, the positive definite matrix P is well-defined. Integrating by parts, we have

$$\begin{aligned} P &= -\int_0^\infty e^{H^T t} S d e^{Ht} H^{-1} \\ &= -e^{H^T t} S e^{Ht} H^{-1}\Big|_0^\infty + H^T \int_0^\infty e^{H^T t} S e^{Ht} dt H^{-1} \\ &= SH^{-1} - H^T P H^{-1}, \end{aligned}$$

which implies

$$H^T P + PH = S. \tag{3.1.9}$$

This means that if H is stable, then for any negative definite matrix S we can find a positive definite matrix P to satisfy equation (3.1.9). This fact is called the Lyapunov theorem and (3.1.9) called the Lyapunov equation. Consequently, we can find $P > 0$ such that

$$PH + H^T P = -2I,$$

where I denotes the identity matrix of compatible dimension.

Since $H_k \longrightarrow H$ as $k \longrightarrow \infty$, there exists k_0 such that for $\forall k \geq k_0$

$$PH_k + H_k^T P \leq -I \tag{3.1.10}$$

Consequently,

$$\begin{aligned} \Phi_{k,j}^T P \Phi_{k,j} &= \Phi_{k-1,j}^T (I + a_k H_k)^T P (I + a_k H_k) \Phi_{k-1,j} \\ &= \Phi_{k-1,j}^T (P + a_k^2 H_k^T P H_k + a_k H_k^T P + a_k P H_k) \Phi_{k-1,j} \\ &\leq \Phi_{k-1,j}^T (P + a_k^2 H_k^T P H_k - a_k I) \Phi_{k-1,j} \\ &= \Phi_{k\ \ 1,j}^T P^{\frac{1}{2}} (I - a_k P^{-1} + a_k^2 P^{-\frac{1}{2}} H_k^T P H_k P^{-\frac{1}{2}}) P^{\frac{1}{2}} \Phi_{k-1,j}. \end{aligned} \tag{3.1.11}$$

Without loss of generality we may assume that k_0 is sufficiently large such that for $\forall k \geq k_0$,

$$\|I - a_k P^{-1} + a_k^2 P^{-\frac{1}{2}} H^T P H P^{-\frac{1}{2}}\| \leq 1 - 2ca_k < e^{-2ca_k} \tag{3.1.12}$$

for some constant $c > 0$, where the first inequality is because $a_k \longrightarrow 0$ as $k \longrightarrow \infty$ and $P^{-1} > 0$, while the second inequality is elementary. Combining (3.1.11) and (3.1.12) leads to

$$\Phi_{k,j}^T P \Phi_{k,j} \leq \big(\exp(-2c\sum_{i=j}^k a_i)\big) I,$$

and hence

$$\|\Phi_{k,j}\| \le \lambda_{\min}^{-\frac{1}{2}}(P)\exp\big(-c\sum_{i=j}^{k} a_i\big), \tag{3.1.13}$$

where $\lambda_{\min}(P)$ denotes the minimum eigenvalue of P.

Paying attention to that

$$\|\Phi_{k_0-1,j}\| \le \prod_{i=j}^{k_0-1}(1+a_i\|H_i\|) \le \prod_{i=1}^{k_0-1}(1+a_i\|H_i\|),$$

from (3.1.13) we derive

$$\|\Phi_{k,j}\| \le \|\Phi_{k,k_0}\| \cdot \|\Phi_{k_0-1,j}\| \le \lambda_{\min}^{-\frac{1}{2}}(P)\exp(-c\sum_{i=k_0}^{k} a_i)\prod_{i=1}^{k_0-1}(1+a_i\|H_i\|)$$

$$\le \lambda_{\min}^{-\frac{1}{2}}(P)\exp(c\sum_{i=j}^{k_0-1} a_i)\prod_{i=0}^{k_0-1}(1+a_i\|H_i\|)\exp(-c\sum_{i=j}^{k} a_i),$$

which verifies (3.1.8).

From (3.1.6) it follows that

$$x_{k+1} = \Phi_{k,0}x_0 + \sum_{j=0}^{k}\Phi_{k,j+1}a_j(e_{j+1}+\nu_{j+1}) \tag{3.1.14}$$

We have to show that the right-hand side of (3.1.14) tends to zero as $k \longrightarrow \infty$.

For any fixed j, $\|\Phi_{k,j}\| \longrightarrow 0$ as $k \longrightarrow \infty$ because of (3.1.1) and (3.1.8). This implies that $\Phi_{k,0}x_0 \longrightarrow 0$ as $k \longrightarrow \infty$ for any initial value x_0.

Since $\nu_k \longrightarrow 0$ as $k \longrightarrow \infty$, for any $\epsilon > 0$, k_1 exists such that $\|\nu_k\| < \epsilon$, $\forall k \ge k_1$. Then by (3.1.8) we have

$$\|\sum_{j=0}^{k}\Phi_{k,j+1}a_j\nu_{j+1}\| \le c_0\sum_{j=0}^{k_1-1}\Big[\exp(-c\sum_{i=j+1}^{k}a_i)\Big]a_j\|\nu_{j+1}\|$$

$$+\epsilon c_0\sum_{j=k_1}^{k}\Big(\exp(-c\sum_{i=j+1}^{k}a_i)\Big)a_j. \tag{3.1.15}$$

The first term at the right-hand side of (3.1.15) tends to zero by A3.1.1, while the second term can be estimated as follows:

$$\epsilon c_0\sum_{j=k_1}^{k}a_j\exp\Big(-c\sum_{i=j+1}^{k}a_i\Big) \le 2\epsilon c_0\sum_{j=k_1}^{k}\Big(a_j-\frac{ca_j^2}{2}\Big)\exp\big(-c\sum_{i=j+1}^{k}a_i\big)$$

$$
\begin{aligned}
&\leq \frac{2\epsilon c_0}{c} \sum_{j=k_1}^{k} (1 - e^{-ca_j}) \cdot \exp(-c \sum_{i=j+1}^{k} a_i) \\
&= \frac{2\epsilon c_0}{c} \sum_{j=k_1}^{k} \Big[\exp(-c \sum_{i=j+1}^{k} a_i) - \exp(-c \sum_{i=j}^{k} a_i) \Big] \\
&\leq \frac{2\epsilon c_0}{c},
\end{aligned} \tag{3.1.16}
$$

where the first inequality is valid for sufficiently large k_1 since $a_j \longrightarrow 0$ as $j \longrightarrow \infty$ and the second inequality is valid when $0 < ca_j < 1$.

Therefore, the right-hand side of (3.1.15) tends to zero as $k \longrightarrow \infty$ and then $\epsilon \longrightarrow 0$.

Let us now estimate $\sum_{j=0}^{k} \Phi_{k,j+1} a_j e_{j+1}$.

Set

$$
s_k = \sum_{j=0}^{k} a_j e_{j+1}, \quad s_{-1} = 0.
$$

By assumption of the lemma $s_n \longrightarrow s < \infty$. Hence, for any $\epsilon > 0$, there exists $k_2 > k_1$ such that $\|s_j - s\| \leq \epsilon$, $\forall j \geq k_2$. By a partial summation, we have

$$
\begin{aligned}
\sum_{j=0}^{k} \Phi_{k,j+1} a_j e_{j+1} &= \sum_{j=0}^{k} \Phi_{k,j+1}(s_j - s_{j-1}) \\
&= s_k - \sum_{j=0}^{k} (\Phi_{k,j+1} - \Phi_{k,j}) s_{j-1} \\
&= s_k - \sum_{j=0}^{k} (\Phi_{k,j+1} - \Phi_{k,j}) s \\
&\quad - \sum_{j=0}^{k} (\Phi_{k,j+1} - \Phi_{k,j})(s_{j-1} - s) \\
&= s_k - s + \Phi_{k,0} s - \sum_{j=0}^{k_2} (\Phi_{k,j+1} - \Phi_{k,j})(s_{j-1} - s) \\
&\quad + \sum_{j=k_2+1}^{k} \Phi_{k,j+1} a_j H_j (s_{j-1} - s),
\end{aligned} \tag{3.1.17}
$$

where except the last term, the sum of remaining terms tends to zero as $k \longrightarrow \infty$ by (3.1.8) and $s_k \longrightarrow s$.

Since $\|s_j - s\| \le \epsilon$, for $j \ge k_2$ and $H_j \longrightarrow H$ as $j \longrightarrow \infty$, by (3.1.8) we have

$$\|\sum_{j=k_2+1}^{k} \Phi_{k,j+1} a_j H_j (s_{j-1} - s)\|$$

$$\le \epsilon \sup_{1 \le j < \infty} \|H_j\| \sum_{j=k_2+1}^{k} c_0 [\exp(-c \sum_{i=j+1}^{k} a_i)] a_j,$$

which tends to zero as $k \longrightarrow \infty$ and $\epsilon \longrightarrow 0$ by (3.1.16) and the fact that $\sup_{1 \le j < \infty} \|H_j\| < \infty$. Thus, the right-hand side of (3.1.17) tends to zero as $k \longrightarrow \infty$, and the proof of the lemma is completed. □

Theorem 3.1.1 *Assume A3.1.1–A3.1.4 hold. Then x_k given by (2.1.1)–(2.1.3) for those sample paths for which (3.1.3) holds converges to x^0 with the following convergence rate:*

$$\|x_k - x^0\| = o(a_k^\delta), \tag{3.1.18}$$

where δ is the one given in (3.1.3).

Proof. We first note that by Theorem 2.4.1 $x_k \longrightarrow x^0$ and there is no truncation after a finite number of steps. Without loss of generality, we may assume $x^0 = 0$.

By (3.1.1), $\frac{a_k - a_{k+1}}{a_{k+1}} \xrightarrow[k \to \infty]{} 0$. Hence, by the Taylor's expansion we have

$$(\frac{a_k}{a_{k+1}})^\delta = (1 + \frac{a_k - a_{k+1}}{a_{k+1}})^\delta = 1 + \delta \frac{a_k - a_{k+1}}{a_{k+1}} + O((\frac{a_k - a_{k+1}}{a_{k+1}})^2). \tag{3.1.19}$$

Write $\delta(x)$ given by (3.1.4) as follows

$$\delta(x) = \Big(\delta(x) \frac{x^T}{\|x\|^2}\Big) x \quad \text{or} \quad \delta(x_k) = D_k x_k \tag{3.1.20}$$

where

$$D_k \triangleq \Big(\delta(x) \frac{x^T}{\|x_k\|^2}\Big) \longrightarrow 0 \quad \text{as} \quad k \longrightarrow \infty. \tag{3.1.21}$$

By (3.1.4) and (3.1.19), for sufficiently large k we have

$$\frac{x_{k+1}}{a_{k+1}^\delta} = (a_k / a_{k+1})^\delta \Big[\frac{1}{a_k^\delta}(x_k + a_k(F x_k + D_k x_k) + a_k \epsilon_{k+1}\Big]$$

$$
\begin{aligned}
&= \Big(1+\delta\frac{a_k-a_{k+1}}{a_{k+1}}+O\big((\frac{a_k-a_{k+1}}{a_{k+1}})^2\big)\Big)\Big(\frac{x_k}{a_k^\delta}+a_kF\frac{x_k}{a_k^\delta} \\
&\quad + a_kD_k\frac{x_k}{a_k^\delta}+a_k^{1-\delta}\epsilon_{k+1}\Big) \\
&= \frac{x_k}{a_k^\delta}+a_k\big(F+\delta\frac{a_k-a_{k+1}}{a_{k+1}}(\frac{1}{a_k}I+F)+C_k\big)\frac{x_k}{a_k^\delta} \\
&\quad + a_k\Big(\frac{\epsilon'_{K+1}}{a_k^\delta}+\frac{\epsilon''_{k+1}}{a_k^\delta}+O(\frac{a_k-a_{k+1}}{a_{k+1}})\frac{\epsilon_{k+1}}{a_k^\delta}\Big),
\end{aligned} \tag{3.1.22}
$$

where

$$
\begin{aligned}
C_k \triangleq &\Big(\delta\frac{a_k-a_{k+1}}{a_{k+1}}+O\big((\frac{a_k-a_{k+1}}{a_{k+1}})^2\big)+1\Big)D_k \\
&+O\big((\frac{a_k-a_{k+1}}{a_{k+1}})^2\big)\Big(\frac{1}{a_k}I+F\Big)\xrightarrow[k\to\infty]{}0.
\end{aligned}
$$

By (3.1.1), (3.1.3) we have

$$
\frac{\epsilon'_{k+1}}{a_k^\delta}\xrightarrow[k\to\infty]{}0,\ \text{and}\ O(\frac{a_k-a_{k+1}}{a_{k+1}})\frac{\epsilon_{k+1}}{a_k^\delta}\xrightarrow[k\to\infty]{}0. \tag{3.1.23}
$$

Denote

$$
\begin{aligned}
&F_k \triangleq F+\delta\frac{a_k-a_{k+1}}{a_{k+1}}(\frac{1}{a_k}I+F)+C_k, \quad z_k\triangleq\frac{x_k}{a_k^\delta}, \\
&e_{k+1}\triangleq\frac{\epsilon'_{k+1}}{a_k^\delta}\ \text{and}\ \nu_{k+1}\triangleq\frac{\epsilon''_{k+1}}{a_k^\delta}+O(\frac{a_k-a_{k+1}}{a_ka_{k+1}})\frac{\epsilon_{k+1}}{a_k^\delta}.
\end{aligned} \tag{3.1.24}
$$

Then (3.1.22) can be rewritten as

$$
z_{k+1}=z_k+a_kF_kz_k+a_k(e_{k+1}+\nu_{k+1}).
$$

Noticing that $F_k\xrightarrow[k\to\infty]{}F+\alpha\delta I$, which is stable by A3.1.4, we see that all conditions of Lemma 3.1.1 are satisfied. Hence, by the lemma $z_k\xrightarrow[k\to\infty]{}0$, which proves the theorem. □

Remark 3.1.2 Consider the dependence of convergence rate on the step size $\{a_k\}$. Take $a_k=\frac{1}{k^\alpha}$, $\alpha\in(\frac{1}{2},1]$ and let $\epsilon''_k\equiv 0$ in (3.1.3). In order to have $\sum_{k=1}^{\infty}a_k^{1-\delta}\epsilon_{k+1}<\infty$ a.s., it suffices to require

$$
\sum_{k=1}^{\infty}E\big(\frac{1}{k^{2\alpha(1-\delta)}}\|\epsilon_{k+1}\|^2/\mathcal{F}_k\big)<\infty, \tag{3.1.25}
$$

if $\{\epsilon_k, \mathcal{F}_k\}$ is a martingale difference sequence with $\sup_k E(\|\epsilon_{k+1}\|^2/\mathcal{F}_k) < \infty$. So, for (3.1.25) it is sufficient to require $2\alpha(1-\delta) > 1$, or $\delta < 1 - \frac{1}{2\alpha}$.

Since $\alpha \in (\frac{1}{2}, 1]$, the best convergence rate $o(\frac{1}{k^\delta})$, $\forall \delta < \frac{1}{2}$, is achieved at $\alpha = 1$. For $\alpha \in (\frac{1}{2}, 1]$, the convergence rate is $o(\frac{1}{k^{\alpha\delta}})$. Since $\alpha\delta < \alpha(1 - \frac{1}{2\alpha}) = \alpha - \frac{1}{2}$, the convergence rate is slowing down as α approaches to $\frac{1}{2}$. When $\alpha = \frac{1}{2}$, (3.1.25) cannot be guaranteed. From this it is seen that the convergence rate depends on how big $\alpha \in (\frac{1}{2}, 1]$ is.

3.2. Convergence Rate: Degenerate Case

In the previous section, for obtaining the convergence rate of $\{x_k\}$, stability and hence nondegeneracy of F is an essential requirement. We now consider what will happen if the linear term vanishes in the Taylor's expansion of $f(\cdot)$. For this we introduce the following set of conditions:

A3.2.1 $a_k > 0$, $a_k \xrightarrow[k\to\infty]{} 0$ *and* $\sum_{k=1}^{\infty} a_k = \infty$;

A3.2.2 *A continuously differentiable function* $v(\cdot) : \mathbb{R}^l \longrightarrow R$ *exists such that*

$$\sup_{\delta \le \|x - x^0\| \le \Delta} f^T(x) v_x(x) < 0 \tag{3.2.1}$$

for any $\Delta > \delta > 0$, *and* $v(x^*) < \inf_{\|x\|=c_0} v(x)$ *for some* $c_0 > 0$ *with* $c_0 > \|x^*\|$, *where* x^* *is used in (2.1.1);*

A3.2.3 *For the observation noise* $\{\epsilon_k\}$ *on the sample path* ω *under consideration the following series converges:*

$$\sum_{k=0}^{\infty} a_k (t_{k+1})^{1/\gamma} \epsilon_{k+1} < \infty, \quad \gamma > 0, \tag{3.2.2}$$

where $t_k = \sum_{i=0}^{k-1} a_i$;

A3.2.4 $f(\cdot)$ *is measurable and locally bounded, and is differentiable at* x^0 *such that as* $x \longrightarrow x^0$

$$f(x) = F(x - x^0)\|x - x^0\|^\gamma + r(x), \quad r(x)/\|x - x^0\|^{1+\gamma} \longrightarrow 0, \tag{3.2.3}$$

where F *is a stable matrix, and* γ *is the one used in A3.2.3.*

We first note that in comparison with A3.1.1–A3.1.4, here we do not require (3.1.1), but A3.2.2 is the same as A3.1.2. From (3.2.3) we see that the Taylor's expansion for $f(\cdot)$ does not contain the linear term. Here F is the coefficient for a term higher than second order in the Taylor's expansion of $f(\cdot)$. The noise condition A3.2.3 is different from A3.1.3, but, as to be shown by the following lemma, it also implies A2.2.3.

Lemma 3.2.1 *If (3.2.2) holds, then* $\sum_{k=0}^{\infty} a_k \epsilon_{k+1} < \infty$, *and hence A2.2.3 is satisfied.*

Proof. We need only to show $\sum_{k=0}^{\infty} a_k \epsilon_{k+1} < \infty$.

Setting

$$s_k \triangleq \sum_{k=0}^{n} a_k (t_{k+1})^{\frac{1}{\gamma}} \epsilon_{k+1}, \quad s_0 = 0,$$

by a partial summation we have

$$\begin{aligned}\sum_{k=m}^{n} a_k \epsilon_{k+1} &= \sum_{k=m}^{n} (s_k - s_{k-1})(t_{k+1})^{-\frac{1}{\gamma}} \\ &= s_k (t_{n+1})^{-\frac{1}{\gamma}} - s_{m-1}(t_{m+1})^{-\frac{1}{\gamma}} \\ &\quad + \sum_{k=m}^{n-1} s_k ((t_{k+1})^{-\frac{1}{\gamma}} - (t_{k+2})^{-\frac{1}{\gamma}}).\end{aligned} \tag{3.2.4}$$

Since $t_k \longrightarrow \infty$ as $k \longrightarrow \infty$ and s_n converges as $n \longrightarrow \infty$, the first two terms on the right-hand side of (3.2.4) tend to zero as $n \longrightarrow \infty$ and $m \longrightarrow \infty$. The last term in (3.2.4) is dominated by

$$\begin{aligned}&\max_{m \le k \le n} |s_k| \sum_{k=m}^{n} \left|(t_{k+1})^{-\frac{1}{\gamma}} - (t_{k+2})^{-\frac{1}{\gamma}}\right| \\ &= \max_{m \le k \le n} |s_k| \sum_{k=m}^{n} (t_{k+2})^{-\frac{1}{\gamma}} \left|(1 - \frac{a_{k+1}}{t_{k+2}})^{-\frac{1}{\gamma}} - 1\right| \\ &= \frac{1}{\gamma} \max_{m \le k \le n} |s_k| \sum_{k=m}^{n} a_{k+1} (t_{k+2})^{-(1+\frac{1}{\gamma})} (1 + o(1)),\end{aligned} \tag{3.2.5}$$

where $o(1) \longrightarrow 0$ as $k \longrightarrow \infty$.

By the following elementary calculation we conclude that the right-hand side of (3.2.5) tends to zero as $n \longrightarrow \infty$ and $m \longrightarrow \infty$:

$$\sum_{k=m}^{n} a_{k+1} (t_{k+2})^{-(1+\frac{1}{\gamma})} = \sum_{k=m}^{n} \int_{t_{k+1}}^{t_{k+2}} dx / t_{k+2}^{1+\frac{1}{\gamma}}$$

$$\leq \sum_{k=m}^{n} \int_{t_{k+1}}^{t_{k+2}} \frac{dx}{x^{1+\frac{1}{\gamma}}} = \gamma[(t_{m+1})^{-\frac{1}{\gamma}} - (t_{n+2})^{-\frac{1}{\gamma}}]$$

which tends to zero as $n \longrightarrow \infty$ and $m \longrightarrow \infty$ because $t_k \longrightarrow \infty$ as $k \longrightarrow \infty$. This combining with (3.2.4) and (3.2.5) shows that $\sum_{k=0}^{\infty} a_k \epsilon_{k+1} < \infty$. □

By the Lyapunov equation (3.1.9), there is a positive definite matrix $P > 0$ such that

$$F^T P + PF = -2I \tag{3.2.6}$$

Denote by $\lambda_{\max}$ and $\lambda_{\min}$ the maximum and minimum eigenvalues of P, respectively, and by K the condition number $\lambda_{\max}/\lambda_{\min}$.

Theorem 3.2.1 *Assume A3.2.1–A3.2.4 hold and $\{x_k\}$ is given by (2.1.1) -(2.1.3). Then for the sample paths where A3.2.3 holds the following convergence rate takes place:*

$$\limsup_{k \longrightarrow \infty} (t_k)^{\frac{1}{\gamma}} \|x_k - x^0\| \leq \sqrt{K} (\frac{\lambda_{\max}}{\gamma})^{\frac{1}{\gamma}}, \tag{3.2.7}$$

where $\lambda_{\max}$ is the maximum eigenvalue of P given by (3.2.6).

We start with lemmas. Note that by Theorems 2.2.1 or 2.4.1 $x_k \longrightarrow x^0$. Therefore, starting from some k, the algorithm has no truncation.

Define

$$z_k = (t_k)^{\frac{1}{\gamma}} (x_k - x^0). \tag{3.2.8}$$

Assuming k is large enough so that there is no truncation, by (3.2.3) we have

$$\begin{aligned}
z_{k+1} &= (t_{k+1})^{\frac{1}{\gamma}} (x_{k+1} - x^0) \\
&= (\frac{t_{k+1}}{t_k})^{\frac{1}{\gamma}} ((t_k)^{\frac{1}{\gamma}} (x_{k+1} - x_k + x_k - x^0)) \\
&= (1 + \frac{a_k}{t_k})^{\frac{1}{\gamma}} (a_k (t_k)^{\frac{1}{\gamma}} (F\|x_k - x^0\|^{\gamma} (x_k - x^0) + r(x_k) + \epsilon_{k+1}) + z_k) \\
&= (1 + \frac{a_k}{t_k})^{\frac{1}{\gamma}} (z_k + \frac{a_k}{t_k} (F\|z_k\|^{\gamma} z_k \\
&\qquad + \frac{r(x_k)}{\|x_k - x^0\|^{1+\gamma}} \|z_k\|^{1+\gamma}) + a_k (t_k)^{\frac{1}{\gamma}} \epsilon_{k+1}) \\
&= z_k + \frac{a_k}{t_k} g_k(z_k) + a_k (t_{k+1})^{\frac{1}{\gamma}} \epsilon_{k+1},
\end{aligned} \tag{3.2.9}$$

where

$$g_k(z) = (1 + o(1))\Big(\frac{z}{\gamma} + F\|z\|^\gamma z + \frac{r(x_k)}{\|x_k - x^0\|^{1+\gamma}} \cdot \|z\|^{1+\gamma}\Big). \quad (3.2.10)$$

Define

$$g(z) = \frac{z}{\gamma} + F\|x\|^\gamma z.$$

Lemma 3.2.2 *Assume A3.2.1–A3.2.4 hold. Then $\{z_k\}$ is bounded.*

Proof. Since $x_k \longrightarrow x^0$, N_0 exists such that $\|x_k - x^0\| < 1$ and

$$\|r(x_k)\|/\|x_k - x^0\|^{1+\gamma} < \frac{1}{2\lambda_{\max}} \wedge 1, \quad \forall k \geq N_0, \quad (3.2.11)$$

where and hereafter, $a \wedge b$ means the smaller one between a and b.

By the definition of z_k, we have $\|z_k\| < (t_k)^{\frac{1}{\gamma}}$ for $k \geq N_0$. By (3.2.2) there exists N_1 such that

$$\|\sum_{k=n}^{m} a_k(t_{k+1})^{\frac{1}{\gamma}}\epsilon_{k+1}\| \leq \frac{1}{16\lambda_{\max}(\frac{1}{\gamma} + 1 + \sup\limits_{\|z\|=1} \|Fz\|^2)} \wedge 1, \ \forall m > n \geq N_1. \quad (3.2.12)$$

Assuming N_1 is large enough such that for $k \geq N_1$, we also have

$$a_k \leq \frac{1}{24\lambda_{\max}(\frac{1}{\gamma^2} + 1 + \sup\limits_{\|z\|=1} \|Fz\|^2)}, \quad t_k > 1, \quad (3.2.13)$$

and $\frac{2}{3} \leq 1 + o(1) \leq \sqrt{2}$.

Define

$$d_0 \triangleq (t_{N_1})^{\frac{1}{\gamma}} \vee (\frac{8\lambda_{\max}}{\gamma})^{\frac{1}{\gamma}} \vee 1 \text{ and } d \triangleq 2K(d_0^2 + 1),$$

where and hereafter $a \vee b = \begin{cases} a, & \text{if } a \geq b, \\ b, & \text{if } b \geq a. \end{cases}$

Since $\|z_k\| < (t_k)^{\frac{1}{\gamma}}$, $\forall k \geq N_0$, we have $\|z_{N_1}\| < d_0$.
Let

$$k_0 + 1 = \inf\{k > N_1, \|z_k\| \geq d_0\}.$$

If $k_0 = \infty$, then $\|z_k\| \leq d_0 \vee \max_{m<N_1} \|z_m\|$, $\forall k$, i.e., $\{z_k\}$ is bounded. Otherwise, let $i_0 + 1 = \inf\{i > k_0 + 1, \|z_i\|^2 > d\}$. We need only to consider the case $i_0 < \infty$, since if it is not true then $\{z_k\}$ is clearly bounded.

Let P be given by (3.2.6). We have

$$\begin{aligned}
z_{i_0+1}^T P z_{i_0+1} =&(z_{i_0+1} - z_{i_0} + z_{i_0})^T P (z_{i_0+1} - z_{i_0} + z_{i_0}) \\
=&\Big[\frac{a_{i_0}}{t_{i_0}} g_{i_0}(z_{i_0}) + a_{i_0}(t_{i_0+1})^{\frac{1}{\gamma}} \epsilon_{i_0+1} + z_{i_0}\Big]^T \\
&\cdot P\Big[\frac{a_{i_0}}{t_{i_0}} g_{i_0}(z_{i_0}) + a_{i_0}(t_{i_0+1})^{\frac{1}{\gamma}} \epsilon_{i_0+1} + z_{i_0}\Big] \\
=&\Phi_1(i_0) + \Phi_2(i_0) + \Phi_3(i_0, a_{i_0}(t_{i_0+1})^{\frac{1}{\gamma}})\epsilon_{i_0+1}) \\
&+ (a_{i_0}(t_{i_0+1})^{\frac{1}{\gamma}} \epsilon_{i_0+1} + z_{i_0})^T P((a_{i_0}(t_{i_0+1})^{\frac{1}{\gamma}} \epsilon_{i_0+1} + z_{i_0})),
\end{aligned} \tag{3.2.14}$$

where

$$\Phi_1(i_0) = \frac{a_{i_0}}{t_{i_0}} (g_{i_0}^T(z_{i_0}) P z_{i_0} + z_{i_0}^T P g_{i_0}(z_{i_0})),$$

$$\Phi_2(i_0) = \frac{a_{i_0}^2}{t_{i_0}^2} g_{i_0}^T(z_{i_0}) P g_{i_0}(z_{i_0}) \le \frac{a_{i_0}^2}{t_{i_0}^2} \lambda_{\max} \|g_{i_0}(z_{i_0})\|^2,$$

$$\Phi_3(i_0, a_{i_0}(t_{i_0+1})^{\frac{1}{\gamma}} \epsilon_{i_0+1}) = \frac{a_{i_0}^2 (t_{i_0+1})^{\frac{1}{\gamma}}}{t_{i_0}} (\epsilon_{i_0+1}^T P g_{i_0}(z_{i_0}) + g_{i_0}^T(z_{i_0}) P \epsilon_{i_0+1}).$$

In what follows we will prove that

$$\Phi_1(i_0) + \Phi_2(i_0) + \Phi_3(i_0, a_{i_0}(t_{i_0+1})^{\frac{1}{\gamma}} \epsilon_{i_0+1}) \le 0. \tag{3.2.15}$$

By (3.2.10) and (3.2.6) it is clear that

$$\begin{aligned}
\Phi_1(i_0) \le& \frac{a_{i_0}}{t_{i_0}}(1 + o(1))\Big[\frac{2}{\gamma} z_{i_0}^T P z_{i_0} - 2\|z_{i_0}\|^{2+\gamma} + 2\lambda_{\max} \frac{\|r(x_{i_0})\|}{\|x_{i_0} - x^0\|^{1+\gamma}} \\
&\cdot \|z_{i_0}\|^{2+\gamma}\Big] \le -\frac{a_{i_0}}{2t_{i_0}} \|z_{i_0}\|^{2+\gamma},
\end{aligned} \tag{3.2.16}$$

where the last inequality follows by the following consideration:

By (3.2.11) $\dfrac{2\lambda_{\max}\|r(x_{i_0})\|}{\|x_{i_0} - x^0\|^{1+\gamma}} < 1$ so for (3.2.16) it suffices to show that

$$\frac{a_{i_0}}{t_{i_0}}(1 + o(1))\Big[\frac{2}{\gamma} z_{i_0}^T P z_{i_0} - \|z_{i_0}\|^{2+\gamma}\Big] \le -\frac{a_{i_0}}{2t_{\epsilon_0}} \|z_{i_0}\|^{2+\gamma}.$$

By definition of i_0, we have $i_0 \ge k + 1$, and hence

$$\|z_{i_0}\| \ge d_0, \text{ or } \|z_{i_0}\|^\gamma \ge d_0^\gamma \ge \frac{8\lambda_{\max}}{\gamma}.$$

Consequently,

$$\frac{2}{\gamma} z_{i_0}^T P z_{i_0} - \|z_{i_0}\|^{2+\gamma} \le \frac{2\lambda_{\max}}{\gamma} \|z_{i_0}\|^{2+\gamma} \cdot \frac{\gamma}{8\lambda_{\max}} - \|z_0\|^{2+\gamma} = -\frac{3}{4} \|z_{i_0}\|^{2+\gamma},$$

and by the agreement $(1 + o(1)) \ge \frac{2}{3}$ for $k \ge N_1$ $(i_0 > N_1)$

$$\begin{aligned} \frac{a_{i_0}}{t_{i_0}} (1 + o(1)) (\frac{2}{\gamma} z_{i_0}^T P z_{i_0} - \|z_{i_0}\|^{2+\gamma}) &\le \frac{a_{i_0}}{t_{i_0}} (1 + o(1)) (-\frac{3}{4} \|z_{i_0}\|^{2+\gamma}) \\ &\le -\frac{a_{i_0}}{2t_{i_0}} \|z_{i_0}\|^{2+\gamma}, \end{aligned}$$

which verifies the last inequality in (3.2.16).

We now estimate $\Phi_2(i_0)$. By (3.2.10) (3.2.11) and the agreement $|1 + o(1)| \le \sqrt{2}$ we have

$$\Phi_2(i_0) \le \frac{a_{i_0}^2}{t_{i_0}^2} \lambda_{\max} \Big(\frac{6}{\gamma^2} \|z_{i_0}\|^2 + 6\|z_{i_0}\|^{2\gamma} \|F z_{i_0}\|^2 + 6\|z_{i_0}\|^{2+2\gamma} \Big). \quad (3.2.17)$$

Noticing that, as agreed, $t_{i_0} > 1$, $\|z_{i_0}\| \ge d_0 \ge 1$ and $\|z_{i_0}\| \le (t_{i_0})^{\frac{1}{\gamma}}$, from (3.2.17) we have

$$\Phi_2(i_0) \le \frac{6a_{i_0}}{t_{i_0}} \lambda_{\max} a_{i_0} (\frac{1}{\gamma^2} + 1 + \sup_{\|z\|=1} \|Fz\|^2) \|z_{i_0}\|^{2+\gamma},$$

and by (3.2.13),

$$\Phi_2(i_0) \le \frac{1}{4} \frac{a_{i_0}}{t_{i_0}} \|z_{i_0}\|^{2+\gamma}. \quad (3.2.18)$$

Again, from (3.2.10) and noticing $\|z_{i_0}\| \ge 1$ we have

$$\begin{aligned} \|g_{i_0}(z_{i_0})\| &\le \sqrt{2} (\frac{\|z_{i_0}\|}{\gamma} + \sup_{\|z\|=1} \|Fz\| \cdot \|z_{i_0}\|^{1+\gamma} + \|z_{i_0}\|^{1+\gamma}) \\ &\le 2(\frac{1}{\gamma} + 1 + \sup_{\|z\|=1} \|Fz\|) \|z_{i_0}\|^{2+\gamma}. \end{aligned} \quad (3.2.19)$$

Consequently, by (3.2.12)

$$\begin{aligned} &\|\Phi_3(i_0, a_{i_0} (t_{i_0+1})^{\frac{1}{\gamma}} \epsilon_{i_0+1}\| \\ &\le 2\frac{a_{i_0}}{t_{i_0}} \|a_{i_0} (t_{i_0+1})^{\frac{1}{\gamma}} \epsilon_{i_0+1}\| \cdot \|P g_{i_0}(z_{i_0})\| \le \frac{1}{4} \frac{a_{i_0}}{t_{i_0}} \|z_{i_0}\|^{2+\gamma}. \end{aligned} \quad (3.2.20)$$

Combining (3.2.14), (3.2.16), (3.2.18), and (3.2.20) yields

$$z_{i_0+1}^T P z_{i_0+1} \le (a_{i_0} (t_{i_0+1})^{\frac{1}{\gamma}} \epsilon_{i_0+1} + z_{i_0})^T P (a_{i_0} (t_{i_0+1})^{\frac{1}{\gamma}} \epsilon_{i_0+1} + z_{i_0}).$$

Similar to (3.2.14) we treat the right-hand side of the above inequality as follows.

$$
\begin{aligned}
z_{i_0+1}^T P z_{i_0+1} \le & (z_{i_0} - z_{i_0-1} + z_{i_0-1} + a_{i_0}(t_{i_0+1})^{\frac{1}{\gamma}} \epsilon_{i_0+1})^T \\
& \cdot P(z_{i_0} - z_{i_0-1} + z_{i_0-1} + a_{i_0}(t_{i_0+1})^{\frac{1}{\gamma}} \epsilon_{i_0+1}) \\
= & \Big[\frac{a_{i_0-1}}{t_{i_0-1}} g_{i_0-1}(z_{i_0-1}) + z_{i_0-1} + \sum_{j=i_0-1}^{i_0} a_j(t_{j+1})^{\frac{1}{\gamma}} \epsilon_{j+1}\Big]^T \\
& \cdot P\Big[\frac{a_{i_0-1}}{t_{i_0-1}} g_{i_0-1}(z_{i_0-1}) + z_{i_0-1} + \sum_{j=i_0-1}^{i_0} a_j(t_{j+1})^{\frac{1}{\gamma}} \epsilon_{j+1}\Big] \\
= & \Phi_1(i_0-1) + \Phi_2(i_0-1) + \Phi_3(i_0-1, \sum_{j=i_0-1}^{i_0} a_j(t_{j+1})^{\frac{1}{\gamma}} \epsilon_{j+1}) \\
& + (z_{i_0-1} + \sum_{j=i_0-1}^{i_0} a_j(t_{j+1})^{\frac{1}{\gamma}} \epsilon_{j+1})^T P(z_{i_0-1} \\
& + \sum_{j=i_0-1}^{i_0} a_j(t_{j+1})^{\frac{1}{\gamma}} \epsilon_{j+1}).
\end{aligned}
$$

By the same argument as that used above, we can show that

$$
\Phi_1(i_0-1) + \Phi_2(i_0-1) + \Phi_3(i_0-1, \sum_{j=i_0-1}^{i_0} a_j(t_{j+1})^{\frac{1}{\gamma}}) \le 0,
$$

and inductively we derive

$$
\begin{aligned}
z_{i_0+1}^T P z_{i_0+1} \le & \big(z_{k_0} + \sum_{j=k_0}^{i_0} a_j(t_{j+1})^{\frac{1}{\gamma}} \epsilon_{j+1}\big)^T P \big(z_{k_0} + \sum_{j=k_0}^{i_0} a_j(t_{j+1})^{\frac{1}{\gamma}} \epsilon_{j+1}\big) \\
\le & 2 z_{k_0}^T P z_{k_0} + 2\lambda_{\max} \| \sum_{j=k_0}^{i_0} a_j(t_{j+1})^{\frac{1}{\gamma}} \epsilon_{j+1} \|^2.
\end{aligned}
$$

Thus, by (3.2.12) and the definition of k_0

$$
\|z_{i_0+1}\|^2 \lambda_{\min} \le 2\lambda_{\max} \|z_{k_0}\|^2 + 2\lambda_{\max} \le 2\lambda_{\max}(d_0^2 + 1),
$$

or

$$
\|z_{i_0+1}\|^2 \le 2K(d_0^2 + 1) = d.
$$

This contradicts with the definition of i_0, and hence i_0 must be infinite. Consequently, $\{z_k\}$ is bounded. □

Proof of Theorem 3.2.1.

By Lemma 3.2.2 and the fact $r(x_k)/\|x_k - x^0\|^{1+\gamma} \longrightarrow 0$ as $k \longrightarrow \infty$, we have

$$\nu_{k+1} \triangleq g_k(z_k) - g(z_k) \longrightarrow 0 \text{ as } k \longrightarrow \infty, \tag{3.2.21}$$

where

$$g(z) = \frac{z}{\gamma} + F\|z\|^\gamma z. \tag{3.2.22}$$

By setting

$$e_{k+1} = t_k (t_{k+1})^{\frac{1}{\gamma}} \epsilon_{k+1},$$

from (3.2.9) it follows that

$$z_{k+1} = z_k + b_k(g(z_k) + \nu_{k+1} + e_{k+1}), \quad b_k = \frac{a_k}{t_k}. \tag{3.2.23}$$

This is nothing else but an RM algorithm. Since by Lemma 3.2.2 $\{z_k\}$ is bounded, no truncation is needed and one may apply Theorem 2.2.1".

First note that

$$\begin{aligned}\sum_{j=1}^{k} b_j &= \sum_{j=1}^{k} \frac{a_j}{t_j} = \sum_{j=1}^{k} \frac{t_{j+1} - t_j}{t_j} \\ &\geq \sum_{j=1}^{k} \int_{t_j}^{t_{j+1}} \frac{dx}{x} = \ln t_{k+1} - \ln t_1 \xrightarrow[k \longrightarrow \infty]{} \infty.\end{aligned}$$

Hence, A2.2.1 is satisfied.

Notice $\sum_{j=1}^{\infty} b_j e_{j+1} = \sum_{j=1}^{\infty} a_j (t_{j+1})^{\frac{1}{\gamma}} \epsilon_{j+1} < \infty$ by (3.2.2), and $\nu_k \longrightarrow 0$ as $k \longrightarrow \infty$. So A2.2.3 holds with a_k replaced by b_k.

A2.2.4 is clearly satisfied, since $g(\cdot)$ is continuous. The key issue is to find a $v(\cdot)$ satisfying A2.2.2".

Take

$$v(z) = z^T P z, \tag{3.2.24}$$

and define $J \triangleq \{z : z^T P z \leq \lambda_{\max} (\frac{\lambda_{\max}}{\gamma})^{\frac{2}{\gamma}}\}$, which is closed.

Notice

$$\begin{aligned}v_z^T(z) g(z) &= \frac{1}{2}(v_z^T(z) g(z) + g^T(z) v_z(z)) \\ &= \frac{1}{\gamma} z^T P z + \frac{1}{2} z^T (PF + F^T P) z \|z\|^\gamma \\ &= \frac{1}{\gamma} z^T P z - \|z\|^{2+\gamma}.\end{aligned}$$

For $z \notin J$, $z^T P z > \lambda_{\max}(\frac{\lambda_{\max}}{\gamma})^{\frac{2}{\gamma}}$, and $\|z\| > (\frac{\lambda_{\max}}{\gamma})^{\frac{1}{\gamma}}$.
Then we have

$$\frac{1}{\gamma} z^T P z - \|z\|^{2+\gamma} \le \frac{1}{\gamma} z^T P z - \frac{z^T P z}{\lambda_{\max}} \|z\|^{\gamma} = \frac{z^T P z}{\lambda_{\max}} (\frac{\lambda_{\max}}{\gamma} - \|z\|^{\gamma}) < 0.$$

This means that

$$v_z^T(z) g(z) < 0, \text{ for } z \notin J,$$

and the condition A2.2.2" holds.

By Theorem 2.2.1", $d(v(z_k), v(J)) \xrightarrow[k \to \infty]{} 0$. This implies

$$\limsup_{k \to \infty} z_k^T P z_k \le \lambda_{\max}(\frac{\lambda_{\max}}{\gamma})^{\frac{2}{\gamma}},$$

which in turn implies (3.2.7) by (3.2.8). □

Imposing some additional conditions on F, we may have more precise than (3.2.7) results by using different Lyapunov functions.

Theorem 3.2.2 *Assume A3.2.1–A3.2.4 hold, in addition, assume F is normal, i.e., $F^T F = F F^T$. Let $\{x_k\}$ be given by (2.1.1)–(2.1.3). Then for those sample paths for which A3.2.3 holds, $(t_k)^{\frac{1}{\gamma}} \|x_k - x^0\|$ converges to either zero or one of $(\frac{1}{-\lambda_j \gamma})^{\frac{1}{\gamma}}$, where λ_j denotes an eigenvalue of $H \triangleq \frac{F+F^T}{2}$. More precisely,*

$$(t_k)^{\frac{1}{\gamma}} (x_k - x^0) \xrightarrow[k \to \infty]{} \begin{cases} 0, & or \\ (\frac{1}{-\lambda_j \gamma})^{\frac{1}{\gamma}} \phi_j, \end{cases} \tag{3.2.25}$$

where ϕ_j is an unit eigenvector of H corresponding to λ_j.

Proof. Since F is stable, the integral

$$Q \triangleq \int_0^{\infty} e^{Ft} e^{F^T t} dt$$

is well defined. Noticing that $F F^T = F^T F$, we have

$$e^{Ft} e^{F^T t} = e^{(F+F^T)t},$$

and

$$Q = \int_0^{\infty} e^{(F+F^T)t} dt = \int_0^{\infty} e^{2Ht} dt < \infty.$$

This means that H is also stable. Therefore, all eigenvalues λ_j are negative. Further, by $FF^T = F^TF$, we find

$$F^{-1}F^T = F^{-1}F^TFF^{-1} = F^{-1}FF^TF^{-1} = F^TF^{-1},$$

and hence

$$(F + F^T)^{-1}F = F(F + F^T)^{-1} \text{ and } H^{-1}F = FH^{-1}. \tag{3.2.26}$$

We consider (3.2.23) and take

$$v(z) = \frac{1}{2\gamma}z^TH^{-1}z + \frac{1}{\gamma+2}\|z\|^{2+\gamma}. \tag{3.2.27}$$

By (3.2.26) we have

$$\begin{aligned}
&v_z^T(z)g(z)\\
&= \frac{1}{2}\big(v_z^T(z)g(z) + g^T(z)v_z(z)\big)\\
&= \frac{1}{2}\big((\frac{1}{\gamma}H^{-1}z + \|z\|^\gamma z)^T(\frac{z}{\gamma} + F\|z\|^\gamma z)\\
&\quad + (\frac{z}{\gamma} + F\|z\|^\gamma z)^T(\frac{1}{\gamma}H^{-1}z + \|z\|^\gamma z)\big)\\
&= \frac{1}{\gamma^2}z^TH^{-1}z + \frac{1}{\gamma}\|z\|^{2+\gamma} + \frac{1}{2\gamma}z^T(H^{-1}F + F^TH^{-1})z\|z\|^\gamma\\
&\quad + \frac{1}{2}z^T(F + F^T)z\|z\|^{2\gamma}\\
&= \frac{1}{\gamma^2}z^TH^{-1}z + \frac{2}{\gamma}\|z\|^{2+\gamma} + z^THz\|z\|^{2\gamma}\\
&= -\|\frac{1}{\gamma}(-H)^{-\frac{1}{2}}z - (-H)^{\frac{1}{2}}\|z\|^\gamma z\|^2.
\end{aligned}$$

Define

$$J = \{z : (\frac{1}{\gamma}(-H)^{-\frac{1}{2}} - (-H)^{\frac{1}{2}}\|z\|^\gamma)z = 0\}.$$

Obviously,

$$v_z^T(z)g(z) < 0 \tag{3.2.28}$$

for any $z \notin J$.

Clearly, $J = \{z : (\frac{1}{\gamma}I + H\|z\|^\gamma)z = 0\} = \{0, (\frac{1}{-\gamma\lambda_j})^{\frac{1}{\gamma}}\phi_j, \ j = 1, \dots, l\}$, where l is the dimension of x_k.

Thus, J is a discrete set, and $v(J)$ is nowhere dense because $v(z)$ is continuous. This together with (3.2.28) shows that A2.2.2' is satisfied.

By Theorem 2.2.1', $d(z_k, J) \xrightarrow[k\to\infty]{} 0$ and (3.2.25) is verified. □

Corollary 3.2.1 *Let* $a_k = \frac{1}{k}$. *Then* $\ln(k-1) < \sum_{j=2}^{k-1} \frac{1}{j} \le t_k = \sum_{j=1}^{k-1} a_j$ $= \sum_{j=1}^{k-1} \frac{1}{j} < \int_1^k \frac{dx}{x} = \ln k$. *In this case,*

$$\left|[(t_k)^{\frac{1}{\gamma}} - (\ln k)^{\frac{1}{\gamma}}](x_k - x^0)\right| \longrightarrow 0 \ as \ k \longrightarrow \infty,$$

and hence (3.2.7) and (3.2.25) are respectively equivalent to

$$\limsup_{k\longrightarrow\infty} (\ln k)^{\frac{1}{\gamma}} \|x_k - x^0\| \le \sqrt{k}\left(\frac{\lambda_{\max}}{r}\right)^{\frac{1}{\gamma}} \tag{3.2.29}$$

and

$$\lim_{k\longrightarrow\infty} (\ln k)^{\frac{1}{\gamma}} (x_k - x^0) = \begin{cases} 0, & or \\ (\frac{1}{-\lambda_j \gamma})^{\frac{1}{\gamma}} \phi_j. \end{cases} \tag{3.2.30}$$

Remark 3.2.1 For $a_k = \frac{1}{k}$, the convergence rate given by (3.1.18) for the nondegenerate case is $o(k^{-\delta})$, while for the degenerate case is $O((\ln k)^{-\frac{1}{\gamma}})$ by (3.2.29), which is much slower than $o(k^{-\delta})$.

3.3. Asymptotic Normality

In Theorem 3.1.1 we have shown that $\|x_k - x^0\| = o(a_k^\delta)$ for $\{x_k\}$ given by (2.1.1)–(2.1.3). As shown in Remark 3.1.2, $\delta < \frac{1}{2}$ if $a_k = \frac{1}{k^\alpha}$, $\alpha \in (\frac{1}{2}, 1]$. This is a path-wise result. Assuming the observation noise $\{\epsilon_k\}$ is a random sequence, we show that $\frac{1}{\sqrt{a_k}}(x_k - x^0)$ is asymptotically normal, i.e., the distribution of $\frac{1}{\sqrt{a_k}}(x_k - x^0)$ converges to a normal distribution as $k \longrightarrow \infty$. This convergence implies that δ in the convergence rate $\|x_k - x^0\| = o(a_k^\delta)$ cannot be improved to $\frac{1}{2}$.

We first consider the linear regression case, i.e., $f(\cdot)$ is a linear function, but may be time-varying.

Let us introduce a central limit theorem on double-indexed random variables. We formulate it as a lemma.

Lemma 3.3.1 *Let* ξ_{ki}, $1 \le i \le k$ *be an array of l-dimensional random vectors. Denote*

$$S_{ki} \triangleq E\xi_{ki}\xi_{ki}^T, \quad R_{ki} = E(\xi_{ki}\xi_{ki}^T | \xi_{k1}, \ldots, \xi_{k,i-1})$$

and

$$S_k = \sum_{i=1}^{k} S_{ki}.$$

Assume

$$E(\xi_{ki}|\xi_{k1},\ldots,\xi_{k,i-1}) = 0, \tag{3.3.1}$$

$$\sup_{k>1} \sum_{i=1}^{k} E\|\xi_{ki}\|^2 < \infty, \quad \lim_{k\longrightarrow\infty} S_k = S, \tag{3.3.2}$$

$$\lim_{k\longrightarrow\infty} \sum_{i=1}^{k} E\|S_{ki} - R_{ki}\| = 0, \tag{3.3.3}$$

and

$$\lim_{k\longrightarrow\infty} \sum_{i=1}^{k} E\|\xi_{ki}\|^2 I_{[\|\xi_{ki}\|>\epsilon]} = 0, \quad \forall\epsilon > 0. \tag{3.3.4}$$

Then

$$\sum_{i=1}^{k} \xi_{ki} \xrightarrow[k\longrightarrow\infty]{d} N(0, S), \tag{3.3.5}$$

where and hereafter $N(\mu, S)$ denotes the normal distribution with mean μ and covariance S.

Let us first consider the linear recursion (3.1.6) and derive its asymptotic normality. We keep the notation $\Phi_{k,j}$ introduced by (3.1.7).

We have obtained estimate (3.1.8) for $\|\Phi_{k,j}\|$ and now derive more properties for it.

Lemma 3.3.2 *Assume $a_k > 0$, $a_k \xrightarrow[k\longrightarrow\infty]{} 0$, $\sum_{k=1}^{\infty} a_k = \infty$, and $H_k \xrightarrow[k\longrightarrow\infty]{} H$ where H is stable. Then for any $r > 0$*

$$\sup_{k} \sum_{i=1}^{k} a_i \|\Phi_{k,i+1}\|^r < \infty. \tag{3.3.6}$$

Proof. By (3.1.8) it follows that

$$\sum_{i=1}^{k} a_i \|\Phi_{k,i+1}\|^r \le c_0^r \sum_{i=1}^{k} a_i \exp(-cr \sum_{j=i+1}^{k} a_j). \tag{3.3.7}$$

We will use the following elementary inequality

$$1 - e^{-x} + \frac{x^2}{2} \geq x, \quad \forall x \geq 0, \tag{3.3.8}$$

which follows from the fact that the function $1 - e^{-x} - x + \frac{x^2}{2}$ equals zero at $x = 0$ and its derivative $e^{-x} - 1 + x \geq 0$. By (3.3.8), we derive

$$\begin{aligned}
\sum_{i=1}^{k} a_i \exp(-cr \sum_{j=i+1}^{k} a_j) &\leq \frac{1}{cr} \sum_{i=1}^{k} (1 - e^{-cra_i} + \frac{(cra_i)^2}{2}) \exp(-cr \sum_{j=i+1}^{k} a_j) \\
&= \frac{1}{cr} \sum_{i=1}^{k} \left[\exp(-cr \sum_{j=i+1}^{k} a_j) - exp(-cr \sum_{j=i}^{k} a_j)\right] \\
&\quad + \frac{cr}{2} \sum_{i=1}^{k} a_i^2 \exp(-cr \sum_{j=i+1}^{k} a_j),
\end{aligned}$$

which implies

$$\sum_{i=1}^{k} (1 - \frac{cr}{2} a_i) a_i \exp(-cr \sum_{j=i+1}^{k} a_j) \leq \frac{1}{cr} [1 - \exp(-cr \sum_{j=1}^{k} a_j)] < \frac{1}{cr}. \tag{3.3.9}$$

Assume k_1 is sufficiently large such that $\frac{cr}{2} a_i < \frac{1}{2}$, $\forall i \geq k_1$. Then

$$\begin{aligned}
\sum_{i=1}^{k} a_i \exp(-cr \sum_{j=i+1}^{k} a_j) &< \sum_{i=1}^{k_1} a_i + \sum_{i=k_1+1}^{k} a_i \exp(-cr \sum_{j=i+1}^{k} a_j) \\
&< \sum_{i=1}^{k_1} a_i + 2 \sum_{i=k_1+1}^{k} (1 - \frac{cr}{2} a_i) a_i \exp(-cr \sum_{j=i+1}^{k} a_j) \\
&< \sum_{i=1}^{k_1} a_i + \frac{2}{cr},
\end{aligned} \tag{3.3.10}$$

where for the last inequality (3.3.9) is invoked.

Combining (3.3.7) and (3.3.10) gives (3.3.6). □

Lemma 3.3.3 *Set*

$$\theta_{k,i} = \exp(H \sum_{j=i}^{k} a_j) - \Phi_{k,i}. \tag{3.3.11}$$

Under conditions of Lemma 3.3.2, $\sup_{k,i} \|\theta_{k,i}\| < \infty$, $\|\theta_{k,i}\| \longrightarrow 0$ *as* $i \longrightarrow \infty$ *uniformly with respect to* $k : k \geq i$, *and* $\|\theta_{k,i}\| \longrightarrow 0$ *as* $k \longrightarrow \infty$ *uniformly with respect to* $i : i \leq k$.

Proof. Expanding $e^{a_k H}$ to the series

$$e^{a_k H} = (I + a_k H + A_k)$$

with $\|A_k\| = O(a_k^2)$, we have

$$\begin{aligned}\theta_{k,i} &= (I + a_k H + A_k)\exp(H\sum_{j=i}^{k-1} a_j) - (I + a_k H_k)\Phi_{k-1,i} \\ &= (I + a_k H)\theta_{k-1,i} + (a_k(H - H_k) + A_k)\exp(H\sum_{j=i}^{k-1} a_j) \\ &= \sum_{s=i}^{k} \Phi_{k,s+1}(a_s(H - H_s) + A_s)\exp(H\sum_{j=i}^{s-1} a_j)\end{aligned} \tag{3.3.12}$$

where by definition $\sum_{j=i}^{i-1} a_j \triangleq 0$.

By stability of H, there exist constants $c_1 > 0$ and $\rho > 0$ such that

$$\|\exp(H\sum_{j=i}^{k} a_j)\| \leq c_1 \exp(-\rho \sum_{j=i}^{k} a_j). \tag{3.3.13}$$

Putting (3.3.13) into (3.3.12) yields that for any $\epsilon > 0$

$$\begin{aligned}\|\theta_{k,i}\| &\leq \sum_{s=i}^{k} \|\Phi_{k,s+1}\| a_s(o(1) + O(a_s))c_1 \exp(-\rho\sum_{j=i}^{s-1} a_j) \\ &\leq c_1 \sum_{s=i}^{k} a_s \|\Phi_{k,s+1}\|(o(1) + O(a_s)) \\ &\leq \frac{\epsilon}{2} c_1 \sum_{s=i_1}^{k} a_s \|\Phi_{k,s+1}\| \\ &\quad + c_0 c_1 \sum_{s=i}^{i_1-1} a_s[\exp(-c\sum_{j=s+1}^{k} a_j)](o(1) + O(a_s)),\end{aligned} \tag{3.3.14}$$

where for the last inequality i_1 is assumed to be sufficiently large such that $(o(1) + O(a_s)) < \frac{\epsilon}{2}$, $\forall s \geq i_1$, and (3.1.8) is used too.

Since $\sum_{j=1}^{\infty} a_j = \infty$ and $\epsilon > 0$ may be arbitrarily small the conclusions of the lemma follow from (3.3.14) by Lemma 3.3.2. □

Lemma 3.3.4 *Assume $a_k > 0$, $a_k \longrightarrow 0$ as $k \longrightarrow \infty$, and $\sum_{k=1}^{\infty} a_k = \infty$. Let A, B, and Q be $l \times l$ matrices and let A and B be stable. Then*

$$\lim_{k\longrightarrow\infty} \sum_{i=1}^{k} a_i (\exp(A \sum_{j=i+1}^{k} a_j))Q \exp(B \sum_{j=i+1}^{k} a_j) = \int_0^{\infty} e^{At} Q e^{Bt} dt. \tag{3.3.15}$$

Proof. For any $T > 0$ define

$$s(k,T) = \min\{s : \sum_{i=s}^{k} a_i \le T\}. \tag{3.3.16}$$

Since $a_i \longrightarrow 0$, $\sum_{i=s(k,T)}^{k} a_i \xrightarrow[k\longrightarrow\infty]{} T$ for fixed T. Denoting $\sum_{j=s(k,T)}^{i} a_j$ by t, we then have $\sum_{j=i+1}^{k} a_j \xrightarrow[k\longrightarrow\infty]{} T - t$. Consequently,

$$\sum_{i=s(k,T)}^{k} a_i \big[\exp(A \sum_{j=i+1}^{k} a_j)\big] Q \exp(B \sum_{j=i+1}^{k} a_j)$$

serves as an integral sum for $\int_0^T e^{A(T-t)} Q e^{B(T-t)} dt$, or equivalently, for $\int_0^T e^{At} Q e^{Bt} dt$, and hence

$$\lim_{k\longrightarrow\infty} \sum_{i=s(k,T)}^{k} a_i \Big[\exp(A \sum_{j=i+1}^{k} a_j)\Big] Q \exp(B \sum_{j=i+1}^{k} a_j) = \int_0^{T} e^{At} Q e^{Bt} dt. \tag{3.3.17}$$

Therefore, for (3.3.15) it suffices to show that

$$\lim_{T\longrightarrow\infty} \lim_{k\longrightarrow\infty} \sum_{i=1}^{s(k,T)-1} a_i [\exp(A \sum_{j=i+1}^{k} a_j)] Q \exp(B \sum_{j=i+1}^{k} a_j) = 0. \tag{3.3.18}$$

Similar to (3.3.10), by stability of A we can show that there is a constant $\eta_0 < \infty$ such that

$$\sup_{k,T} \sum_{i=1}^{s(k,T)-1} a_i \| \exp(A \sum_{j=i+1}^{s(k,T)-1} a_j) \| < \eta_0.$$

By stability of A and B, constants $\eta_1 > 0$, $\eta_2 > 0$, and $\eta > 0$ exist such that

$$\|Q\exp(B\sum_{j=i+1}^{k} a_j)\| \le \eta_1, \quad \forall k \ge i+1, \quad \|e^{At}\| \le \eta_2 e^{-\eta t}, \quad \forall t \ge 0.$$

Consequently, we have

$$\limsup_{k\longrightarrow\infty} \sum_{i=1}^{s(k,T)-1} a_i \|[\exp(A\sum_{j=i+1}^{k} a_j)]Q(B\sum_{j=i+1}^{k} a_j)\|$$

$$\le \eta_1 \limsup_{k\longrightarrow\infty} \|\exp(A\sum_{j=s(k,T)}^{k} a_j)\| \sum_{i=1}^{s(k,T)-1} a_i\|\exp(A\sum_{j=i+1}^{s(k,T)-1} a_j)\|$$

$$\le \eta_0\eta_1\eta_2 \limsup_{k\longrightarrow\infty} e^{-\eta\sum_{j=s(k,T)}^{k} a_j} = \eta_0\eta_1\eta_2 e^{-\eta T} \xrightarrow[T\longrightarrow\infty]{} 0,$$

which verifies (3.3.18) and completes the proof of the lemma. □

Theorem 3.3.1 *Let $\{x_k\}$ be given by (3.1.6) with an arbitrarily given initial value. Assume the following conditions holds:*

i) $a_k > 0$, $a_k \longrightarrow 0$ as $k \longrightarrow \infty$, $\sum_{k=1}^{\infty} a_k = \infty$, and

$$a_{k+1}^{-1} - a_k^{-1} \longrightarrow \alpha \ge 0 \text{ as } k \longrightarrow \infty; \tag{3.3.19}$$

ii) $H_k \longrightarrow H$ and $H + \frac{\alpha}{2}I$ is stable;
iii)

$$e_k = \sum_{i=0}^{\infty} C_i w_{k-i}, \quad w_i = 0 \quad for \quad i < 0, \quad \nu_k = o(\sqrt{a_k}), \tag{3.3.20}$$

where C_i are $l \times l$ constant matrices with $\sum_{i=0}^{\infty} \|C_i\| < \infty$ and $\{w_k, \mathcal{F}_k\}$ is a martingale difference sequence of l-dimension satisfying the following conditions:

$$E(w_k|\mathcal{F}_{k-1}) = 0, \quad \sup_k E(\|w_k\|^2|\mathcal{F}_{k-1}) \le \sigma \text{ with } \sigma \text{ being a constant}, \tag{3.3.21}$$

$$\lim_{k\longrightarrow\infty} E(w_k w_k^T|\mathcal{F}_{k-1}) = \lim_{k\longrightarrow\infty} Ew_k w_k^T \triangleq S_0 \text{ a.s.}, \tag{3.3.22}$$

and

$$\lim_{N \longrightarrow \infty} \sup_k E\|w_k\|^2 I_{[\|w_k\|>N]} = 0. \tag{3.3.23}$$

Then $\frac{1}{\sqrt{a_k}} x_k$ *is asymptotically normal:*

$$\frac{1}{\sqrt{a_k}} x_k \xrightarrow[k \longrightarrow \infty]{d} N(0, S), \tag{3.3.24}$$

where

$$S = \int_0^\infty e^{(H + \frac{\alpha}{2} I)t} \sum_{i=0}^\infty C_i S_0 \sum_{i=0}^\infty C_i^T e^{(H^T + \frac{\alpha}{2} I)t} dt. \tag{3.3.25}$$

Proof. Define $\{u_k\}$ by the following recursion

$$u_{k+1} = (I + a_k H_k) u_k + a_k e_{k+1}, \quad u_0 = 0. \tag{3.3.26}$$

By (3.1.6) it follows that

$$x_{k+1} - u_{k+1} = (I + a_k H_k)(x_k - u_k) + a_k \nu_{k+1}. \tag{3.3.27}$$

Using (3.3.19) we have

$$\begin{aligned}
(\frac{a_k}{a_{k+1}})^{\frac{1}{2}} &= (\frac{a_k - a_{k+1}}{a_{k+1}} + 1)^{\frac{1}{2}} \\
&= 1 + \frac{1}{2} a_k (a_{k+1}^{-1} - a_k^{-1}) + O((\frac{a_k - a_{k+1}}{a_{k+1}})^2) \\
&= 1 + \frac{1}{2} \alpha a_k + o(a_k).
\end{aligned} \tag{3.3.28}$$

Consequently,

$$\begin{aligned}
\frac{x_{k+1} - u_{k+1}}{\sqrt{a_{k+1}}} &= (1 + \frac{1}{2} \alpha a_k + o(a_k)) \big[(I + a_k H_k) \frac{(x_k - u_k)}{\sqrt{a_k}} + \sqrt{a_k} \nu_{k+1} \big] \\
&= (I + a_k A_k) \frac{x_k - u_k}{\sqrt{a_k}} + \sqrt{a_k} \nu'_{k+1},
\end{aligned} \tag{3.3.29}$$

where

$$\begin{aligned}
A_k =& H_k + \frac{\alpha}{2} I + \frac{1}{2} a_k \alpha H_k + o(1)(I + a_k H_k) \\
& \xrightarrow[k \longrightarrow \infty]{} A \triangleq H + \frac{\alpha}{2} I,
\end{aligned} \tag{3.3.30}$$

and

$$\nu'_{k+1} \triangleq (1 + \frac{1}{2}\alpha a_k + o(a_k))\nu_{k+1} = o(\sqrt{a_{k+1}}) \tag{3.3.31}$$

by (3.3.20).

Define

$$\Psi_{k,i} \triangleq (I + a_k A_k) \cdots (I + a_i A_i), \quad \Psi_{i-1,i} \triangleq I. \tag{3.3.32}$$

By (3.3.30) and stability of A, from (3.1.8) it follows that constants $b_0 > 0$ and $b > 0$ exist such that

$$\|\Psi_{k,i}\| \leq b_0 \exp(-b \sum_{j=i}^{k} a_j), \quad k \geq i. \tag{3.3.33}$$

Consequently, from (3.3.29) we have

$$\frac{x_{k+1} - u_{k+1}}{\sqrt{a_{k+1}}} = \Psi_{k,1}(\frac{x_1 - u_1}{\sqrt{a_1}}) + \sum_{i=1}^{k} \Psi_{k,i+1}\sqrt{a_i}\nu'_{i+1}. \tag{3.3.34}$$

The first term on the right-hand side of (3.3.34) tends to zero as $k \longrightarrow \infty$ by (3.3.33), while the second term is estimated as follows. By (3.3.31)

$$\begin{aligned}\lim_{k\longrightarrow\infty} \|\sum_{i=1}^{k} \Psi_{k,i+1}\sqrt{a_i}\nu'_{i+1}\| &\leq \lim_{k\longrightarrow\infty} \sum_{i=1}^{k} a_i \|\Psi_{k,i+1}\| o(1) \\ &= \lim_{k\longrightarrow\infty} \sum_{i=1}^{k} a_i \|\Psi_{k,i+1}\|^{\frac{1}{2}} \cdot \|\Psi_{k,i+1}\|^{\frac{1}{2}} o(1) = 0,\end{aligned}$$

where for the last equality, Lemma 3.3.2 and (3.3.33) are used. This means that $\frac{x_k}{\sqrt{a_k}}$ and $\frac{u_k}{\sqrt{a_k}}$ have the same limit distribution if exists. Consequently, for the theorem it suffices to show

$$\frac{u_k}{\sqrt{a_k}} \xrightarrow[k\longrightarrow\infty]{d} N(0, S). \tag{3.3.35}$$

Similar to (3.3.29) and (3.3.31), by (3.3.28) we have

$$\begin{aligned}\frac{1}{\sqrt{a_{k+1}}} u_{k+1} = &\sum_{i=1}^{k} \Psi_{k,i+1}\sqrt{a_i} \sum_{j=0}^{\infty} C_j w_{i+1-j} \\ &+ \sum_{i=1}^{k} \Psi_{k,i+1}\sqrt{a_i} e_{i+1}(\frac{1}{2}\alpha a_i + o(a_i)).\end{aligned} \tag{3.3.36}$$

By (3.3.21) and $\sum_{i=1}^{\infty} \|C_i\| < \infty$, we see that $\sup_i E\|e_i\|^2 \triangleq \mu^2 < \infty$.

Noticing

$$E\|\sum_{i=1}^{k} \Psi_{k,i+1}\sqrt{a_i}e_{i+1}(\frac{1}{2}\alpha a_i + o(a_i))\|$$
$$\leq \sum_{i=1}^{k} \|\Psi_{k,i+1}\|a_i\mu(\frac{\alpha}{2}\sqrt{a_i} + o(\sqrt{a_i})) \xrightarrow[k\longrightarrow\infty]{} 0$$

by Lemma 3.3.2 and (3.1.8), we find that the last term of (3.3.36) tends to zero in probability. Therefore, for (3.3.24) it suffices to show

$$\sum_{i=1}^{k} \Psi_{k,i+1}\sqrt{a_i}\sum_{j=0}^{\infty} C_j w_{i+1-j} \xrightarrow[k\longrightarrow\infty]{d} N(0,S). \tag{3.3.37}$$

We now show that for (3.3.37) it is sufficient to prove

$$\sum_{i=1}^{k} \Psi_{k,i+1}\sqrt{a_i}(\sum_{j=0}^{\infty} C_j) w_{i+1} \xrightarrow[k\longrightarrow\infty]{d} N(0,S). \tag{3.3.38}$$

For any fixed $r > 0$, we have

$$\sum_{i=1}^{k} \Psi_{k,i+1}\sqrt{a_i}\sum_{j=0}^{\infty} C_j (w_{i+1-j} - w_{i+1})$$
$$= \sum_{j=0}^{r}\sum_{i=1}^{k} \Psi_{k,i+1}\sqrt{a_i}C_j(w_{i+1-j} - w_{i+1})$$
$$+ \sum_{i=1}^{k} \Psi_{k,i+1}\sqrt{a_i}\sum_{j=r+1}^{\infty} C_j(w_{i+1-j} - w_{i+1}). \tag{3.3.39}$$

By (3.3.21) we have

$$E\|\sum_{i=1}^{k} \Psi_{k,i+1}\sqrt{a_i}\sum_{j=r+1}^{\infty} C_j w_{i+1-j}\| \leq \sum_{j=r+1}^{\infty} \|C_j\| E\|\sum_{i=1}^{k} \Psi_{k,i+1}\sqrt{a_i}w_{i+1-j}\|$$
$$\leq \sum_{j=r+1}^{\infty} \|C_j\|(\sum_{i=1}^{k} \|\Psi_{k,i+1}\|^2 a_i\sigma)^{\frac{1}{2}} \xrightarrow[r\longrightarrow\infty]{} 0,$$

where convergence to zero follows from $\sum_{i=1}^{\infty} \|C_i\| < \infty$ and Lemma 3.3.2. It is worth noting that the convergence is uniform with respect to k. This implies that the second term on the right-hand side of (3.3.39) tends to zero in probability. The first term on the right-hand side of (3.3.39) can be rewritten as

$$\sum_{j=0}^{r}\sum_{i=1}^{j}\sqrt{a_i}\Psi_{k,i+1}C_j w_{i+1-j} - \sum_{j=0}^{r}\sum_{i=k-j+1}^{k}\sqrt{a_i}\Psi_{k,i+1}C_j w_{i+1}$$
$$-\sum_{j=0}^{r}\sum_{i=1}^{k-j}(\sqrt{a_i}\Psi_{k,i+1}-\sqrt{a_{i+j}}\Psi_{k,i+j+1})C_j w_{i+1}. \quad (3.3.40)$$

By (3.3.33) for any fixed r we estimate the first term of (3.3.40) as follows

$$\|\sum_{j=0}^{r}\sum_{i=1}^{j}\sqrt{a_i}\Psi_{k,i+1}C_j w_{i+1-j}\|$$
$$\leq b_0\sum_{j=0}^{r}\sum_{i=1}^{j}\sqrt{a_i}\exp(-b\sum_{s=i+1}^{k}a_s)\|C_j w_{i+1-j}\| \xrightarrow[k\to\infty]{} 0, \quad (3.3.41)$$

while for the second term we have

$$E\|\sum_{j=0}^{r}\sum_{i=k-j+1}^{k}\sqrt{a_i}\Psi_{k,i+1}C_j w_{i+1}\|$$
$$\leq b_0\sum_{j=0}^{r}(\sum_{i=k-j+1}^{k}a_i\exp(-2b\sum_{s=i+1}^{k}a_s)\|C_j\|^2\sigma)^{\frac{1}{2}} \xrightarrow[k\to\infty]{} 0, \quad (3.3.42)$$

since $a_i \longrightarrow 0$ and $\sum_{s=i+1}^{k} a_s \longrightarrow \infty$ as $k \longrightarrow \infty$.

We now show that the last term of (3.3.40) also converges to zero in probability as $k \longrightarrow \infty$.

Notice that by (3.3.28), $\frac{a_k}{a_{k+j}} \xrightarrow[k\to\infty]{} 1$ for any fixed $j \geq 1$, and $\left(\frac{a_k}{a_{k+j}}\right)^{\frac{1}{2}} = 1 + O(a_k)$. Therefore, for a fixed r, there exist constants m_1 and m_0 such that

$$0 \leq \left(\frac{a_k}{a_{k+j}}\right)^{\frac{1}{2}} - 1 \leq m_1 a_k = m_1\frac{a_k}{a_{k+j}}\cdot a_{k+j} \leq m_1(1+m_1a_k)^2 a_{k+j}$$
$$\leq m_0 a_{k+j}, \quad \forall k \geq 1, \forall j \leq r. \quad (3.3.43)$$

Then the last term of (3.3.40) is estimated as follows:

$$\sum_{j=0}^{r}\sum_{i=1}^{k-j}(\sqrt{a_i}\Psi_{k,i+1}-\sqrt{a_{i+j}}\Psi_{k,i+j+1})C_jw_{i+1}$$

$$=\sum_{j=0}^{r}\sum_{i=1}^{k-j}\sqrt{a_i}(\Psi_{k,i+1}-\Psi_{k,i+j+1})C_jw_{i+1}$$

$$+\sum_{j=0}^{r}\sum_{i=1}^{k-j}((\frac{a_i}{a_{i+j}})^{\frac{1}{2}}-1)\sqrt{a_{i+j}}\Psi_{k,i+j+1}C_jw_{i+1}$$

$$=\sum_{j=0}^{r}\sum_{i=1}^{k-j}\sqrt{a_i}\sum_{s=1}^{j}\Psi_{k,i+s+1}a_{i+s}A_{i+s}C_jw_{i+1}$$

$$+\sum_{j=0}^{r}\sum_{i=1}^{k-j}((\frac{a_i}{a_{i+j}})^{\frac{1}{2}}-1)\sqrt{a_{i+j}}\Psi_{k,i+j+1}C_jw_{i+1}. \tag{3.3.44}$$

For the first term on the right-hand side of (3.3.44) we have

$$\sum_{j=0}^{r}E\|\sum_{i=1}^{k-j}\sqrt{a_i}\sum_{s=1}^{j}\Psi_{k,i+s+1}a_{i+s}A_{i+s}C_jw_{i+1}\|$$

$$\leq\sum_{j=0}^{r}(\sum_{i=1}^{k-j}a_i\|\sum_{s=1}^{j}\Psi_{k,i+s+1}a_{i+s}A_{i+s}C_j\|^2\sigma)^{\frac{1}{2}}$$

$$\leq\sum_{j=0}^{r}(\sum_{i=1}^{k-j}a_i\sum_{s=1}^{j}\|\Psi_{k,i+s+1}\|^2a_{i+s}^2\sum_{s=1}^{j}\|A_{i+s}C_j\|^2\sigma)^{\frac{1}{2}}$$

$$\leq\sum_{j=0}^{r}(m_2\sum_{s=1}^{j}\sum_{i=1}^{k-j}a_i\|\Psi_{k,i+s+1}\|^2a_{i+s}^2)^{\frac{1}{2}}, \tag{3.3.45}$$

where the last inequality is obtained because $\sum_{s=1}^{j}\|A_{i+s}C_j\|^2\sigma$ is bounded by some constant m_2, $\forall j\leq r$, by (3.3.30). Since r is fixed, in order to prove that the right-hand side of (3.3.45) tends to zero as $k\longrightarrow\infty$, it suffices to show

$$\sum_{i=1}^{k-j}a_i\|\Psi_{k,i+s+1}\|^2a_{i+s}^2\xrightarrow[k\longrightarrow\infty]{}0,\quad\forall s\leq j,\quad\forall j\leq r. \tag{3.3.46}$$

By (3.3.33), for any fixed i_1,

$$\sum_{i=1}^{i_1} a_i \|\Psi_{k,i+s+1}\|^2 a_{i+s}^2 \xrightarrow[k\longrightarrow\infty]{} 0, \qquad (3.3.47)$$

while for any given $\epsilon > 0$, we may take i_1 sufficiently large such that $a_i < \epsilon$, $\forall i \geq i_1$. Therefore,

$$\begin{aligned}\sum_{i=i_1+1}^{k-j} a_i \|\Psi_{k,i+s+1}\|^2 a_{i+s}^2 &\leq \epsilon \sum_{i=i_1+1}^{k-j} \|\Psi_{k,i+s+1}\|^2 a_{i+s}^2 \\ &\leq \epsilon^2 \sum_{i=i_1+1}^{k-j} \|\Psi_{k,i+s+1}\|^2 a_{i+s} \\ &\leq \epsilon^2 \sup_k \sum_{i=1}^{k} a_i \|\Psi_{k,i+1}\|^2 \xrightarrow[\epsilon\longrightarrow 0]{} 0 \qquad (3.3.48)\end{aligned}$$

by Lemma 3.3.2.

Incorporating (3.3.47) with (3.3.48) proves (3.3.46). Therefore, the right-hand side of (3.3.45) tends to zero as $k \longrightarrow \infty$. This implies that the first term on the right-hand side of (3.3.44) tends to zero in probability.

By (3.3.43), for the last term of (3.3.44) we have

$$\begin{aligned}&E\|\sum_{j=0}^{r}\sum_{i=1}^{k-j}((\frac{a_i}{a_{i+j}})^{\frac{1}{2}} - 1)\sqrt{a_{i+j}}\Psi_{k,i+j+1}C_j w_{i+1}\| \\ &\leq \sum_{j=0}^{r}(\sum_{i=1}^{k-j} m_0^2 a_{i+j}^3 \|\Psi_{k,i+j+1}\|^2 \|C_j\|^2 \sigma)^{\frac{1}{2}},\end{aligned}$$

which tends to zero as $k \longrightarrow \infty$ as can be shown by an argument similar to that used for (3.3.45).

In summary we conclude that the right-hand side of (3.3.44) tends to zero in probability, and hence all terms in (3.3.40) tend to zero in probability. This implies that the right-hand side of (3.3.39) tends to zero in probability as $k \longrightarrow \infty$ and then $r \longrightarrow \infty$. Thus, we have shown that for (3.3.37) it suffices to show (3.3.38).

We now intend to apply Lemma 3.3.1, identifying $\Psi_{k,i+1}\sqrt{a_i}(\sum_{j=0}^{\infty} C_j) \cdot w_{i+1}$ to $\xi_{k,i}$ in that lemma. We have to check conditions of the lemma.

Since $\{w_i\}$ is a martingale difference sequence, (3.3.1) is obviously satisfied.

By (3.3.22) and Lemma 3.3.2,

$$\lim_{k\longrightarrow\infty}\sum_{i=1}^{k} E\|S_{ki}-R_{ki}\| = \lim_{k\longrightarrow\infty}\sum_{i=1}^{k} E\|a_i\Psi_{k,i+1}(\sum_{j=0}^{\infty} C_j)[Ew_{i+1}w_{i+1}^T$$
$$- E(w_{i+1}w_{i+1}^T|\mathcal{F}_i)](\sum_{j=0}^{\infty} C_j^T)\Psi_{k,i+1}^T\| = 0.$$

This verifies (3.3.3). We now verify (3.3.2). We have

$$\begin{aligned} S_k &= \sum_{i=1}^{k} a_i\Psi_{k,i+1}(\sum_{j=0}^{\infty} C_j)Ew_{i+1}w_{i+1}^T(\sum_{j=0}^{\infty} C_j^T)\Psi_{k,i+1}^T \\ &= \sum_{i=1}^{k} a_i\Psi_{k,i+1}(\sum_{j=0}^{\infty} C_j)S_0(\sum_{j=0}^{\infty} C_j^T)\Psi_{k,i+1}^T \\ &\quad + \sum_{i=1}^{k} a_i\Psi_{k,i+1}(\sum_{j=0}^{\infty} C_j)(Ew_{i+1}w_{i+1}^T - S_0)(\sum_{j=0}^{\infty} C_j^T)\Psi_{k,i+1}^T, \end{aligned} \tag{3.3.49}$$

where the last term tends to zero by (3.3.22) and Lemma 3.3.2.

We show that the first term on the right-hand side of (3.3.49) tends to (3.3.25).

With A and $\Psi_{k,i}$ respectively identified to H and $\Phi_{k,i}$ in Lemma 3.3.3, by Lemmas 3.3.2 and 3.3.3 we have

$$\sum_{i=1}^{k} a_i\|\exp(A\sum_{j=i+1}^{k} a_j) - \Psi_{k,i+1}\|\|S_0\|\|\sum_{j=0}^{\infty} C_j\|^2\|\Psi_{k,i+1}\| \xrightarrow[k\longrightarrow\infty]{} 0.$$

This incorporating with (3.3.49) leads to

$$\|S_k - \sum_{i=1}^{k} a_i[\exp(A\sum_{j=i+1}^{k} a_j)](\sum_{j=0}^{\infty} C_j)S_0(\sum_{j=0}^{\infty} C_j)^T\exp(A^T\sum_{j=i+1}^{k} a_j)\|$$
$$\xrightarrow[k\longrightarrow\infty]{} 0.$$

By Lemma 3.3.4 we conclude

$$\|S_k - S\| \xrightarrow[k\longrightarrow\infty]{} 0.$$

Finally, we have to verify (3.3.4).

By (3.3.33) we have

$$[\|\Psi_{k,i+1}\sqrt{a_i}(\sum_{j=0}^{\infty} C_j)w_{i+1}\| > \epsilon]$$

$$\subset [\|w_{i+1}\| > b_0^{-1}\|\sum_{j=0}^{\infty} C_j\|^{-1}\frac{\epsilon}{\sqrt{a_i}}\exp(b\sum_{j=i+1}^{k} a_j)].$$

Noticing that $\sqrt{a_i}\exp(-b\sum_{j=i+1}^{k} a_j) \xrightarrow[k\to\infty]{} 0$ uniformly with respect to $i : i \geq 1$ since $a_i \xrightarrow[i\to\infty]{} 0$, or equivalently, $f(k,i) \triangleq \frac{1}{\sqrt{a_i}}\exp(b\sum_{j=i+1}^{k} a_j)$ $\xrightarrow[k\to\infty]{} \infty$ uniformly with respect to $i : i \geq 1$, by (3.3.23) we have

$$\sup_i E\Big(\|w_{i+1}\|^2 I_{[b_0\|\sum_{j=0}^{\infty} C_j\|\|w_{i+1}\| > \epsilon f(k,i)]}\Big) \xrightarrow[k\to\infty]{} 0.$$

Consequently, for any $\epsilon > 0$ by Lemma 3.3.2

$$\sum_{i=1}^{k} E(\|\Psi_{k,i+1}\sqrt{a_i}(\sum_{j=0}^{\infty} C_j)w_{i+1})w_{i+1}\|^2 I_{[\|\Psi_{k,i+1}\sqrt{a_i}(\sum_{j=0}^{\infty} C_j)w_{i+1}\| > \epsilon]}$$

$$\leq \sum_{i=1}^{k} a_i\|\Psi_{k,i+1}\|^2\|\sum_{j=0}^{\infty} C_j\|^2 E\|w_{i+1}\|^2 I_{[b_0\|\sum_{j=0}^{\infty} C_j\|\cdot\|w_{i+1}\| > \epsilon f(k,i)]}$$

$$\xrightarrow[k\to\infty]{} 0.$$

Thus, all conditions of Lemma 3.3.1 hold, and by this lemma we conclude (3.3.38). The proof is completed. □

Remark 3.3.1 Under the conditions of Theorem 3.3.1, if integers $k_i > k_{i-1} > \cdots > k_1 > a_k^{-1}$ are such that $\lim_{k\to\infty}(a_k k_i) = t_i$, then it can be shown that $(\frac{1}{\sqrt{a_{k_1}}}x_{k_1}, \ldots, \frac{1}{\sqrt{a_{k_i}}}x_{k_i})$ converges in distribution to $(x(t_1), \ldots, x(t_i))$, where $x(t)$ is a stationary Gaussian Markov process satisfying the following stochastic differential equation

$$dx(t) = Ax(t)dt + \sum_{j=0}^{\infty} C_j S_0^{\frac{1}{2}} dw_t,$$

where w_t is the l-dimensional standard Wiener process.

Corollary 3.3.1 From (3.1.7) and (3.3.28), similar to (3.3.29)–(3.3.31) we have

$$\frac{x_{k+1}}{\sqrt{a_{k+1}}} = (1+\frac{\alpha}{2}a_k + o(a_k))(I + a_k H_k)\frac{x_k}{\sqrt{a_k}} + \frac{a_k}{\sqrt{a_{k+1}}}(\nu_{k+1} + e_{k+1})$$
$$= (I + a_k A_k)\frac{x_k}{\sqrt{a_k}} + \frac{a_k}{\sqrt{a_{k+1}}}(\nu_{k+1} + e_{k+1}),$$

and

$$\frac{x_{k+1}}{\sqrt{a_{k+1}}} = \Psi_{k+1,m}\frac{x_m}{\sqrt{a_m}} + \sum_{i=m}^{k}\Psi_{k,i+1}\frac{a_i}{\sqrt{a_{i+1}}}(\nu_{i+1} + e_{i+1}). \tag{3.3.50}$$

By (3.3.33), the first term on the right-hand side of (3.3.50) tends to zero as $k \longrightarrow \infty$. Note that the last term in (3.3.34) has been proved to vanish as $k \longrightarrow \infty$, and it is just a different writing of $\sum_{i=1}^{\infty}\Psi_{k,i+1}\frac{a_i}{\sqrt{a_{i+1}}}\nu_{i+1}$. Therefore, from (3.3.50) by Theorem 3.3.1, it follows that for any fixed $m \geq 1$

$$\sum_{i=m}^{k}\Psi_{k,i+1}\frac{a_i}{\sqrt{a_{i+1}}}e_{i+1} \xrightarrow[k\longrightarrow\infty]{d} N(0,S). \tag{3.3.51}$$

We have discussed the asymptotic normality of $\frac{x_k - x^0}{\sqrt{a_k}}$ for the case where $f(\cdot)$ is linear. We now consider the general $f(\cdot)$. Let us first introduce conditions to be used.

A3.3.1 $a_k > 0,\ a_k \longrightarrow 0,\ \sum_{k=1}^{\infty} a_k = \infty,$

$$a_{k+1}^{-1} - a_k^{-1} \longrightarrow \alpha \geq 0, \tag{3.3.52}$$

and

$$\sum_{k=1}^{\infty} a_k^{1+\eta} < \infty \text{ for some } \eta \in (0,1). \tag{3.3.53}$$

A3.3.2 *A continuously differentiable function $v(\cdot)$ exists such that*

$$\sup_{\delta\leq\|x-x^0\|\leq\Delta} f^T(x)v_x(x) < 0$$

for any $0 < \delta < \Delta$, and $v(x^*) < \inf_{\|x\|=c_0} v(x)$ for some $c_0 > 0$ with $c_0 > \|x^*\|$, where x^* is used in (2.1.1).

A3.3.3

$$\epsilon_k = e_k + \nu_k, \quad e_k = \sum_{j=0}^{r} C_j w_{k-j}, \quad r < \infty, \quad \nu_{k+1} = o(\sqrt{a_k}) \ a.s. \tag{3.3.54}$$

where $\{w_k, \mathcal{F}_k\}$ *is a martingale difference sequence satisfying (3.3.21)–(3.3.23).*

A3.3.4 $f(\cdot)$ *is measurable and locally bounded. As* $x \longrightarrow x^0$,

$$\|f(x) - F(x - x^0)\| \le c\|x - x^0\|^{1+\beta}, \tag{3.3.55}$$

where $c > 0$, $F + \frac{\alpha}{2}I$ *with* α *specified in (3.3.52) is stable and* $\beta \in (0, 1]$ *satisfying* $\frac{\beta}{1+\beta} > \eta$, *which is specified in (3.3.53).*

Theorem 3.3.2 *Let* $\{x_k\}$ *be given by (2.1.1)–(2.1.3) and let A3.3.1–A3.3.4 be held. Then*

$$\frac{x_k - x^0}{\sqrt{a_k}} \xrightarrow[k\longrightarrow\infty]{d} N(0, S), \tag{3.3.56}$$

where

$$S = \int_0^\infty e^{(F+\frac{\alpha}{2}I)t} \sum_{i=0}^{r} C_i S_0 \sum_{i=0}^{r} C_i^T e^{(F^T+\frac{\alpha}{2}I)t} dt. \tag{3.3.57}$$

Proof. Since $1 - \eta > \frac{1}{1+\beta}$, there exists δ such that

$$1 - \eta > 2\delta > \frac{1}{1+\beta}, \tag{3.3.58}$$

which implies $2 - 2\delta > 1 + \eta$. From (3.3.53) it follows that

$$\sum_{k=1}^{\infty} a_k^{2-2\delta} < \sum_{k=1}^{\infty} a_k^{1+\eta} < \infty.$$

This together with the convergence theorem for martingale difference sequences yields

$$\sum_{k=1}^{\infty} a_k^{1-\delta} w_{k+1-j} < \infty \quad \text{a.s.,} \quad \forall j = 0, 1, \ldots, r,$$

which implies

$$\begin{aligned}\sum_{k=1}^{\infty} a_k^{1-\delta} e_{k+1} &= \sum_{k=1}^{\infty} a_k^{1-\delta} \sum_{j=0}^{r} C_j w_{k+1-j} \\ &= \sum_{j=0}^{r} C_j \sum_{k=1}^{\infty} a_k^{1-\delta} w_{k+1-j} < \infty, \quad \text{a.s.}\end{aligned}$$

Since $\delta < \frac{1}{2}$ $(\delta < \frac{1-\eta}{2})$, from $\nu_{k+1} = o(\sqrt{a_k})$ it follows that $\nu_{k+1} = o(a_k^\delta)$. Stability of $F + \alpha\delta I$ is implied by stability of $F + \frac{\alpha}{2}I$, which is a part of A3.3.4. Then by Theorem 3.1.1

$$\|x_k - x^0\| = o(a_k^\delta) \text{ a.s.}$$

By (3.3.55) and (3.3.58) we have

$$\begin{aligned}\|f(x_k) - F(x_k - x^0)\| &\le c\|x_k - x^0\|^{1+\beta} \\ &= o(a_k^{\delta(1+\beta)}) < o(\sqrt{a_k}) \text{ a.s.}\end{aligned} \quad (3.3.59)$$

From Theorem 3.1.1 we also know that there is an integer-valued $N < \infty$ (possibly depending on sample paths) such that $\sigma_k \equiv \sigma_N$, $\forall k \ge N$ and there is no truncation in (2.1.1) for $k \ge N$. Consequently, for $k \ge N$ we have

$$\begin{aligned}x_{k+1} - x^0 =&(I + a_k F)(x_k - x^0) + a_k(f(x_k) - F(x_k - x^0)) \\ &+ a_k(e_{k+1} + \nu_{k+1}).\end{aligned} \quad (3.3.60)$$

Denoting

$$\nu'_{k+1} = f(x_k) - F(x_k - x^0) + \nu_{k+1},$$

by (3.3.59) and (3.3.54) we see $\nu'_{k+1} = o(\sqrt{a_k})$ a.s.

Then (3.3.60) is written as

$$x_{k+1} - x^0 = (I + a_k F)(x_k - x^0) + a_k(e_{k+1} + \nu'_{k+1}).$$

By (3.3.28) it follows that

$$\frac{x_{k+1} - x^0}{\sqrt{a_{k+1}}} = (I + a_k A_k)\frac{x_k - x^0}{\sqrt{a_k}} + \frac{a_k}{\sqrt{a_{k+1}}}(e_{k+1} + \nu'_{k+1}), \quad k \ge N, \quad (3.3.61)$$

where

$$A_k \triangleq F + \frac{\alpha}{2}I + \frac{\alpha a_k}{2}F + o(\sqrt{a_k})(I + a_k F) \longrightarrow F + \frac{\alpha}{2}I \triangleq A. \quad (3.3.62)$$

Using $\Psi_{k,i}$ introduced by (3.3.32), we find

$$\frac{x_{k+1}-x^0}{\sqrt{a_{k+1}}} = \Psi_{k,N}\frac{x_N - x^0}{\sqrt{a_N}} + \sum_{i=N}^{k} \Psi_{k,i+1}\frac{a_i}{\sqrt{a_{i+1}}}(e_{i+1} + \nu'_{i+1}). \quad (3.3.63)$$

By the argument similar to that used in Corollary 3.3.1, we have $\Psi_{k,N}\frac{x_N-x^0}{\sqrt{a_N}} \longrightarrow 0$ and $\sum_{i=N}^{k} \Psi_{k,i+1}\frac{a_i}{\sqrt{a_{i+1}}}\nu'_{i+1} \longrightarrow 0$ as $k \longrightarrow \infty$.

Then by (3.3.51) from (3.3.63) we conclude (3.3.56). □

Corollary 3.3.2 Let D be an $l \times l$ matrix and let a_k in (2.1.1)–(2.1.2) be replaced by $a_k D_k$. In other words, in stead of (2.1.1) and (2.1.2) if we consider

$$\begin{aligned} x_{k+1} =& (x_k + a_k D y_{k+1}) I_{[\|x_k + a_k D y_{k+1}\| \le M_{\sigma_k}]} \\ &+ x^* I_{[\|x_k + a_k D y_{k+1}\| > M_{\sigma_k}]}, \end{aligned} \quad (3.3.64)$$

$$\sigma_k = \sum_{i=0}^{k-1} I_{[\|x_i + a_i D y_{i+1}\| > M_{\sigma_i}]}, \quad \sigma_0 = 0, \quad (3.3.65)$$

$$y_{k+1} = f(x_k) + \epsilon_{k+1}, \quad (3.3.66)$$

then this is equivalent to replacing $f(\cdot)$ and ϵ_{k+1} by $Df(\cdot)$ and $D\epsilon_{k+1}$, respectively.

In this case the only modification should be made in conditions of Theorem 3.3.2 consists in that stability of $F + \frac{\alpha}{2}I$ in A3.3.4 should be replaced by stability of $DF + \frac{\alpha}{2}I$. The conclusion of Theorem 3.3.2 remains valid with only modification that $\sum_{i=0}^{r} C_i S_0 \sum_{i=0}^{r} C_i^T$ and F in (3.3.57) should be replaced by $D\sum_{i=0}^{r} C_i S_0 \sum_{i=0}^{r} C_i^T D^T$ and DF, respectively.

3.4. Asymptotic Efficiency

In Corollary 3.3.2 we have mentioned that the limiting covariance matrix $S(D)$ for $\frac{x_k - x^0}{\sqrt{a_k}}$ depends on D, if $\{a_k\}$ in (2.1.1)–(2.1.3) is replaced by $\{Da_k\}$. By efficiency we mean that $S(D)$ reaches its minimum with respect to D.

Denote

$$Q = \sum_{i=0}^{r} C_i S_0 \sum_{i=0}^{r} C_i^T. \quad (3.4.1)$$

By Corollary 3.3.2, the limiting covariance matrix for $\frac{x_k - x^0}{\sqrt{a_k}}$ with $\{x_k\}$ given by (3.3.64)–(3.3.66) is expressed by

$$S(D) \stackrel{\Delta}{=\!=} \int_0^\infty e^{(DF+\frac{\alpha}{2}I)t} DQDe^{(DF+\frac{\alpha}{2}I)^T t} dt. \tag{3.4.2}$$

Theorem 3.4.1 *Assume $DF + \frac{\alpha}{2}I$ is stable. i) If $\alpha > 0$, then $S(D)$ reaches its minimum at $D = -\alpha F^{-1}$ and $S(-\alpha F^{-1}) = \alpha F^{-1}QF^{-T}$, where $F^{-T} = (F^{-1})^T$. ii) If $\alpha = 0$, then $S(-\epsilon F^{-1}) = \frac{\epsilon}{2}F^{-1}QF^{-1} \longrightarrow 0$ as $\epsilon \longrightarrow 0$.*

Proof. i) Integrating by parts, we have

$$\begin{aligned}(DF + \frac{\alpha}{2}I)S(D) &= \int_0^\infty de^{(DF+\frac{\alpha}{2}I)t} DQD^T e^{(DF+\frac{\alpha}{2}I)^T t} \\ &= -DQD^T - S(D)(F^T D^T + \frac{\alpha}{2}I).\end{aligned}$$

This means that $S(D)$ satisfies the following algebraic Riccati equation

$$(DF + \frac{\alpha}{2}I)S(D) + S(D)(DF + \frac{\alpha}{2}I)^T + DQD^T = 0 \tag{3.4.3}$$

By stability of $DF + \frac{\alpha}{2}I$ and $\alpha > 0$, DF is nondegenerate. Thus, (3.4.3) is equivalent to

$$(F + \frac{\alpha D^{-1}}{2})S(D)D^{-T} + D^{-1}S(D)(F^T + \frac{\alpha D^{-T}}{2}) + Q = 0,$$

or

$$(\frac{F}{\sqrt{\alpha}} + \sqrt{\alpha}D^{-1})S(D)(\frac{F}{\sqrt{\alpha}} + \sqrt{\alpha}D^{-T}) - \frac{1}{\alpha}FS(D)F^T + Q = 0,$$

or

$$S(D) = \alpha F^{-1}\big[(\frac{F}{\sqrt{\alpha}} + \sqrt{\alpha}D^{-1})S(D)(\frac{F^T}{\sqrt{\alpha}} + \sqrt{\alpha}D^{-T}) + Q\big]F^{-T}. \tag{3.4.4}$$

From (3.4.4) it follows that

$$S(D) \geq \alpha F^{-1}QF^{-T}$$

and the equality is achieved at $D = -\alpha F^{-1}$.

ii) If $\alpha = 0$, then

$$S(D) = \int_0^\infty e^{DFt} DQD^T e^{F^T D^T t} dt.$$

When $D = -\epsilon F^{-1}$, $\epsilon > 0$, then

$$S(D) = \int_0^\infty e^{-2\epsilon t}\epsilon^2 F^{-1}QF^{-T}dt = \frac{\epsilon}{2}F^{-1}QF^{-T} \xrightarrow[\epsilon \longrightarrow 0]{} 0.$$

□

For the commonly used step size $a_k = \frac{a}{k}$, $a_{k+1}^{-1} - a_k^{-1} = \frac{1}{a}$, i.e., α specified in (3.3.52) equals $\frac{1}{a}$. By Theorem 3.4.1 the optimal $D = -\alpha F^{-1} = \frac{-F^{-1}}{a}$ and the optimal step size is $a_k D = -\frac{F^{-1}}{k}$. For $\frac{x_k - x^0}{\sqrt{a_k}} = \frac{\sqrt{k}(x_k - x^0)}{\sqrt{a}}$ the limiting covariance matrix is $\frac{1}{a}F^{-1}QF^{-T}$. Therefore, the optimal limiting covariance matrix for $\sqrt{k}(x_k - x^0)$ is $F^{-1}QF^{-T}$ no matter what a is taken in $a_k = \frac{a}{k}$.

Let us take $a_k = \frac{1}{k}$. Then $\alpha = 1$ and the optimal $D = -F^{-1}$. In this case $\sqrt{k}(x_k - x^0) \xrightarrow[k \longrightarrow \infty]{d} N(0, F^{-1}QF^{-T})$ and $F^{-1}QF^{-T}$ is the minimum of the limiting covariance matrix. However, $f(\cdot)$ is unknown and F^{-1} is unknown too. Hence, F^{-1} cannot be directly used in the algorithm. To achieve asymptotic efficiency, one way is to estimate F, and replace the optimal step size $-\frac{F^{-1}}{k}$ by its estimate $-\frac{-F_k^{-1}}{k}$. This is the so-called adaptive SA. But, to guarantee its convergence and optimality, rather restrictive conditions are needed.

Let $\{x_k\}$ be estimates for x^0 being the root of $f(\cdot)$ satisfying

$$\|f(x) - Fx^0\| \le c\|x - x^0\|^{1+\beta} \text{ as } x \longrightarrow x^0,$$

where F is stable and $\beta > 0$. The estimates are obtained on the basis of observations

$$y_{k+1} = f(x_k) + \epsilon_{k+1}$$

with $\frac{1}{k}\sum_{i=1}^{k} E\epsilon_i\epsilon_i^T \longrightarrow Q$.

If $\sqrt{k}(x_k - x^0) \xrightarrow[k \longrightarrow \infty]{d} N(0, S)$, $S = F^{-1}QF^{-T}$, then we call x_k asymptotically efficient for x^0.

To achieve asymptotic efficiency we apply the averaging technique that is different from adaptive SA.

For $\{a_k\}$ satisfying A3.3.1, if α in (3.3.52) equals zero, then $\{a_k\}$ is called slowly decreasing step size. As a typical example of slowly decreasing step sizes, one may take $a_k = \frac{1}{k^\nu}$, $\nu \in (0, 1)$.

Let $\{x_k\}$ be generated by (2.1.1)–(2.1.3) with slowly decreasing $\{a_k\}$. Define

$$\overline{x}_k = \frac{1}{k}\sum_{i=1}^{k} x_i. \tag{3.4.5}$$

In what follows we will show that $\sqrt{k}(\overline{x}_k - x^0)$ is asymptotically normal and $\overline{x}_k$ is asymptotically efficient.

We list the conditions to be used.

A3.4.1 *$a_k > 0$, a_k nonincreasingly converges to zero, $\sum_{k=1}^{\infty} a_k = \infty$,*

$$a_k k \xrightarrow[k\to\infty]{} \infty, \quad a_{k+1}^{-1} - a_k^{-1} \xrightarrow[k\to\infty]{} 0, \tag{3.4.6}$$

and for some $\delta \in (0,1)$

$$\sum_{i=1}^{\infty} \frac{a_i^{\frac{1+\delta}{2}}}{i^{\frac{1}{2}}} < \infty. \tag{3.4.7}$$

A3.4.2 *A continuously differentiable function $v(\cdot)$ exists such that*

$$\sup_{\delta \le \|x - x^0\| \le \Delta} f^T(x) v_x(x) < 0$$

for any $0 \le \delta < \Delta$, and $v(x^*) < \inf_{\|x\|=c_0} v(x)$ for some $c_0 > 0$ with $c_0 > \|x^*\|$, where x^* is used in (2.1.1).

A3.4.3 *The observation noise $\{\epsilon_k\}$ is such that*

$$\epsilon_k = e_k + \nu_k,$$

$$a_k \sum_{i=0}^{k} e_{i+1} \longrightarrow 0, \quad \frac{1}{\sqrt{k}} \sum_{i=1}^{k} e_i \xrightarrow[k\to\infty]{d} N(0, Q), \tag{3.4.8}$$

$$E\|e_k\|^2 < \infty, \quad \sum_{s \ge -i}^{\infty} \|E e_i e_{i+s}^T\| \le c_2 \tag{3.4.9}$$

with c_2 being a constant independent of i, and

$$E\|\nu_k\|^2 = O(a_k^{1+\delta}), \tag{3.4.10}$$

where δ is specified in (3.4.7).

A3.4.4 *$f(\cdot)$ is measurable and locally bounded. There exist a stable matrix F, $\gamma > 0$, and $\beta \in (\delta, 1]$ such that*

$$\|f(x) - F(x - x^0)\| \le c_1 \|x - x^0\|^{1+\beta}, \quad \forall x \in \{x : \|x - x^0\| \le \gamma\} \tag{3.4.11}$$

where $c_1 > 0$ is a constant.

Remark 3.4.1 It is clear that $a_k = \frac{1}{k^\gamma}$, $\forall \gamma \in (0,1)$, satisfies A3.4.1. From (3.4.7) it follows that

$$\sum_{i=[\frac{k}{2}]}^{k} \frac{a_i^{\frac{1+\delta}{2}}}{\sqrt{i}} = o(1), \tag{3.4.12}$$

where $[x]$ denotes the integer part of x.

Since a_k is nonincreasing, from (3.4.12) we have

$$a_k^{\frac{1+\delta}{2}} \cdot \sum_{i=[\frac{k}{2}]}^{k} \frac{1}{\sqrt{i}} = o(1),$$

which implies

$$a_k^{\frac{1+\delta}{2}} \cdot \sqrt{k} = o(1),$$

or

$$a_k = o(k^{-\mu}), \quad \mu \triangleq \frac{1}{1+\delta} \in (\frac{1}{2}, 1). \tag{3.4.13}$$

Remark 3.4.2 If $e_i = \sum_{j=0}^{r} C_j w_{i-j}$ with $\{w_k, \mathcal{F}_k\}$ being a martingale difference sequence satisfying (3.3.21)–(3.3.23), then identifying $\frac{1}{\sqrt{k}} e_i$ to ξ_{ki} in Lemma 3.3.1, by this lemma we have

$$\frac{1}{\sqrt{k}} \sum_{i=1}^{k} e_i \xrightarrow[k \to \infty]{d} N(0, Q),$$

where Q is given by (3.4.1). Thus, in this case the second condition in (3.4.8) holds.

We now show that the first condition in (3.4.8) holds too.

By the estimate for the weighted sum of martingale difference sequences (See Appendix B) we have

$$\sum_{i=1}^{n} e_i = \sum_{i=r}^{n} (\sum_{j=0}^{r} C_j) w_i = o(\sqrt{n} (\log n)^{\frac{1}{2}+\epsilon}) \text{ a.s. } \forall \epsilon > 0,$$

which incorporating with (3.4.13) yields

$$a_k \sum_{i=0}^{k} e_{i+1} \xrightarrow[k \to \infty]{} 0 \text{ a.s.}$$

It is clear that (3.4.9) is implied by (3.3.21). Therefore, in the present case all requirements in A3.4.3 are satisfied.

Theorem 3.4.2 *Assume A3.4.1–A3.4.4 hold. Let* $\{x_k\}$ *be given by (2.1.1)–(2.1.3) and* $\{\overline{x}_k\}$ *be given by (3.4.5). Then* $\overline{x}_k$ *is asymptotically efficient:*

$$\sqrt{k}(\overline{x}_k - x^0) \xrightarrow[k\longrightarrow\infty]{d} N(0,S), \quad S = F^{-1}QF^{-T}.$$

Prior to proving the theorem we establish some properties of slowly decreasing step size.

Set

$$\Phi_{k,j} = \prod_{i=j}^{k}(I + a_i F), \quad k \geq j, \quad \Phi_{j,j+1} = I. \tag{3.4.14}$$

By (3.1.8) we have

$$\|\Phi_{k,j}\| \leq c_0 \exp(-c\sum_{i=j}^{k} a_i), \quad \forall k \geq j, \quad \forall j \geq 0, \tag{3.4.15}$$

where $c_0 > 0$ and $c > 0$ are constants.

Set

$$G_{k,j} = \sum_{i=j}^{k}(a_{j-1} - a_i)\Phi_{i-1,j} + F^{-1}\Phi_{k,j}. \tag{3.4.16}$$

Lemma 3.4.1 *i) The following estimate takes place*

$$\frac{a_j}{a_k} \leq \exp(o(1)\sum_{i=j}^{k} a_i), \quad \forall k \geq j, \quad \forall j \geq 1, \tag{3.4.17}$$

where $o(1)$ *denotes a magnitude that tends to zero as* $j \longrightarrow \infty$.

ii) $G_{k,j}$ *is uniformly bounded with respect to both* k *and* $j : 1 \leq j \leq k$ *and*

$$\frac{1}{k}\sum_{j=1}^{k}\|G_{k,j}\| \xrightarrow[k\longrightarrow\infty]{} 0. \tag{3.4.18}$$

Proof. i) By (3.4.6) we know that

$$\frac{a_j}{a_{j+1}} = 1 + o(1)a_j, \quad o(1) \xrightarrow[j\longrightarrow\infty]{} 0,$$

and

$$\frac{a_j}{a_k} = \frac{a_j}{a_{j+1}} \cdots \frac{a_{k-1}}{a_k} = \prod_{i=j}^{k-1} (1 + o(1)a_i)$$
$$\leq \prod_{i=j}^{k-1} e^{o(1)a_k} = \exp(o(1) \sum_{i=j}^{k-1} a_i), \tag{3.4.19}$$

which implies (3.4.17) since $a_k \longrightarrow 0$ as $k \longrightarrow \infty$.

ii) By (3.4.6) $ka_k \longrightarrow \infty$ as $k \longrightarrow \infty$, and hence for any $\epsilon > 0$ we have

$$\sum_{i=[(1-\epsilon)k]}^{k} a_i = \sum_{i=[(1-\epsilon)k]}^{k} \frac{ia_i}{i} \xrightarrow[k \longrightarrow \infty]{} \infty, \tag{3.4.20}$$

where $[x]$ denotes the integer part of x.

Using (3.4.15) we have

$$\frac{1}{k} \sum_{j=0}^{k} \|\Phi_{k,j}\| \leq \frac{1}{k} (\sum_{j=0}^{[(1-\epsilon)k]} + \sum_{j=[(1-\epsilon)k]+1}^{k}) c_0 \exp(-c \sum_{i=j}^{k} a_i)$$
$$\leq (1-\epsilon) c_0 (-c \sum_{i=[(1-\epsilon)k]}^{k} a_i) + c_0 \epsilon$$

for any $\epsilon > 0$, where the first term at the right-hand side tends to zero as $k \longrightarrow \infty$ by (3.4.20), and the last term tends to zero as $\epsilon \longrightarrow 0$. Therefore, for (3.4.18) it suffices to show

$$\frac{1}{k} \sum_{j=1}^{k} \| \sum_{i=j}^{k} (a_{j-1} - a_i) \Phi_{i-1,j} \| \xrightarrow[k \longrightarrow \infty]{} 0. \tag{3.4.21}$$

Noticing that (3.4.13) implies $\sum_{i=1}^{\infty} a_i^2 < \infty$, for any $\lambda > 0$ we have

$$\sum_{i=j}^{k} a_i \exp(-\lambda \sum_{s=j}^{i-1} a_s) \leq \frac{1}{\lambda} \sum_{i=j}^{k} (1 - e^{-\lambda a_i} + \frac{\lambda^2 a_i^2}{2!}) \exp(-\lambda \sum_{s=j}^{i-1} a_s)$$
$$= \frac{1}{\lambda} \sum_{i=j}^{k} (\exp(-\lambda \sum_{s=j}^{i-1} a_s) - \exp(-\lambda \sum_{s=j}^{i} a_s)) + \frac{\lambda}{2} \sum_{i=j}^{k} a_i^2 \exp(-\lambda \sum_{s=j}^{i-1} a_s)$$
$$\leq \frac{1}{\lambda} + \frac{\lambda}{2} \sum_{i=1}^{\infty} a_i^2 < \infty, \tag{3.4.22}$$

and hence

$$\sum_{i=j}^{k} a_i \sum_{s=j}^{i-1} a_s \exp(-\lambda \sum_{t=j}^{i-1} a_t) < \sum_{i=j}^{k} a_i \frac{2}{\lambda}\{\exp(\frac{\lambda}{2}\sum_{s=j}^{i-1} a_s)\} \exp(-\lambda \sum_{t=j}^{i-1} a_t)$$

$$= \frac{2}{\lambda}\sum_{i=j}^{k} a_i \exp(-\frac{\lambda}{2}\sum_{s=j}^{i-1} a_s) \le \frac{4}{\lambda^2} + \frac{1}{2}\sum_{i=1}^{\infty} a_i^2. \tag{3.4.23}$$

By (3.4.6) $a_k - a_{k+1} = o(1)a_k a_{k+1}$, where $o(1) \longrightarrow 0$ as $k \longrightarrow \infty$. Taking this into account, by (3.4.15) and (3.4.17) we have

$$\begin{aligned}
&\|\sum_{i=j}^{k}(a_{j-1} - a_i)\Phi_{i-1,j}\| \le \sum_{i=j}^{k}\sum_{s=j}^{i}(a_{s-1} - a_s)\|\Phi_{i-1,j}\| \\
&\le \sum_{i=j}^{k} o(a_{j-1}) \sum_{s=j}^{i} a_s c_0 \exp(-c\sum_{t=j}^{i-1} a_t) \\
&= o(1)\sum_{i=j}^{k} a_j \sum_{s=j}^{i-1} a_s c_0 \exp(-c\sum_{t=j}^{i-1} a_t) \\
&\le o(1)\sum_{i=j}^{k} a_i \exp(o(1)\sum_{p=j}^{i} a_p) \sum_{s=j}^{i-1} a_s c_0 \exp(-c\sum_{t=j}^{i-1} a_t) \\
&\le o(1)c_0 \sum_{i=j}^{k} a_i \sum_{s=j}^{i-1} a_s \exp(-\frac{c}{2}\sum_{t=j}^{i-1} a_t),
\end{aligned}$$

where $o(1) \longrightarrow 0$ as $j \longrightarrow \infty$.

Thus, by (3.4.23) we have

$$\|\sum_{i=j}^{k}(a_{j-1} - a_i)\Phi_{i-1,j}\| \le o(1)(\frac{16}{c^2} + \frac{1}{2}\sum_{i=1}^{\infty} a_i^2) \xrightarrow[j\longrightarrow\infty]{} 0.$$

This implies (3.4.21), and together with (3.4.15) shows that $G_{k,j}$ is uniformly bounded with respect to both k and j. □

We now express $\{x_k\}$ given by (2.1.1)–(2.1.3) in a different form by introducing a sequence of stopping times and a sequence of processes $\{x_k^{(i)}\}$ $i = 1, 2, \ldots$. To be precise, define $\alpha_0 = 0$,

$$\alpha_1 \triangleq \begin{cases} \min\{i : i \ge \alpha_0\}, & \|x_i^{(1)}\| > M_0, \\ \infty, & \text{if } \sup_i \|x_i^{(1)}\| \le M_0, \end{cases} \tag{3.4.24}$$

where by definition

$$x_{k+1}^{(1)} \triangleq x_0 + \sum_{i=0}^{k} a_i y_{i+1}^{(1)}, \quad \forall k \geq 0, \quad y_{i+1}^{(1)} \triangleq f(x_i^{(1)}) + \epsilon_{i+1}. \tag{3.4.25}$$

Remind that $\{M_i\}$ is the sequence used in (2.1.1)–(2.1.3).

It is clear that $y_{i+1}^{(1)} = y_{i+1}, \forall i < \alpha_1$.

Similarly, define

$$\alpha_2 = \begin{cases} \min\{i : i > \alpha_1, \|x_i^{(2)}\| > M_1\}, \\ \infty, \quad \text{if } \sup_i \|x_i^{(2)}\| \leq M_1, \end{cases} \tag{3.4.26}$$

where

$$x_{k+1}^{(2)} \triangleq x^* + \sum_{i=0}^{k} a_i y_{i+1}^{(2)} I_{[\alpha_1 \leq i]}, \quad \forall k \geq 0, \tag{3.4.27}$$

$$y_{i+1}^{(2)} \triangleq f(x_i^{(2)}) + \epsilon_{i+1}, \quad \forall i \in [\alpha_1, \alpha_2). \tag{3.4.28}$$

Recursively define

$$\alpha_j = \begin{cases} \min\{i : i > \alpha_{j-1}, \|x_i^{(j)}\| > M_{j-1}\}, \\ \infty, \quad \text{if } \sup_i \|x_i^{(j)}\| \leq M_{j-1}, \end{cases} \tag{3.4.29}$$

where

$$x_{k+1}^{(j)} \triangleq x^* + \sum_{i=0}^{k} a_i y_{i+1}^{(j)} I_{[\alpha_{j-1} \leq i]}, \quad \forall k \geq 0, \tag{3.4.30}$$

$$y_{i+1}^{(j)} \triangleq f(x_i^{(j)}) + \epsilon_{i+1}, \quad \forall i \in [\alpha_{j-1}, \alpha_j). \tag{3.4.31}$$

As a matter of fact, α_j is the first exit time of $x_k^{(j)}$ from the sphere with radius M_{j-1} after time α_{j-1}, and during the time period $[\alpha_{j-1}, \alpha_j)$, $x_k^{(j)}$ evolves as same as x_k and is recursively defined as an RM process. Therefore, $\{x_k\}$ given by (2.1.1)–(2.1.3) can be expressed as

$$x_k = \sum_{j=1}^{\infty} x_k^{(j)} I_{[\alpha_{j-1} \leq k < \alpha_j]}. \tag{3.4.32}$$

Lemma 3.4.2 *Under Conditions A3.4.1–A3.4.4, there exists an integer-valued i_0 such that $\alpha_{i_0} < \infty$ a.s., $\alpha_{i_0+1} = \infty$ a.s., and $\{x_k\}$ given by (2.1.1)–(2.1.3) has no truncation for $k \geq \alpha_{i_0}$, i.e.,*

$$x_{k+1} = x_k + a_k y_{k+1}, \quad \forall k \geq \alpha_{i_0}, \tag{3.4.33}$$

and $x_k \longrightarrow x^0$ a.s.

Proof. If we can show that A2.2.3 is implied by A3.4.3, then all conditions of Theorem 2.2.1 are fulfilled a.s., and the conclusions of the lemma follow from Theorem 2.2.1.

Since $\nu_k \xrightarrow[k\longrightarrow\infty]{} 0$, we have

$$\frac{1}{T}\|\sum_{i=n_k}^{m(n_k,T_k)} a_i \nu_{i+1}\| \leq \max_{i\geq n_k} \|\nu_{i+1}\| \xrightarrow[k\longrightarrow\infty]{} 0, \quad \forall T_k \in [0,T],$$

which means that (2.2.2) is satisfied for $\{\nu_i\}$.

We now check (2.2.2) for $\{e_i\}$. By a partial summation we have

$$\begin{aligned}\|\sum_{i=n}^{m(n,T)} a_i e_{i+1}\| =& \|a_m(n,T) \sum_{i=1}^{m(n,T)} e_{i+1} - a_n \sum_{i=1}^{n-1} e_{i+1} \\ &+ \sum_{i=n}^{m(n,T)} (a_i - a_{i+1}) \sum_{j=1}^{i} e_{j+1}\| \\ \leq& \|a_{m(n,T)} \sum_{i=1}^{m(n,T)} e_{i+1}\| + \|a_n \sum_{i=1}^{n-1} e_{i+1}\| \\ &+ o(1) \sum_{i=n}^{m(n,T)} a_{i+1} \|a_i \sum_{j=1}^{i} e_{j+1}\|,\end{aligned} \tag{3.4.34}$$

where (3.4.6) is used and $o(1) \longrightarrow 0$ as $n \longrightarrow \infty$.

By (3.4.8) the first two terms on the right-hand side of (3.4.34) tend to zero as $n \longrightarrow \infty$, by the same reason and by the fact $\sum_{i=n}^{m(n,T)} a_{i+1} \leq T$, the last term of (3.4.34) also tends to zero as $n \longrightarrow \infty$. This means that $\{e_i\}$ satisfies (2.2.2), and the lemma follows. □

By Lemma 3.4.2 we have

$$x_{k+1} - x^0 = (I + a_k F)(x_k - x^0) + a_k(f(x_k) - F(x_k - x^0)) + a_k \epsilon_{k+1}, \quad \forall k \geq \alpha_{i_0} \tag{3.4.35}$$

and by (3.4.14)

$$
\begin{aligned}
x_{k+1} - x^0 =& \Phi_{k,\alpha_{i_0}}(x_{\alpha_{i_0}} - x^0) + \sum_{j=\alpha_{i_0}}^{k} \Phi_{k,j+1} a_j \epsilon_{j+1} \\
&+ \sum_{j=\alpha_{i_0}}^{k} \Phi_{k,j+1} a_j (f(x_j) - F(x_j - x^0)), \quad \forall k \geq \alpha_{i_0}.
\end{aligned}
\tag{3.4.36}
$$

For γ specified in (3.4.11) and a deterministic integer k_0, define the stopping time μ as follows

$$
\mu = \begin{cases} \min\{j : j > k_0, \|x_j - x^0\| \geq \gamma\}, \\ 0, \quad \text{if} \quad \|x_{k_0} - x^0\| \geq \gamma. \end{cases} \tag{3.4.37}
$$

From (3.4.35) we have

$$
\begin{aligned}
(x_{k+1} - x^0) I_{[\alpha_{i_0} < k_0]} =& \Phi_{k,k_0}(x_{k_0} - x^0) I_{[\alpha_{i_0} < k_0]} \\
&+ \sum_{j=k_0}^{k} \Phi_{k,j+1} a_j \epsilon_{j+1} I_{[\alpha_{i_0} < k_0]} \\
&+ \sum_{j=k_0}^{k} \Phi_{k,j+1} a_j (f(x_j) - F(x_j - x^0)) I_{[\alpha_{i_0} < k_0]}, \quad \forall k \geq k_0,
\end{aligned}
\tag{3.4.38}
$$

and

$$
\begin{aligned}
(x_{k+1} - x^0) I_{[\mu > k+1,\ \alpha_{i_0} < k_0]} =& \Phi_{k,k_0}(x_{k_0} - x^0) I_{[\mu > k+1,\ \alpha_{i_0} < k_0]} \\
&+ \sum_{j=k_0}^{k} \Phi_{k,j+1} a_j \epsilon_{j+1} I_{[\mu > k+1,\ \alpha_{i_0} < k_0]} \\
&+ \sum_{j=k_0}^{k} \Phi_{k,j+1} a_j (f(x_j) - F(x_j - x^0)) I_{[\mu > j,\ \alpha_{i_0} < k_0]} \\
&\cdot I_{[\mu > k+1,\ \alpha_{i_0} < k_0]}, \quad \forall k \geq k_0.
\end{aligned}
\tag{3.4.39}
$$

Lemma 3.4.3 *If A 3.4.1-A3.4.4 hold, then*

$$
\frac{1}{a_{k+1}} E\|(x_{k+1} - x^0) I_{[\mu > k+1,\ \alpha_{i_0} < k_0]}\|^2
$$

is uniformly bounded with respect to k.

Proof. By (3.4.11) and (3.4.15) from (3.4.39) we have

$$\frac{1}{a_{k+1}} E\|x_{k+1}-x^0\|^2 I_{[\mu>k+1,\, \alpha_{i_0}<k_0]}$$

$$\leq \frac{4c_0^2\gamma^2}{a_{k+1}} \exp(-2c\sum_{i=k_0}^{k} a_i) + \frac{4c_0^2}{a_{k+1}} \sum_{i=k_0}^{k}\sum_{j=k_0}^{k} [\exp(-c\sum_{s=j+1}^{k} a_s)]$$

$$\cdot\, a_j [\exp(-c\sum_{s=i+1}^{k} a_s)] a_i \|E e_{i+1} e_{j+1}^T\|$$

$$+ \frac{4c_0^2}{a_{k+1}} \sum_{i=k_0}^{k}\sum_{j=k_0}^{k} [\exp(-c\sum_{s=j+1}^{k} a_s)] a_j [\exp(-c\sum_{s=i+1}^{k} a_s)] a_i E\|\nu_{i+1}\nu_{j+1}^T\|$$

$$+ \frac{4}{a_{k+1}} E(\sum_{j=k_0}^{k} c_0 c_1 [\exp(-c\sum_{s=j+1}^{k} a_s)] a_j \|x_j - x^0\|^{1+\beta} I_{[\mu>j,\, \alpha_{i_0}<k_0]})^2$$

$$\triangleq I_1 + I_2 + I_3 + I_4, \tag{3.4.40}$$

where $I_i, i=1,\cdots,4$, respectively denote the terms on the right-hand side of the inequality in (3.4.40).

By (3.4.19) we see

$$I_1 \leq \frac{4c_0^2\gamma^2}{a_{k_0}} \exp(o(1)\sum_{i=k_0}^{k+1} a_i) \cdot \exp(-2c\sum_{i=k_0}^{k} a_i), \tag{3.4.41}$$

where $o(1) \longrightarrow 0$ as $k_0 \longrightarrow \infty$. From this we find that I_1 is bounded in k if k_0 is large enough so that $o(1) - 2c < 0$.

By (3.4.19) we estimate I_2 as follows:

$$I_2 \leq 4c_0^2 \sum_{i=k_0}^{k}\sum_{j=k_0}^{k} [\exp(-\frac{c}{2}\sum_{s=j+1}^{k} a_s)] a_i [\exp(-c\sum_{s=i+1}^{k} a_s)] \|E e_{i+1} e_{j+1}^T\|$$

where k_0 is assumed to be large enough such that

$$\frac{a_j}{a_{k+1}} \leq \exp(\frac{c}{2}\sum_{s=j+1}^{k} a_s), \quad \forall j \geq k_0.$$

Thus, by (3.4.9)

$$I_2 \leq 4c_0^2 \sum_{i=k_0}^{k}\sum_{j=k_0}^{k} a_i \|E e_{i+1} e_{j+1}^T\| \exp(-c\sum_{s=i+1}^{k} a_s)$$

$$= 4c_0^2 \sum_{i=k_0}^{k} \sum_{s=k_0-i}^{k-i} a_i \| E e_{i+1} e_{i+s+1}^T \| \exp(-c \sum_{s=i+1}^{k} a_s)$$

$$\leq 4c_0^2 c_2 \sum_{i=k_0}^{k} a_i \exp(-c \sum_{s=i+1}^{k} a_s). \tag{3.4.42}$$

We now pay attention to (3.3.10) in the proof of Lemma 3.3.2 and find that the right-hand side of (3.4.42) is bounded with respect to k.

For I_3 by (3.4.19) and (3.4.10) we have

$$I_3 \leq 4c_0 \sum_{i=k_0}^{k} \sum_{j=k_0}^{k} [\exp(-\frac{c}{2} \sum_{s=j+1}^{k} a_s)] a_j^{\frac{1}{2}} [\exp(-\frac{c}{2} \sum_{j=i+1}^{k} a_s)] a_j^{\frac{1}{2}} O(a_i^{\frac{1+\delta}{2}} a_j^{\frac{1+\delta}{2}})$$

$$\leq c_4 \sum_{i=k_0}^{k} a_i (\exp -\frac{c}{2} \sum_{s=i+1}^{k} a_s) \sum_{j=k_0}^{k} a_j (\exp -\frac{c}{2} \sum_{s=j+1}^{k} a_s), \tag{3.4.43}$$

where $c_4 > 0$ is a constant. Again, by (3.3.10), I_3 is bounded in k.

It remains to estimate I_4. By Schwarz inequality we have

$$I_4 \leq \frac{4c_0^2 c_1^2}{a_{k+1}} \sum_{j=k_0}^{k} a_j^2 \exp(-c \sum_{s=j+1}^{k} a_s) \sum_{j=k_0}^{k} [\exp(-c \sum_{s=j+1}^{k} a_s)]$$
$$\cdot E(\|x_j - x^0\|^{2(1+\beta)} I_{[\mu > j,\, \alpha_{i_0} < k_0]}).$$

By (3.4.19), for large enough k_0

$$\sum_{j=k_0}^{k} \frac{a_j^2}{a_{k+1}} \exp(-c \sum_{s=j+1}^{k} a_s) \leq \sum_{j=k_0}^{k} a_j \exp(-\frac{c}{2} \sum_{s=j+1}^{k} a_s),$$

which, as shown by (3.3.11), is bounded in k, we then by (3.4.37) have

$$I_4 \leq c_5 \gamma^{2\beta} \sum_{j=k_0}^{k} \exp(-c \sum_{s=j+1}^{k} a_s) a_j E(\frac{\|x_j - x^0\|^2}{a_j} I_{[\mu > j,\, \alpha_{i_0} < k_0]}) \tag{3.4.44}$$

where c_5 is a constant.

Combing (3.4.40)-(3.4.44) we find that there exists a constant $c_6 > 0$ such that

$$\frac{1}{a_{k+1}} E(\|x_{k+1} - x^0\|^2 I_{[\mu > k+1,\, \alpha_{i_0} < k_0]})$$
$$\leq c_6 + c_5 \gamma^{2\beta} \sum_{j=k_0}^{k} a_j [\exp(-c \sum_{s=j+1}^{k} a_s)] E(\frac{\|x_j - x^0\|^2}{a_j} I_{[\mu > j,\, \alpha_{i_0} < k_0]}). \tag{3.4.45}$$

Setting

$$g_k = [\exp(c\sum_{s=1}^{k} a_s)]\frac{1}{a_k}E(\|x_j - x^0\|^2 I_{[\mu>k,\ \alpha_{i_0}<k_0]}), \tag{3.4.46}$$

and

$$p_{k+1} = c_6 \exp(c\sum_{s=1}^{k+1} a_s), \tag{3.4.47}$$

from (3.4.45) we have

$$\begin{aligned} g_{k+1} &\le p_{k+1} + c_5\gamma^{2\beta}\sum_{j=k_0}^{k} a_j e^{ca_{k+1}} g_j \\ &\le p_{k+1} + c_7\gamma^{2\beta}\sum_{j=k_0}^{k} a_j g_j, \end{aligned} \tag{3.4.48}$$

where c_7 is a constant.

Denoting

$$f_k = \sum_{j=k_0}^{k} a_j g_j, \tag{3.4.49}$$

from (3.4.48) we find

$$\begin{aligned} f_{k+1} =& a_{k+1}g_{k+1} + f_k \le (1 + c_7\gamma^{2\beta}a_{k+1})f_k + a_{k+1}p_{k+1} \\ &\le \cdots \le \sum_{i=k_0}^{k+1}\prod_{j=i+1}^{k+1}(1 + c_7\gamma^{2\beta}a_j)a_i p_i, \end{aligned} \tag{3.4.50}$$

where $\prod_{j=k+2}^{k+1}(1 + c_7\gamma^{2\beta}a_j)$ is set to equal to 1.

From (3.4.48) and (3.4.50) it then follows that

$$g_{k+1} \le p_{k+1} + c_7\gamma^{2\beta}\sum_{i=k_0}^{k}[\exp(c_7\gamma^{2\beta}\sum_{j=i+1}^{k} a_j)]a_i p_i,$$

which combining (3.4.46) leads to

$$\begin{aligned} &\frac{1}{a_{k+1}}E(\|x_{k+1} - x^0\|^2 I_{[\mu>k+1,\ \alpha_{i_0}<k_0]}) \\ &\le [\exp(-c\sum_{i=1}^{k+1} a_i)[p_{k+1} + c_7\gamma^{2\beta}\sum_{i=k_0}^{k}[\exp(c_7\gamma^{2\beta}\sum_{j=i+1}^{k} a_j)]a_i p_i] \end{aligned}$$

$$= c_6 + c_6 c_7 \gamma^{2\beta} \exp(-c \sum_{i=1}^{k+1} a_i) \sum_{i=k_0}^{k} a_i [\exp(c_7 \gamma^{2\beta} \sum_{j=i+1}^{k} a_j)][\exp(c \sum_{s=1}^{i} a_s)], \tag{3.4.51}$$

where for the last equality we have used (3.4.47).

Choosing γ sufficiently small so that

$$-c + c_7 \gamma^{2\beta} \triangleq -c_8 < 0,$$

from (3.4.51) we then have

$$\frac{1}{a_{k+1}} E(\|x_{k+1} - x^0\|^2 I_{[\mu > k+1, \, \alpha_{i_0} < k_0]})$$
$$\leq c_6 + c_6 c_7 \gamma^{2\beta} \sum_{i=k_0}^{k} a_i \exp(-c_8 \sum_{j=i+1}^{k} a_j),$$

which is bounded with respect to k as shown by (3.3.10). □

Lemma 3.4.4 *If A3.4.1-A3.4.4 hold, then*

$$\frac{1}{\sqrt{k}} \sum_{i=1}^{k} \|f(x_i) - F(x_i - x^0)\| \xrightarrow[k \to \infty]{} 0 \quad \text{a.s.} \tag{3.4.52}$$

Proof. It suffices to prove

$$\sum_{i=k_0}^{\infty} \frac{1}{\sqrt{i}} \|f(x_i) - F(x_i - x^0)\| < \infty \quad \text{a.s.} \tag{3.4.53}$$

Then the lemma follows from (3.4.53) by using the Kronecker lemma.

By (3.4.11) and (3.4.37) we have

$$E \sum_{i=k_0}^{\infty} \frac{1}{\sqrt{i}} \|f(x_i) - F(x_i - x^0)\| I_{[\mu > i, \, \alpha_{i_0} < k_0]}$$
$$\leq E \sum_{i=k_0}^{\infty} \frac{1}{\sqrt{i}} c_1 \|x_i - x^0\|^{1+\beta} I_{[\mu > i, \, \alpha_{i_0} < k_0]}$$
$$\leq \sum_{i=k_0}^{\infty} \frac{1}{\sqrt{i}} c_1 [E(\|x_i - x^0\|^2 I_{[\mu > i, \, \alpha_{i_0} < k_0]})]^{\frac{1+\beta}{2}},$$

where the last inequality follows by using the Lyapunov inequality.

Applying Lemma 3.4.3, from the above estimate we derive

$$E\sum_{i=k_0}^{\infty}\frac{1}{\sqrt{i}}\|f(x_i)-F(x_i-x^0)\|I_{[\mu>i,\ \alpha_{i_0}<k_0]}$$

$$\leq \sum_{i=k_0}^{\infty}\frac{1}{\sqrt{i}}c_1(E(\|x_i-x^0\|^2 I_{[\mu>i,\ \alpha_{i_0}<k_0]}))^{\frac{1+\beta}{2}}\cdot\frac{a_i^{\frac{1+\beta}{2}}}{a_i^{\frac{1+\beta}{2}}}$$

$$\leq c_9\sum_{i=k_0}^{\infty}\frac{1}{\sqrt{i}}a_i^{\frac{1+\beta}{2}}<\infty, \tag{3.4.54}$$

where c_9 is a constant and the convergence of the series follows from (3.4.13).

From (3.4.54) it follows that

$$\sum_{i=k_0}^{\infty}\frac{1}{\sqrt{i}}\|f(x_i)-F(x_i-x^0)\|I_{[\mu>i,\ \alpha_{i_0}<k_0]}<\infty \quad \text{a.s.},$$

which means that

$$\{\sum_{i=k_0}^{\infty}\frac{1}{\sqrt{i}}\|f(x_i)-F(x_i-x^0)\|<\infty\} \supset \{\sup_{k_0\leq i<\infty}\|x_i-x^0\|<\gamma, \alpha_{i_0}<k_0\}. \tag{3.4.55}$$

By Lemma 3.4.2, for any given $\epsilon>0$,

$$P\{\sup_{k_0\leq i<\infty}\|x_i-x^0\|<\gamma,\quad \alpha_{i_0}<k_0\}>1-\epsilon,$$

if k_0 is sufficiently large. This together with (3.4.55) shows that

$$P\{\sum_{i=k_0}^{\infty}\frac{1}{\sqrt{i}}\|f(x_i)-F(x_i-x^0)\|<\infty\}>1-\epsilon,$$

or equivalently,

$$P\{\sum_{i=1}^{\infty}\frac{1}{\sqrt{i}}\|f(x_i)-F(x_i-x^0)\|<\infty\}>1-\epsilon.$$

This verifies (3.4.53) because $\epsilon>0$ can be arbitrarily small. The proof of the lemma is completed. □

Proof of Theorem 3.4.2.

By Lemma 3.4.2, $x_k \xrightarrow[k\to\infty]{} x^0$ a.s. and

$$x_{k+1} = x_k + a_k y_{k+1}, \quad \forall k \geq \alpha_{i_0}.$$

Consequently,

$$\begin{aligned}
\sqrt{k}(\bar{x}_k - x^0) &= \frac{1}{\sqrt{k}}\sum_{i=1}^{k}(x_i - x^0) = o(1) + \frac{1}{\sqrt{k}}\sum_{i=\alpha_{i_0}}^{k}(x_i - x^0) \\
&= \frac{1}{\sqrt{k}}\sum_{i=\alpha_{i_0}}^{k}\Phi_{i-1,\alpha_{i_0}}(x_{\alpha_{i_0}} - x^0) \\
&\quad + \frac{1}{\sqrt{k}}\sum_{i=\alpha_{i_0}}^{k}\sum_{j=\alpha_{i_0}}^{i-1}\Phi_{i-1,j+1}a_j\epsilon_{j+1} \\
&\quad + \frac{1}{\sqrt{k}}\sum_{i=\alpha_{i_0}}^{k}\sum_{j=\alpha_{i_0}}^{i-1}\Phi_{i-1,j+1}a_j[f(x_j) - F(x_j - x^0)] + o(1),
\end{aligned} \tag{3.4.56}$$

where $o(1) \longrightarrow 0$ as $k \longrightarrow \infty$.

Noticing $\Phi_{k,j} = \Phi_{k-1,j} + a_k F \Phi_{k-1,j}$, we have

$$\Phi_{k,j} = I + \sum_{i=j}^{k} a_i F \Phi_{i-1,j}, \quad \text{and} \quad F^{-1}\Phi_{k,j} = F^{-1} + \sum_{i=j}^{k} a_i \Phi_{i-1,j}, \tag{3.4.57}$$

and hence

$$a_{j-1}\sum_{i=j}^{k}\Phi_{i-1,j} = \sum_{i=j}^{k}(a_{j-1} - a_i)\Phi_{i-1,j} + \sum_{i=j}^{k} a_i \Phi_{i-1,j}.$$

By (3.4.16) and (3.4.57), from here we derive

$$a_{j-1}\sum_{i=j}^{k}\Phi_{i-1,j} = -F^{-1} + G_{k,j}. \tag{3.4.58}$$

By Lemma 3.4.1, $G_{k,j}$ is bounded. Then with the help of (3.4.58) we have

$$\begin{aligned}
&\frac{1}{\sqrt{k}}\sum_{i=\alpha_{i_0}}^{k}\Phi_{i-1,\alpha_{i_0}}(x_{\alpha_{i_0}} - x^0) \\
&= \frac{1}{\sqrt{k}a_{\alpha_{i_0}-1}}(-F^{-1} + G_{k,\alpha_{i_0}})(x_{\alpha_{i_0}} - x^0) \xrightarrow[k\to\infty]{} 0 \quad \text{a.s.}
\end{aligned} \tag{3.4.59}$$

From (3.4.58) and the boundedness of $G_{k,j}$, there exists a constant $c_{10} > 0$ such that

$$\|\sum_{i=j+1}^{k} \Phi_{i-1,j+1}a_j\| < c_{10}, \quad \forall k, \quad \forall j < k. \tag{3.4.60}$$

Then, we have

$$\begin{aligned}
&\frac{1}{\sqrt{k}}\|\sum_{i=\alpha_{i_0}}^{k}\sum_{j=\alpha_{i_0}}^{i-1} \Phi_{i-1,j+1}a_j[f(x_j)+F(x_j-x^0)]\| \\
&=\frac{1}{\sqrt{k}}\|\sum_{j=\alpha_{i_0}}^{k}\sum_{i=j+1}^{k} \Phi_{i-1,j+1}a_j[f(x_j)-F(x_j-x^0)]\| \\
&\leq \frac{c_{10}}{\sqrt{k}}\sum_{j=\alpha_{i_0}}^{k} \|f(x_j)-F(x_j-x^0)\| \xrightarrow[k\to\infty]{} 0, \quad \text{a.s.,}
\end{aligned} \tag{3.4.61}$$

where the convergence to zero a.s. follows from Lemma 3.4.4.
Putting (3.4.59), (3.4.61) into (3.4.56) leads to

$$\sqrt{k}(\bar{x}_k - x^0) = \frac{1}{\sqrt{k}}\sum_{j=\alpha_{i_0}}^{k}\sum_{i=j+1}^{k} \Phi_{i-1,j+1}a_j\epsilon_{j+1} + o(1).$$

By (3.4.58) we then have

$$\begin{aligned}
\sqrt{k}(\bar{x}_k - x^0) = &-\frac{F^{-1}}{\sqrt{k}}\sum_{j=\alpha_{i_0}}^{k} e_{j+1} + \frac{1}{\sqrt{k}}\sum_{j=\alpha_{i_0}}^{k} G_{k,j+1}e_{j+1} \\
&+\frac{1}{\sqrt{k}}\sum_{j=\alpha_{i_0}}^{k}(-F^{-1}+G_{k,j+1})\nu_{j+1} + o(1).
\end{aligned} \tag{3.4.62}$$

Notice that

$$E\|\frac{1}{\sqrt{k}}\sum_{j=0}^{k} G_{k,j+1}e_{j+1}\|^2 \leq \frac{1}{k}\sum_{j=0}^{k}\|G_{k,j+1}\|\sum_{s=-j}^{k-j}\|G_{k,j+s}\|Ee_{j+s+1}e_{j+1}^T\|.$$

Let us denote by c_{11} the upper bound for $\|G_{k,j}\|$, where the existence of c_{11} is guaranteed by Lemma 3.4.1. Then using (3.4.9) and (3.4.18) we have

$$E\|\frac{1}{\sqrt{k}}\sum_{j=0}^{k} G_{k,j+1}e_{j+1}\|^2 \leq \frac{c_2c_{11}}{k}\sum_{j=0}^{k}\|G_{k,j+1}\| \xrightarrow[k\to\infty]{} 0,$$

which implies that

$$\frac{1}{\sqrt{k}}\sum_{j=0}^{k} G_{k,j+1}e_{j+1} \xrightarrow[k\longrightarrow\infty]{P} 0,$$

and hence

$$\frac{1}{\sqrt{k}}\sum_{i=\alpha_{i_0}}^{k} G_{k,j+1}e_{j+1} \xrightarrow[k\longrightarrow\infty]{P} 0, \tag{3.4.63}$$

because $\{G_{k,j}\}$ is bounded.

By (3.4.10) we see that

$$\sum_{j=1}^{k}\frac{1}{\sqrt{j}}E\|\nu_{j+1}\| \le \sum_{j=1}^{k}\frac{1}{\sqrt{j}}O(a_j^{\frac{1+\delta}{2}}) < \infty,$$

where the convergence follows from (3.4.13).

From this by the Kronecker lemma it follows that

$$\frac{1}{\sqrt{k}}\sum_{j+1}^{k} E\|\nu_{j+1}\| \xrightarrow[k\longrightarrow\infty]{} 0.$$

Therefore, we have

$$E\frac{1}{\sqrt{k}}\|\sum_{j=\alpha_{i_0}}^{k}(-F^{-1}+G_{k,j+1})\nu_{j+1}\|$$

$$\le (\|F^{-1}\|+c_{11})\frac{1}{\sqrt{k}}\sum_{j=0}^{k}E\|\nu_{j+1}\| \xrightarrow[k\longrightarrow\infty]{} 0,$$

and hence

$$\frac{1}{\sqrt{k}}\sum_{j=\alpha_{i_0}}^{k}(F^{-1}+G_{k,j+1}) \xrightarrow[k\longrightarrow\infty]{P} 0. \tag{3.4.64}$$

Combining (3.4.62)–(3.4.64) we arrive at

$$\sqrt{k}(\bar{x}_k - x^0) + \frac{F^{-1}}{\sqrt{k}}\sum_{j=\alpha_{i_0}}^{k} e_{j+1} \xrightarrow[k\longrightarrow\infty]{P} 0,$$

or

$$\sqrt{k}(\bar{x}_k - x^0) + \frac{F^{-1}}{\sqrt{k}}\sum_{j=0}^{k} e_{j+1} \xrightarrow[k\longrightarrow\infty]{P} 0.$$

This incorporating with (3.4.8) implies the conclusion of the theorem. □

This theorem tells us that if in (2.1.1)-(2.1.3) we apply the slowly decreasing step size, then the averaged estimate $\bar{x}_k$ leads to the minimal covariance matrix of the limit distribution.

3.5. Notes and References

Convergence rates and asymptotic normality can be found in [28, 68, 78] for the nondegenerate case. The rate of convergence for the degenerate case was first considered by Pflug in [74]. The results presented in Section 3.2 are given in [15, 47].

For the proof of central limit theorem (Lemma 3.3.1) we refer to [6, 56, 78], while for Remark 3.3.1 refer to [78]. The proof of Theorem 3.3.1 and 3.3.2 can be found in [28].

Asymptotic normality of stochastic approximation algorithm was first considered in [44].

For asymptotic efficiency the averaging technique was introduced in [80, 83], and further considered in [35, 59, 66, 67, 74, 98]. Theorems given in Section 3.4 can be found in [13]. For adaptive stochastic approximation refer to [92, 95].

Chapter 4

OPTIMIZATION BY STOCHASTIC APPROXIMATION

Up-to now we have been concerned with finding roots of an unknown function $f(\cdot)$ observed with noise. In applications, however, one often faces to the optimization problem, i.e., to finding the minimizer or maxinizer of an unknown function $L(\cdot)$. It is well know that $L(\cdot)$ achieves its maximum or minimum values at the root set of its gradient, i.e., at $J = \{x : \nabla L(x) \triangleq f(x) = 0\}$, although it may be only in the local sense. The gradient $\nabla L(x)$ is also written as $L_x(x)$.

If the gradient $L_x(\cdot)$ can be observed with or without noise, then the optimization problem is reduced to the SA problem we have discussed in previous chapters. Here, we are considering the optimization problem for the case where the function $L(\cdot)$ itself rather than its gradient is observed and the observations are corrupted by noise. This problem was solved by the classical Kiefer-Wolfowitz (KW) algorithm which took the finite differences to serve as estimates for the partial derivatives. To be precise, let x_k be the estimate at time k for the minimizer (maximizer) of $L(\cdot)$, and let

$$y_{k+1}^{i+} = L(x_k^{i+}) + \xi_{k+1}^{i+}, \quad y_{k+1}^{i-} = L(x_k^{i-}) + \xi_{k+1}^{i-}$$

be two observations on $L(\cdot)$ at time $k+1$ with noises ξ_{k+1}^{i+} and ξ_{k+1}^{i-}, respectively, where

$$\begin{aligned} x_k^{i+} &= [x_k^1, \ldots, x_k^{i-1}, x_k^i + c_k, x_k^{i+1}, \ldots, x_k^l]^T \\ x_k^{i-} &= [x_k^1, \ldots, x_k^{i-1}, x_k^i - c_k, x_k^{i+1}, \ldots, x_k^l]^T \end{aligned}$$

are two vectors perturbed from the estimate x_k by $+c_k$ and $-c_k$, respectively, on the ith component of x_k. The KW algorithm suggests taking the finite difference

$$y_{k+1}^i \triangleq \frac{y_{k+1}^{i+} - y_{k+1}^{i-}}{2c_k}$$

as the observation of $f_i(x_k)$, the ith component of the gradient $L_x(x)(\triangleq f(x))$. It is clear that

$$y_{k+1} \triangleq [y_{k+1}^1, \dots, y_{k+1}^l]^T = f(x_k) + \epsilon_{k+1}$$

where the ith component of ϵ_{k+1} equals

$$\frac{L(x_k^{i+}) - L(x_k^{i-})}{2c_k} - f_i(x_k) + \frac{\xi_{k+1}^{i+} - \xi_{k+1}^{i-}}{2c_k}.$$

The RM algorithm

$$x_{k+1} = x_k + a_k y_{k+1}.$$

with y_{k+1} defined above is called the KW algorithm.

It is understandable that in the classical theory for convergence of the KW algorithm rather restrictive conditions are imposed not only on $L(\cdot)$ but also on ξ_{k+1}^{i+} and ξ_{k+1}^{i-}. Besides, at each iteration to form finite differences, $2l$ observations are needed, where l is the dimension of x_k. In some problems l may be very large, for example, in the problem of optimizing weights in a neuro-network l corresponds to the number of nodes, which may be large. Therefore, it is of interest not only to weaken conditions required for convergence of the optimizing algorithm but also to reduce the number of observations per iteration.

In Section 4.1 the KW algorithm with expanding truncations using randomized differences is considered. As to be shown, because of replacing finite differences by randomized differences, the number of observations is reduced from $2l$ to 2 for each iteration, and because of involving expanding truncations in the algorithm and applying TS method for convergence analysis, the conditions needed for $L(\cdot)$ have been weakened significantly and the conditions imposed on the noise have been improved to the weakest possible. The convergence rate and asymptotic normality for the KW algorithm with randomized differences and expanding truncations are given in Section 4.2.

The KW algorithm as other gradient-based optimization algorithms may be stuck at a local minimizer (or maximizer). How to approach to the global optimizer is one of the important issues in optimization theory. Especially, how to pathwisely reach the global optimizer is a difficult and challenging problem. In Section 4.3 the KW algorithm is combined with searching initial values, and it is shown that the resulting algorithm a.s. converges to the global optimizer of the unknown function

$L(\cdot)$. The obtained results are then applied to some practical problems in Section 4.4.

4.1. Kiefer-Wolfowitz Algorithm with Randomized Differences

There is a fairly long history of random search or approximation ideas in SA. Different random versions of KW algorithm were introduced: for example, in one version a sequence of random unit vectors that are independent and uniformly distributed on the unit sphere or unit cube was used; and in another version the KW algorithm with random directions was introduced and was called a simultaneous perturbation stochastic approximation algorithm.

Here, we consider the expandingly truncated KW algorithm with randomized differences. Conditions needed for convergence of the proposed algorithm are considerably weaker than existing ones.

Conditions on $\{\Delta_k\}$

Let $(\Delta_k^i, i = 1, \dots, l, k = 1, 2, \dots)$ be a sequence of independent and identically distributed (iid) random variables such that

$$|\Delta_k^i| < a, \quad |\frac{1}{\Delta_k^i}| < b, \quad E(\frac{1}{\Delta_k^i}) = 0, \tag{4.1.1}$$
$$\forall i = 1, \dots, l, \quad \forall k = 1, 2, \dots$$

Furthermore, let $\Delta_k \triangleq [\Delta_k^1, \dots, \Delta_k^l]^T$ be independent of $\mathcal{F}_k^\xi \triangleq \sigma(\xi_i^+, \xi_i^0, \xi_i^-, i = 0, \dots, k)$, the σ-algebra generated by $(\xi_i^+, \xi_i^0, \xi_i^-, i = 0, \dots, k), \forall k$. $(\xi_i^+, \xi_i^0, \xi_i^-)$ is the observation noise to be explained later.

For convenience of writing let us denote

$$\Delta_k^{-1} \triangleq [\frac{1}{\Delta_k^1}, \dots, \frac{1}{\Delta_k^l}]^T. \tag{4.1.2}$$

It should be emphasized that Δ_k^{-1} is a vector and is irrelevant to inverse.

At each time two observations are taken: either

$$y_{k+1}^+ = L(x_k + c_k\Delta_k) + \xi_{k+1}^+, \tag{4.1.3}$$
$$y_{k+1}^- = L(x_k - c_k\Delta_k) + \xi_{k+1}^-, \tag{4.1.4}$$

or

$$y_{k+1}^+ = L(x_k + c_k\Delta_k) + \xi_{k+1}^+, \tag{4.1.5}$$
$$y_{k+1}^0 = L(x_k) + \xi_{k+1}^0, \tag{4.1.6}$$

where x_k is the estimate for the sought-for minimizer (maximizer) of $L(\cdot)$, $\xi^+_{k+1}, \xi^-_{k+1}, \xi^0_{k+1}$ denote the observation noises, and $c_k > 0$ is a real number.

The randomized differences are defined as

$$\frac{L(x_k + c_k\Delta_k) - L(x_k - c_k\Delta_k)}{2c_k}\Delta_k^{-1} \text{ or } \frac{L(x_k + c_k\Delta_k) - L(x_k)}{c_k}\Delta_k^{-1}, \tag{4.1.7}$$

and

$$y_{k+1} = \frac{y^+_{k+1} - y^-_{k+1}}{2c_k}\Delta_k^{-1} \quad \text{or} \quad y_{k+1} = \frac{y^+_{k+1} - y^0_{k+1}}{c_k}\Delta_k^{-1} \tag{4.1.8}$$

may serve as observations of randomized differences.

To be fixed, let us consider observations defined by (4.1.3) and (4.1.4). The convergence analysis, however, can analogously be done for observations (4.1.5) and (4.1.6).

Thus, the observations considered in the sequel are

$$\begin{aligned} y_{k+1} &= \frac{y^+_{k+1} - y^-_{k+1}}{2c_k}\Delta_k^{-1} \\ &= \frac{L(x_k + c_k\Delta_k) - L(x_k - c_k\Delta_k)}{2c_k}\Delta_k^{-1} + \frac{\xi_{k+1}\Delta_k^{-1}}{2c_k}, \end{aligned} \tag{4.1.9}$$

where

$$\xi_{k+1} = \xi^+_{k+1} - \xi^-_{k+1}. \tag{4.1.10}$$

We now define the KW algorithm with expanding truncations and randomized differences. Let $\{M_k\}$ be a sequence of positive numbers increasingly diverging to infinity, and let x^* be a fixed point in $\mathbb{R}^l$. Given any initial value x_0, the algorithm is defined by:

$$x_{k+1} = (x_k + a_k y_{k+1}) I_{[\|x_k + a_k y_{k+1}\| \le M_{\sigma_k}]} + x^* I_{[\|x_k + a_k y_{k+1}\| > M_{\sigma_k}]}, \tag{4.1.11}$$

$$\sigma_k = \sum_{i=0}^{k-1} I_{[\|x_i + a_i y_{i+1}\| > M_{\sigma_i}]}, \quad \sigma_0 = 0, \tag{4.1.12}$$

where $\{y_k\}$ is given by (4.1.9) and (4.1.10).

It is worth noting that the algorithm (4.1.9)-(4.1.12) differs from (2.1.1)- (2.1.3) only by observations $\{y_k\}$. As a matter of fact, (4.1.11) and (4.1.12) are exactly the same as (2.1.1) and (2.1.2), but (4.1.9) and

(4.1.10) are different from (2.1.3). As before, σ_k is the number of truncations that have occurred before time k. Clearly the random vector x_k is measurable with respect to $\mathcal{F}_k \triangleq \mathcal{F}_k^\xi \vee \mathcal{F}_{k-1}^\Delta$, the minimal σ-algebra containing both $\mathcal{F}_k^\xi$ and $\mathcal{F}_{k-1}^\Delta$, where $\mathcal{F}_k^\Delta = \sigma(\Delta_i, i = 0, \dots, k)$. Thus the random vector Δ_k is independent of $\sigma(x_i, i \le k)$.

Let $L_x(x) \triangleq f(x)$.

The observation (4.1.9) can be written in the standard form of RM algorithm. In fact, we can rewrite y_{k+1} as follows:

$$y_{k+1} = f(x_k) + \epsilon_{k+1}, \tag{4.1.13}$$

where

$$\epsilon_{k+1} = \frac{L(x_k + c_k\Delta_k) - L(x_k - c_k\Delta_k)}{2c_k}\Delta_k^{-1} - f(x_k) + \frac{\xi_{k+1}^+ - \xi_{k+1}^-}{2c_k}\Delta_k^{-1} \tag{4.1.14}$$

Thus, the KW algorithm (4.1.9)-(4.1.12) turns to be a standard RM algorithm with expanding truncations (4.1.11)-(4.1.14) considered in Chapter 2. Of course, the observation noise ϵ_{k+1} expressed by (4.1.14) is quite complicated: it is composed of the structural error

$$\frac{L(x_k + c_k\Delta_k) - L(x_k - c_k\Delta_k)}{2c_k}\Delta_k^{-1} - f(x_k) \tag{4.1.15}$$

and the random noise $\frac{\xi_{k+1}^+ - \xi_{k+1}^-}{2c_k}\Delta_k^{-1}$ caused by inaccuracy of observations.

We now list conditions to be used.

A4.1.1 *i)* $a_k > 0, a_k \longrightarrow 0$ as $k \longrightarrow \infty$, $\sum_{k=1}^{\infty} a_k = \infty$ *and a* $p \in (1, 2]$ *exists such that* $\sum_{k=1}^{\infty} a_k^p < \infty$*;*

ii) $c_k > 0$ *and* $c_k \longrightarrow 0$ *as* $k \longrightarrow \infty$.

A4.1.2 $f(x) \triangleq L_x(x)$ *is locally Lipschitz continuous. There is an unique maximum of* $L(\cdot)$ *at* x^0 *that is the only root for* $f(\cdot)$*:* $f(x^0) = 0$ *and* $f(x) \neq 0$ *for* $x \neq x^0$*. Further,* x^* *used in (4.1.11) is such that* $L(x^*) > \sup_{\|x\|=c} L(x)$ *for some* c *and* $\|x^*\| < c$.

Remark 4.1.1 If $L(\cdot)$ is twice continuously differentiable, then $f(\cdot)$ is locally Lipschitz continuous.

Remark 4.1.2 If x^0 is the unique minimizer of $L(\cdot)$, then in (4.1.11) and (4.1.12) a_k should be replaced by $-a_k$.

Theorem 4.1.1 *Assume A4.1.1, A4.1.2, and Conditions on $\{\Delta_k\}$ hold. Let $\{x_k\}$ be given by (4.1.9)-(4.1.12) (or (4.1.11)-(4.1.14)) with any initial value. Then*

$$\lim_{k\longrightarrow\infty} x_k = x^0 \quad \text{a.s.}$$

if and only if for each k the random noise ξ_{k+1} given by (4.1.10) can be decomposed into the sum of two terms in l ways such that

$$\xi_k = e_k^j + \nu_k^j, \quad j = 1, \dots, l \tag{4.1.16}$$

with

$$\sum_{k=1}^{\infty} \frac{a_k e_{k+1}^j}{c_k \Delta_k^j} < \infty \quad \text{a.s.} \tag{4.1.17}$$

and

$$\lim_{k\longrightarrow\infty} \frac{\nu_{k+1}^j}{c_k \Delta_k^j} = 0 \quad \text{a.s.}, \tag{4.1.18}$$

where Δ_k^j is given in Conditions on $\{\Delta_k\}$.

Proof. We will apply Theorem 2.2.1 for sufficiency and Theorem 2.4.1 for necessity.

Let us first check Conditions A2.2.1–A2.2.4. Condition A2.2.1 is a part of A4.1.1. Condition A2.2.2 is automatically satisfied if we take $v(x) = -L(x)$ noticing that $J = \{x^0\}$ in the presented case. Condition A2.2.4 is contained in A4.1.2. So, the key issue is to verify that $\{\epsilon_k\}$ given by (4.1.14) satisfies the requirements.

Let $\Delta_k(\cdot)$ and $\Delta_k^c(\cdot)$ be l-dimensional vector functions obtained from Δ_k with some of its components replaced by zero:

$$\Delta_k(s) = [\Delta_k^1, \dots, \Delta_k^s, 0, \dots, 0]^T, \qquad \Delta_k(0) = 0, \tag{4.1.19}$$

$$\Delta_k^c(s) = [0, \dots, 0, \Delta_k^{s+1}, \dots, \Delta_k^l,]^T, \qquad \Delta_k^c(l) = 0. \tag{4.1.20}$$

It is clear that

$$\Delta_k = \Delta_k^c(0) = \Delta_k(l), \tag{4.1.21}$$

and

$$\Delta_k(i-1) + \Delta_k^c(i) = [\Delta_k^1, \dots, \Delta_k^{i-1}, 0, \Delta_k^{i+1}, \dots, \Delta_k^l]^T. \tag{4.1.22}$$

For notational convenience, let $\delta_k(i)$ denote a generic l-dimensional random vector such that

$$\delta_k(i) = [\underbrace{0, \dots, 0}_{i-1}, \delta_k^i, \underbrace{0, \dots, 0}_{l-i}]^T, \tag{4.1.23}$$

where $|\delta_k^i| \le c_k a$, a is specified in (4.1.1), and δ_k^i may vary for different applications.

We express ϵ_{k+1} given by (4.1.14) in an appropriate form to be dealt with. We mainly use the local Lipschitz-continuity to treat the structural error (4.1.15) in ϵ_{k+1}.

Rewrite the ith component of the structural error as follows

$$\begin{aligned} &\frac{L(x_k + c_k\Delta_k) - L(x_k - c_k\Delta_k)}{2c_k\Delta_k^i} - f_i(x_k) \\ &= \frac{L(x_k + c_k\Delta_k) - L(x_k)}{2c_k\Delta_k^i} + \frac{L(x_k) - L(x_k - c_k\Delta_k)}{2c_k\Delta_k^i} - f_i(x_k) \end{aligned} \tag{4.1.24}$$

and for any $1 \le i \le l$ express

$$\begin{aligned} &L(x_k + c_k\Delta_k) - L(x_k) \\ &= L(x_k + c_k\Delta_k) - L(x_k + c_k(\Delta_k(i-1) + \Delta_k^c(i))) \\ &\quad + L(x_k + c_k(\Delta_k(i-1) + \Delta_k^c(i))) - L(x_k + c_k(\Delta_k(i-1) + \Delta_k^c(i+1))) \\ &\quad + \cdots\cdots \\ &\quad + L(x_k + c_k(\Delta_k(i-1) + \Delta_k^c(l-1))) - L(x_k + c_k\Delta_k(i-1)) \\ &\quad + L(x_k + c_k\Delta_k(i-1)) - L(x_k - c_k\Delta_k(i-2)) \\ &\quad + L(x_k + c_k\Delta_k(i-2)) - L(x_k - c_k\Delta_k(i-3)) \\ &\quad + \cdots\cdots \\ &\quad + L(x_k + c_k\Delta_k(1)) - L(x_k), \end{aligned} \tag{4.1.25}$$

where on the right-hand side of the equality all terms are cancelled except the first and the last terms, and in each difference of L, the arguments of L differ from each other only by one $\Delta_k^i, 1 \le i \le l$.

We write (4.1.25) in the compact from:

$$\begin{aligned} &L(x_k + c_k\Delta_k) - L(x_k) \\ &\quad = L(x_k + c_k\Delta_k) - L\Big(x_k + c_k\big(\Delta_k(i-1) + \Delta_k^c(i)\big)\Big) \end{aligned}$$

$$+\sum_{j=i+1}^{l}\Big[L\Big(x_k+c_k\big(\Delta_k(i-1)+\Delta_k^c(j-1)\big)\Big)$$
$$-L\Big(x_k+c_k\big(\Delta_k(i-1)+\Delta_k^c(j)\big)\Big)\Big]$$
$$+\sum_{j=1}^{i-1}\Big[L(x_k+c_k\Delta_k(j))-L(x_k+c_k\Delta_k(j-1))\Big],\quad i=1,\dots,l. \tag{4.1.26}$$

Applying the Taylor's expansion to (4.1.26) we derive

$$\frac{L(x_k+c_k\Delta_k)-L(x_k)}{2c_k\Delta_k^i}-\frac{1}{2}f_i(x_k)=w_{ki}+\sum_{j\neq i}^{l}h_{kj}^i/\Delta_k^i,\ i=1,\dots,l, \tag{4.1.27}$$

where

$$w_{ki}=\frac{1}{2}[f_i(x_k+c_k\Delta_k+\delta_k(i))-f_i(x_k)], \tag{4.1.28}$$

$$h_{kj}^i=\begin{cases}\frac{1}{2}f_j(x_k+c_k\Delta_k(j-1)+\delta_k(j))\Delta_k^j, & j=1,\dots,i-1,\\ \frac{1}{2}f_j(x_k+c_k(\Delta_k(i-1)+\Delta_k^c(j))+\delta_k(j))\Delta_k^j, & j=i+1,\dots,l.\end{cases}$$

Similarly, we have

$$L(x_k)-L(x_k-c_k\Delta_k)=L(x_k-c_k(\Delta_k(i-1)+\Delta_k^c(i)))-L(x_k-c_k\Delta_k)$$
$$+\sum_{j=1}^{i-1}[L(x_k-c_k\Delta_k(j-1))-L(x_k-c_k\Delta_k(j))]$$
$$+\sum_{j=i+1}^{l}\Big[L\big(x_k-c_k(\Delta_k(i-1)+\Delta_k^c(j))\big)$$
$$-L\big(x_k-c_k(\Delta_k(i-1)+\Delta_k^c(j-1))\big)\Big] \tag{4.1.29}$$

and

$$\frac{L(x_k)-L(x_k-c_k\Delta_k)}{2c_k\Delta_k^i}-\frac{1}{2}f_i(x_k)=u_{ki}+\sum_{j\neq i}^{l}g_{kj}^i/\Delta_k^i, \tag{4.1.30}$$

where

$$u_{ki}=\frac{1}{2}[f_i(x_k-c_k\Delta_k+\delta_k(i))-f_i(x_k)], \tag{4.1.31}$$

$$g_{kj}^i = \begin{cases} \frac{1}{2} f_j(x_k - c_k\Delta_k(j-1) + \delta_k(j))\Delta_k^j, & j = 1, \dots, i-1, \\ \frac{1}{2} f_j(x_k - c_k(\Delta_k(i-1) + \Delta_k^c(j)) + \delta_k(j))\Delta_k^j, & j = i+1, \dots, l. \end{cases} \tag{4.1.32}$$

Define the following vectors:

$$w_k = [w_{k1}, \dots, w_{kl}]^T, \quad u_k = [u_{k1}, \dots, u_{kl}]^T, \tag{4.1.33}$$

$$h_k = \Big[\frac{1}{\Delta_k^1}\sum_{j\neq 1}^{l} h_{kj}^1, \frac{1}{\Delta_k^2}\sum_{j\neq 2}^{l} h_{kj}^2, \dots, \frac{1}{\Delta_k^l}\sum_{j\neq l}^{l} h_{kj}^l\Big]^T, \tag{4.1.34}$$

$$g_k = \Big[\frac{1}{\Delta_k^1}\sum_{j\neq 1}^{l} g_{kj}^1, \frac{1}{\Delta_k^2}\sum_{j\neq 2}^{l} g_{kj}^2, \dots, \frac{1}{\Delta_k^l}\sum_{j\neq l}^{l} g_{kj}^l\Big]^T. \tag{4.1.35}$$

Finally, putting (4.1.27)-(4.1.35) into (4.1.14) we obtain the following expression for ϵ_{k+1}:

$$\epsilon_{k+1} = w_k + u_k + h_k + g_k + \frac{\xi_{k+1}}{2c_k}\Delta_k^{-1}, \quad \xi_{k+1} = \xi_{k+1}^+ - \xi_{k+1}^-. \tag{4.1.36}$$

It is worth noting that each component of h_k and g_k is a martingale difference sequence, because both $\sum_{j\neq i}^{l} h_{kj}^i$ and $\sum_{j\neq i}^{l} g_{kj}^i$ are independent of Δ_k^i and $E(\frac{1}{\Delta_k^i}) = 0, i = 1, \dots, l$.

For the sufficiency part we have to show that (2.2.2) is satisfied a.s. Let us show that (2.2.2) is satisfied by all components of w_k, u_k, h_k, and g_k. For components w_{ji} of w_j we have for any $i = 1, \dots, l$

$$|w_{ji}|I_{[\|x_j\|\leq N]} \leq \frac{1}{2}\big|f_i(x_j + c_j\Delta_j + \delta_j(i)) - f_i(x_j)\big|I_{[\|x_j\|\leq N]} \xrightarrow[j\to\infty]{} 0,$$

since $\|\Delta_j\| < a\sqrt{l}$ by (4.1.1), $\|\delta_j(i)\| \leq ac_j$ and $c_j \longrightarrow 0$ as $j \longrightarrow \infty$. Therefore, for any integer N

$$\lim_{T\to 0}\limsup_{k\to\infty}\frac{1}{T}\Big\|\sum_{j=n_k}^{m(n_k,T_k)} a_j w_{ji} I_{[\|x_j\|\leq N]}\Big\| = 0, \quad \forall T_k \in [0,T], \quad \forall w \in \Omega \tag{4.1.37}$$

for any $\{n_k\}$ such that x_{n_k} converges.

Thus, all sample paths of components of $\{w_k\}$ satisfy (2.2.2). Completely the same situation takes place for the components of $\{u_k\}$.

By the convergence theorem for martingale difference sequences, we find that for any integer N

$$\sum_{k=1}^{\infty} a_k \frac{1}{\Delta_k^i}\Big(\sum_{j\neq i}^{l} h_{kj}^i\Big) I_{[\|x_k\|\leq N]}$$
$$= \frac{1}{2}\sum_{k=1}^{\infty} a_k\Big(\sum_{j\neq i}^{l} f_j(x_k + c_k\Delta_k(j-1)$$
$$+ \delta_k(j))\Delta_k^j\Big) I_{[\|x_k\|\leq N]}\frac{1}{\Delta_k^i} < \infty \quad \text{a.s.} \tag{4.1.38}$$

This is because $\Big(\sum_{j\neq i}^{l} f_j(x_k + c_k\Delta_k(j-1) + \delta_k(j))\Delta_k^j\Big) I_{[\|x_k\|\leq N]}$ is independent of $\frac{1}{\Delta_k^i}$ and is bounded by a constant uniformly with respect to k by Lipschitz-continuity of $f(\cdot)$. Then the martingale convergence theorem applies since $\sum_{k=1}^{\infty} a_k^p < \infty$ for some $p \in (1, 2]$ by A4.1.1.

Similar argument can be applied to components of g_k. Since for any integer N (4.1.38) holds outside an exceptional set with probability zero, there is an ω-set Ω' with $P\Omega' = 1$ such that for any $\omega \in \Omega'$

$$\sum_{k=1}^{\infty} a_k \frac{1}{\Delta_k^i}\Big(\sum_{j\neq i}^{l} h_{kj}^i\Big) I_{[\|x_k\|\leq N]} < \infty \tag{4.1.39}$$

and

$$\sum_{k=1}^{\infty} a_k \frac{1}{\Delta_k^i}\Big(\sum_{j\neq i}^{l} g_{kj}^i\Big) I_{[\|x_k\|\leq N]} < \infty \tag{4.1.40}$$

for all $i = 1, \ldots, l$ and $N = 1, 2, \ldots$.

Therefore, for all $\omega \in \Omega'$ and any integer N

$$\lim_{T\longrightarrow 0}\limsup_{n\longrightarrow\infty}\frac{1}{T}\left\|\sum_{k=n}^{m(n,t)} a_k(w_k + u_k + h_k + g_k) I_{[\|x_k\|\leq N]}\right\| = 0,\ \forall t \in [0, T], \tag{4.1.41}$$

where $m(n, t)$ is given by (1.3.2).

From (4.1.17) and (4.1.18) it follows that there exists $\Omega_0 \subset \Omega'$ such that $P\Omega_0 = 1$ and for each $\omega \in \Omega_0$

$$\sum_{k=1}^{\infty} \frac{a_k e_{k+1}^j}{c_k \Delta_k^j} < \infty, \quad \lim_{k \longrightarrow \infty} \frac{\nu_{k+1}^j}{c_k \Delta_k^j} = 0, \quad \forall j = 1, \ldots, l,$$

and hence

$$\lim_{T \longrightarrow 0} \limsup_{n \longrightarrow \infty} \frac{1}{T} \| \sum_{k=n}^{m(n,t)} \frac{a_k}{2c_k} \xi_{k+1} \Delta_k^{-1} \| = 0, \quad \forall t \in [0, T]. \tag{4.1.42}$$

Combining (4.1.41) and (4.1.42), we find for each $\omega \in \Omega_0$,

$$\lim_{T \longrightarrow 0} \limsup_{n \longrightarrow \infty} \frac{1}{T} \| \sum_{k=n}^{m(n,t)} a_k \epsilon_{k+1} I_{[\|x_k\| \le N]} \| = 0, \quad \forall t \in [0, T], \quad \forall N.$$

This means that for the algorithm (4.1.11)-(4.1.14), Condition A2.2.3 is satisfied on Ω_0. Thus by Theorem 2.2.1, $x_k \longrightarrow x^0$ on Ω_0. This proves the sufficiency part of the theorem.

Under the assumption $x_k \longrightarrow x^0$ a.s. it is clear that both w_k and u_k converge to zero a.s. and (4.1.39) and (4.1.40) turn to be

$$\sum_{k=1}^{\infty} a_k \frac{1}{\Delta_k^i} \sum_{j \neq i}^{l} h_{kj}^i < \infty \quad \text{a.s.}$$

and

$$\sum_{k=1}^{\infty} a_k \frac{1}{\Delta_k^i} \sum_{j \neq i}^{l} g_{kj}^i < \infty \quad \text{a.s.}$$

Then the necessity part of the theorem follows from Theorem 2.4.1. We show this. By Theorem 2.4.1, ϵ_{k+1} can be decomposed into two parts ϵ_{k+1}^1 and ϵ_{k+1}^2 such that $\sum_{k=1}^{\infty} a_k \epsilon_{k+1}^{(1)} < \infty$ and $\epsilon_{k+1}^{(2)} \xrightarrow[k \longrightarrow \infty]{} 0$ a.s.. Let us denote by $(x)_j$ the jth component of a vector x. Define

$$\nu_{k+1}^j = (\epsilon_{k+1}^{(2)} - w_k - u_k)_j 2c_k \Delta_k^j, \quad e_{k+1}^j = (\epsilon_{k+1}^{(1)} - h_k - g_k)_j 2c_k \Delta_k^j \tag{4.1.43}$$

Then for $j = 1, \ldots, l$,

$$\frac{\nu_{k+1}^j}{c_k \Delta_k^j} = 2(\epsilon_{k+1}^{(2)} - w_k - u_k)_j \xrightarrow[k \longrightarrow \infty]{} 0, \tag{4.1.44}$$

and

$$\sum_{k=1}^{\infty} a_k \frac{e_{k+1}^j}{c_k \Delta_k^j} = 2\sum_{k=1}^{\infty} a_k (\epsilon_{k+1}^{(1)} - h_k - g_k)_j < \infty \quad \text{a.s.} \tag{4.1.45}$$

From (4.1.43) and (4.1.36) it follows that

$$\begin{aligned} e_{k+1}^j + \nu_{k+1}^j &= 2c_k \Delta_k^j (\epsilon_{k+1}^{(1)} + \epsilon_{k+1}^{(2)} - w_k - u_k - h_k - g_k)_j \\ &= 2c_k \Delta_k^j (\epsilon_{k+1} - w_k - u_k - h_k - g_k)_j \\ &= \xi_{k+1}, \qquad j = 1, \ldots, l. \end{aligned}$$

This together with (4.1.44) and (4.1.45) proves the necessity of the theorem. □

Theorem 2.4.1 gives necessary and sufficient condition on the observation noise in order the KW algorithm with expanding truncations and randomized differences converges to the unique maximizer of a function L. We now give some simple sufficient conditions on $\{\xi_k\}$.

Theorem 4.1.2 *Assume A4.1.1 and A4.1.2 hold. Further, assume that* $\sum_{k=1}^{\infty} a_k^2/c_k^2 < \infty$, ξ_{k+1} *is independent of* $(\Delta_j, j = 1, 2, \ldots, k)$, $\forall k = 1, 2, \ldots$, *and* $\{\xi_k\}$ *satisfies one of the following two conditions:*

i) $\sup_k |\xi_k| \le \xi$ *a.s., where* ξ *is a random variable;*

ii) $\sup_k E\xi_k^2 < \infty$. *Then*

$$\lim_{k \longrightarrow \infty} x_k = x^0 \quad \text{a.s.,}$$

where $\{x_k\}$ *is given by (4.1.9)-(4.1.12).*

Proof. It suffices to prove (4.1.16)-(4.1.18). Assume i) holds. Let $\mathcal{F}_k^j$ be given by

$$\mathcal{F}_k^j = \sigma(\Delta_i^j, i = 0, \ldots, k, \ \xi_s, s = 0, 1, \ldots, k+2).$$

By definition, Δ_k is independent of ξ_{k+1} and $\mathcal{F}_{k-1}^j$, so

$$E(\frac{\xi_{k+1}}{\Delta_k^j} | \mathcal{F}_{k-1}^j) = \xi_{k+1} E(\frac{1}{\Delta_k^j} | \mathcal{F}_{k-1}^j) = 0 \quad \text{a.s.}$$

and

$$\sum_{k=1}^{\infty} E(\frac{a_k^2}{c_k^2} \frac{\xi_{k+1}^2}{(\Delta_k^j)^2} | \mathcal{F}_{k-1}^j) \le \sum_{k=1}^{\infty} \frac{a_k^2}{c_k^2} b\xi^2 < \infty \quad \text{a.s.,}$$

where b is an upper bound for $\frac{1}{\Delta_k^j}$.

By the convergence theorem for martingale difference sequences, it follows that

$$\sum_{k=1}^{\infty} \frac{a_k}{c_k} \frac{\xi_{k+1}}{\Delta_k^j} < \infty \quad \text{a.s.}, \quad \forall j = 1, \dots, l.$$

Thus in (4.1.16) it can be assumed that $\nu_k^j \equiv 0, \forall j = 1, \dots, l, \forall k = 1, 2, \dots$ and $e_k^j = \xi_k, \forall j = 1, \dots, l, k = 1, 2, \dots$, and the conclusion of the theorem follows from Theorem 4.1.1.

Assume now ii) holds.

By the independence assumption it follows that for $j < k, \Delta_k$ is independent of $(\Delta_j, \xi_{j+1}, \xi_{k+1})$ so that

$$E\Big(\frac{\xi_{j+1}\xi_{k+1}}{\Delta_j^i \Delta_k^i}\Big) = E\Big(\frac{\xi_{j+1}\xi_{k+1}}{\Delta_j^i}\Big) E\Big(\frac{1}{\Delta_k^i}\Big) = 0, \quad \forall i = 1, \dots, l.$$

Then, we have

$$E\Big(\sum_{k=1}^{\infty} \frac{a_k}{c_k} \frac{\xi_{k+1}}{\Delta_k^j}\Big)^2 \le b^2 \sup_k E\xi_{k+1}^2 \sum_{k=1}^{\infty} a_k^2/c_k^2 < \infty.$$

It directly follows that

$$\sum_{k=1}^{\infty} \frac{a_k}{c_k} \frac{\xi_{k+1}}{\Delta_k^j} < \infty \quad \text{a.s.}, \quad \forall j = 1, \dots, l.$$

Again, it suffices to takes $\nu_k^j \equiv 0, \forall j = 1, \dots, l, \forall k = 1, 2, \dots$. □

We now extend the results to the case of multi-extremes. For this A 4.1.2 is replaced by A4.1.2'.

A4.1.2' *$f(x) \triangleq L_x(x)$ is locally Lipschitz continuous, $L(J)$ is nowhere dense, where $J \triangleq \{x : f(x) = 0\}$, the set where L takes extremes, and x^* used in (4.1.11) is such that $L(x^*) > \sup_{\|x\|=c} L(x)$ for some c and $\|x^*\| < c$.*

Theorem 4.1.3 *Let $\{x_k\}$ be given by (4.1.9)-(4.1.12) with a given initial value x_0. Assume A 4.1.1 and A 4.1.2' hold. Then $d(x_k, J^*) \xrightarrow[k\to\infty]{} 0$ on an ω-set Ω_0 with $P\Omega_0 = 1$, if $\{\xi_k\}$ satisfies (4.1.16)- (4.1.18), or $\{\xi_k\}$ satisfies conditions given in Theorem 4.1.2, where J^* is a connected set contained in $\bar{J}$, the closure of J.*

Proof. Condition A2.2.2 is implied by A4.1.2' with $v(\cdot) = -L(\cdot)$, and A2.2.1 and A2.2.4 are implied by A4.1.1 and A4.1.2, respectively, while

A2.2.3 is satisfied as shown in Theorems 4.1.1 and 4.1.2. Then the conclusion of the theorem follows from Theorem 2.2.2.

Remark 4.1.3 In the multi-extreme case, the necessary conditions on $\{\xi_k\}$ for convergence $d(x_k, J^*) \xrightarrow[k\to\infty]{} 0$ can also be obtained on the analogy of Theorem 2.4.2.

Remark 4.1.4 Conditions i) or ii) used in Theorem 4.1.2 are simple indeed. However, ξ_{k+1} in Theorem 4.1.2 is required to be independent of $(\Delta_j, j = 1, \dots, k)$. This may not be satisfied if the observation noise is state-dependent. Taking into account that $\xi_{k+1} \triangleq \xi^+_{k+1} - \xi^-_{k+1}$ is the observation noise when observing $L(\cdot)$ at $x_k + c_k\Delta_k$ and $x_k - c_k\Delta_k$, we see that ξ_{k+1} depends on $x_k + c_k\Delta_k$ and $x_k - c_k\Delta_k$ if the observation noise is state-dependent. In this case, ξ_{k+1} does depend on Δ_k. This violates the assumption about independence made in Theorem 4.1.2.

Consider the case where the observation noise may depend on locations of measurement, i.e., in lieu of (4.1.3) and (4.1.4) consider

$$y^+_{k+1} = L(x_k + c_k\Delta_k) + \xi^+_{k+1}(x_k + c_k\Delta_k, \omega), \tag{4.1.46}$$
$$y^-_{k+1} = L(x_k - c_k\Delta_k) + \xi^-_{k+1}(x_k - c_k\Delta_k, \omega), \tag{4.1.47}$$
$$y_{k+1} = \frac{y^+_{k+1} - y^-_{k+1}}{2c_k}, \tag{4.1.48}$$
$$\xi_{k+1} = \xi^+_{k+1}(x_k + c_k\Delta_k, \omega) - \xi^-_{k+1}(x_k - c_k\Delta_k, \omega). \tag{4.1.49}$$

Introduce the following condition.

A4.1.3 *Both $\xi^+_k(x,\omega)$ and $\xi^-_k(x,\omega)$ are measurable functions $(I\!R^l \times \Omega, \mathcal{B}^l \times \mathcal{F}) \longrightarrow (R, \mathcal{B})$ $\forall k$, $(\xi^+_k(x,\omega), \mathcal{F}_k)$ and $(\xi^-_k(x,\omega), \mathcal{F}_k)$ are martingale difference sequences for any $x \in I\!R^l$, and*

$$E[(\|\xi^+_{k+1}(x,\omega)\|^p + \|\xi^-_{k+1}(x,\omega)\|^p)|\mathcal{F}_k] \triangleq \sigma_{k+1}(x) < \infty, \quad \forall x \ a.s. \tag{4.1.50}$$

for p specified in A4.1.1 with

$$\sup_k \sup_{\|x\|\le N} \sigma_{k+1}(x) \triangleq \sigma(N) < \infty, \quad \forall N, \tag{4.1.51}$$

where $\{\mathcal{F}_k\}$ is a family of nondecreasing σ-algebras independent of both x and $\{\Delta^i_k, i = 1, \dots, l, k = 1, 2, \dots\}$.

Theorem 4.1.4 *Let $\{x_k\}$ be given by (4.1.9)–(4.1.12) with a given initial value x_0. Assume A4.1.1, A4.1.2', and A4.1.3 hold. Then $d(x_k, J^*) \xrightarrow[k\to\infty]{} 0$ a.s., where J^* is a connected subset of $\overline{J}$.*

Proof. Introduce the σ-algebra $\mathcal{F}'_k$ generated by $\{\Delta_i, i \le k\}$ and $\{\mathcal{F}_j, j \le k\}$, i.e., $\mathcal{F}'_k = \sigma\{\Delta_i, 1 \le i \le k, \mathcal{F}_j, 0 \le j \le k\}$.

It is clear that x_k is measurable with respect to $\sigma\{\Delta_j, 1 \le j \le k-1, \mathcal{F}_j, 0 \le j \le k\}$, and hence $x_k \pm c_k\Delta_k$ are $\mathcal{F}'_k$-measurable. Both $\xi^+_{k+1}(x_k + c_k\Delta_k, \omega)$ and $\xi^-_{k+1}(x_k - c_k\Delta_k, \omega)$ are $\mathcal{F}'_{k+1}$-measurable. Approximating $\xi^+_{k+1}(x, \omega)$ and $\xi^-_{k+1}(x, \omega)$ by simple functions, it is seen that

$$E(\xi^+_{k+1}(x_k + c_k\Delta_k, \omega)|\mathcal{F}'_k) = E(\xi^+_{k+1}(x, \omega)|\mathcal{F}'_k)|_{x=x_k+c_k\Delta_k} = 0$$
$$E(\xi^-_{k+1}(x_k - c_k\Delta_k, \omega)|\mathcal{F}'_k) = E(\xi^-_{k+1}(x, \omega)|\mathcal{F}'_k)|_{x=x_k-c_k\Delta_k} = 0.$$

Therefore, $(\xi^+_{k+1}(x_k + c_k\Delta_k, \omega), \mathcal{F}'_{k+1})$ and $(\xi^-_{k+1}(x_k - c_k\Delta_k, \omega), \mathcal{F}'_{k+1})$ are martingale difference sequences, and

$$\begin{aligned} E\Big[\big(\|\xi^+_{k+1}(x_k + c_k\Delta_k, \omega)\Delta_k^{-1}\|^p + \|\xi^-_{k+1}(x_k - c_k\Delta_k, \omega)\Delta_k^{-1}\|^p\big)|\mathcal{F}'_k\Big] \\ \le \gamma(\sigma_{k+1}(x_k + c_k\Delta_k) + \sigma_{k+1}(x_k - c_k\Delta_k)), \end{aligned}$$

where

$$\gamma \triangleq (b\sqrt{l})^p \ge \|\Delta_k^{-1}\|^p.$$

Hence, $(\xi_{k+1}\Delta_k^{-1}I_{[\|x_k\| \le N]}, \mathcal{F}'_{k+1})$ is a martingale difference sequence with

$$\begin{aligned} E\Big(\|\xi_{k+1}\Delta_k^{-1}\|^p I_{[\|x_k\|\le N]}\big|\mathcal{F}'_k\Big) \\ \le 2^{p-1}\gamma I_{[\|x_k\|\le N]}\big(\sigma_{k+1}(x_k + c_k\Delta_k) + \sigma_{k+1}(x_k - c_k\Delta_k)\big). \end{aligned}$$

Noticing $\{\|\Delta_k\|\}$ is bounded and $c_k \longrightarrow 0$ as $k \longrightarrow \infty$, by (4.1.50) and (4.1.51) and the convergence theorem for martingale difference sequences we have, for any integer $N > 0$

$$\sum_{k=1}^{\infty} \frac{a_k}{c_k}\xi_{k+1}\Delta_k^{-1}I_{[\|x_k\|\le N]} < \infty \text{ a.s.} \tag{4.1.52}$$

This together with (4.1.37) with w_{ji} replaced by u_{ji}, (4.1.39), and (4.1.40) verifies that $\{\epsilon_{k+1}\}$ expressed by (4.1.36) satisfies A2.2.3. Then the conclusion of the theorem follows from Theorem 2.2.2. □

Remark 4.1.5 If J consists of a singleton x^0, then Theorems 4.1.3 and 4.1.4 ensure $x_k \longrightarrow x^0$ a.s. If J is composed of isolated points, then theorems ensure that x_k converges to some point in J. However, the limit is not guaranteed to be a global minimizer of $L(\cdot)$. Depending on initial value, x_k may converge to a local minimizer. We will return back to this issue in Section 4.3.

4.2. Asymptotic Properties of KW Algorithm

We now present results on convergence rate and asymptotic normality of the KW algorithm with randomized differences.

Theorem 4.2.1 *Assume hypotheses of Theorem 4.1.2 or Theorem 4.1.4 with* $J = \{x^0\}$ *and that*

$$\lim_{k \longrightarrow \infty} (a_{k+1}^{-1} - a_k^{-1}) = \alpha \geq 0, \tag{4.2.1}$$

$$c_k = o(a_k^\delta) \quad \text{and} \quad \sum_{j=1}^{\infty} a_j^{2(1-\delta)}/c_j^2 < \infty \tag{4.2.2}$$

for some $\delta \in (0, 1)$ *and as* $x \longrightarrow x^0$

$$f(x) = F(x - x^0) + r(x), \quad r(x^0) = 0, \quad r(x) = o(\|x - x^0\|), \tag{4.2.3}$$

where $F + \alpha\delta I$ *is stable and* α *and* δ *are specified in (4.2.1) and (4.2.2), respectively.*

Then $\{x_k\}$ *given by (4.1.9)–(4.1.12) satisfies*

$$\|x_k - x^0\| = O(a_k^\delta) \qquad \text{a.s.}$$

Proof. First of all, under conditions of Theorems 4.1.2 or 4.1.4, $x_k \longrightarrow x^0$ a.s. By Theorem 3.1.1 it suffices to show that $\{\epsilon_{k+1}\}$ given by (4.1.36) can be represented as

$$\epsilon_k = e_k + \nu_k,$$

where

$$\nu_k = o(a_k^\delta) \quad \text{as} \quad k \longrightarrow \infty \text{ and } \sum_{k=1}^{\infty} a_k^{1-\delta} e_{k+1} < \infty \quad \text{a.s.} \tag{4.2.4}$$

From (4.1.28) and (4.1.31) by the local Lipschitz continuity of f it follows that

$$\|w_k\| = O(c_k) = o(a_k^\delta) \quad \text{and} \quad \|u_k\| = O(c_k) = o(a_k^\delta)$$

by (4.2.2). Since $c_k = o(a_k^\delta)$ it follows that

$$a_k^{1-\delta} = o(\frac{a_k}{c_k}) \quad \text{and} \quad \sum_{k=1}^{\infty} a_k^{2(1-\delta)} < \infty.$$

Since $x_k \longrightarrow x^0$ a.s., $\{h_{kj}^i\}$ and $\{g_{kj}^i\}$ given by (4.1.27) and (4.1.32) are uniformly bounded for $i = 1, \dots, l$, $j = 1, \dots, l$, $k = 1, 2, \dots$, for each ω where $\{x_k\}$ converges. By the convergence theorem for martingale difference sequences it follows that

$$\sum_{j=1}^{\infty} a_j^{1-\delta} h_j < \infty \text{ a.s.}, \qquad \sum_{j=1}^{\infty} a_j^{1-\delta} g_j < \infty \text{ a.s.},$$

where h_k and g_k are are given by (4.1.35).

In the proof of Theorem 4.1.2, replacing $\frac{a_k}{c_k}$ by $\frac{a_k^{1-\delta}}{c_k}$ and using (4.2.2), the same argument leads to

$$\sum_{k=1}^{\infty} \frac{a_k^{1-\delta}}{c_k} \frac{\xi_{k+1}}{\Delta_k^j} < \infty \text{ a.s.}, \quad \forall j = 1, \dots, l. \tag{4.2.5}$$

Then by defining

$$e_{k+1} \triangleq h_k + g_k + \frac{\xi_{k+1}}{2c_k} \Delta_k^{-1}, \quad \nu_{k+1} \triangleq w_k + u_k$$

we have shown (4.2.4) under the hypotheses of Theorem 4.1.2.

Under the hypotheses of Theorem 4.1.4 we have the same conclusions about w_k, u_k, g_k and h_k as before. We need only to show (4.2.5). But this follows from (4.1.52) with a_k replaced by $a_k^{1-\delta}$ and the convergence $x_k \longrightarrow x^0$ a.s. □

Remark 4.2.1 Let $\{x_k\}$ be given by (4.1.9)–(4.1.12). If $a_k = \frac{1}{k}$, and $c_k = \frac{1}{k^\nu}$ with $\nu \in (0, \frac{1}{2})$, then conditions (4.2.1) and (4.2.2) are satisfied.

Theorem 4.2.2 *Assume A4.1.1 and A4.1.2 hold and that*

i) $\lim_{k\longrightarrow\infty} (a_{k+1}^{-1} - a_k^{-1}) = \alpha > 0$ *and* $c_k = a_k^\gamma$ *for some* $\gamma \in (\frac{1}{4}, \frac{1}{2})$*;*

ii) $\|f(x) - F(x - x^0)\| \le c\|x - x^0\|^{1+\beta}$ *for some* $c > 0$ *and* $\beta > 0$*;*

iii) $F + \alpha\gamma I$ *is stable and* $\sum_{j=1}^{\infty} a_j^{2(1-\gamma-\delta)} < \infty$ *for some* $\delta \in (\frac{\gamma}{1+\beta}, \gamma)$*;*

iv) $\{\xi_k\}$ *given by (4.1.10) is an MA process:*

$$\xi_k = 2\sum_{i=0}^{r} b_i w_{k-i}, \quad w_i = 0 \textit{ for } i < 0,$$

where $b_i, i = 0, 1, \dots, r$, are real numbers and $\{w_i, \mathcal{F}_i\}$ is a martingale difference sequence which is independent of $\{\Delta_i\}$ and satisfies

$$E(w_k^2|\mathcal{F}_{k-1}) \le \sigma_0, \quad \forall i \ge 0, \tag{4.2.6}$$

$$\lim_{k\longrightarrow\infty} E(w_k^2|\mathcal{F}_{k-1}) = \sigma^2 > 0, \tag{4.2.7}$$

$$\lim_{K\longrightarrow\infty} \sup_i E(w_i^2 I_{[|w_i|>K]}) = 0. \tag{4.2.8}$$

Then

$$a_k^{-\mu}(x_k - x^0) \xrightarrow[k\longrightarrow\infty]{d} N(0, S), \tag{4.2.9}$$

where $\mu = \frac{1}{2} - \gamma$ and

$$\begin{aligned} S =&\sigma^2\sigma_1^2(\sum_{i=0}^r b_i)^2 \int_0^\infty e^{t(F+\alpha\mu I)} e^{t(F+\alpha\mu I)^T} dt, \qquad (4.2.10)\\ \sigma_1^2 =&E(\frac{1}{(\Delta_i^1)^2}). \end{aligned}$$

Proof. Since $\gamma > \delta$, it follows that $c_k = a_k^\gamma = o(a_k^\delta)$ and

$$\sum_{k=1}^\infty \frac{a_k^{2(1-\delta)}}{c_k^2} = \sum_{k=1}^\infty a_k^{2(1-\gamma-\delta)} < \infty. \tag{4.2.11}$$

By assumption $\{w_i\}$ is independent of $\{\Delta_i\}$, and hence $\{\xi_i\}$ is independent of $\{\Delta_k\}$. Then by (4.2.11) and the convergence theorem for martingale difference sequences we obtain (4.2.5). By Theorem 4.2.1 we have as $k \longrightarrow \infty$

$$\|x_k - x^0\| = o(a_k^\delta) \tag{4.2.12}$$

and after a finite number of iterations of (4.1.11), say, for $k \ge N$ there are no more truncations.

Since $\mu = \frac{1}{2} - \gamma$, $\gamma \in (\frac{1}{4}, \frac{1}{2})$, $\mu \in (0, \frac{1}{4})$, $\gamma > \mu$, and $F + \alpha\mu I$ is stable, it follows that

$$\begin{aligned} \frac{a_{k+1}^{-\mu}}{a_k^{-\mu}} =&(1 + \frac{a_{k+1}^{-1} - a_k^{-1}}{a_k^{-1}})^\mu \\ =&(1 + \alpha a_k + a_k(a_{k+1}^{-1} - a_k^{-1} - \alpha))^\mu \\ =&1 + \alpha\mu a_k + o(a_k). \qquad (4.2.13) \end{aligned}$$

Let z_k be given by

$$z_k = a_k^{-\mu}(x_k - x^0). \tag{4.2.14}$$

By (4.1.11), (4.1.13), (4.1.36), and condition ii) it follows that for $k \geq N$

$$\begin{aligned} z_{k+1} =&\big(1+\alpha\mu a_k + o(a_k)\big)\Big[z_k + a_k a_k^{-\mu}\big(F(x_k - x^0) + f(x_k) \\ & - F(x_k - x^0) + w_k + u_k + h_k + g_k + \frac{\xi_{k+1}}{2c_k}\Delta_k^{-1}\big)\Big] \qquad (4.2.15) \\ =&\Big[I + a_k\big(F + \alpha\mu I + o(1)\big)\Big] z_k \\ & + (1+O(a_k))a_k^{(\frac{1}{2}+\gamma)} c\|x_k - x^0\|^{1+\beta} \\ & + (1+O(a_k))a_k^{\frac{1}{2}+\gamma}(w_k + u_k + h_k + g_k) \\ & + (1+O(a_k))a_k^{\frac{1}{2}}\xi_{k+1}\Delta_k^{-1}. \qquad (4.2.16) \end{aligned}$$

Let Φ_{ki} be given by

$$\Phi_{k,i} = \prod_{j=i}^{k}\Big(I + a_j\big(F + \alpha\mu I + o(1)\big)\Big), \quad \Phi_{i,i+1} = I, \qquad (4.2.17)$$

where $i < k$.

Since $F + \alpha\mu I$ is stable, by (3.1.8) it follows that there are constants $\lambda_1 > 0$ and $\lambda > 0$ such that

$$\|\Phi_{k,i}\| \leq \lambda_1 \exp\Big(-\lambda\sum_{j=i}^{k} a_j\Big). \qquad (4.2.18)$$

Noticing $\|x_k - x^0\|^{1+\beta} = o(a_k^{\delta(1+\beta)}) = o(a_k^{(\gamma+e)})$, where $e > 0$ because $\delta \in (\frac{\gamma}{1+\beta}, \gamma)$ by condition iii), we have

$$\begin{aligned} z_{k+1} =&\Phi_{k,N}z_N + \sum_{j=N}^{k}\Phi_{k,j+1}\big(1+O(a_j)\big)a_j^{1+e+(2\gamma-\frac{1}{2})} \\ & + \sum_{j=N}^{k}\Phi_{k,j+1}\big(1+O(a_j)\big)a_j^{\frac{1}{2}+\gamma}(w_j + u_j) \\ & + \sum_{j=N}^{k}\Phi_{k,j+1}\big(1+O(a_j)\big)a_j^{\frac{1}{2}+\gamma}(h_j + g_j) \\ & + \sum_{j=N}^{k}\Phi_{k,j+1}\big(1+O(a_j)\big)a_j^{\frac{1}{2}}\xi_{j+1}\Delta_j^{-1} \\ \triangleq & I_1 + I_2 + I_3 + I_4 + I_5 \qquad (4.2.19) \end{aligned}$$

where $I_i, i = 1,2,3,4,5$, respectively denote the five terms on the right-hand side of the first equality of (4.2.19).

By (4.2.18), $I_1 \triangleq \Phi_{k,N} z_N \xrightarrow[k\to\infty]{} 0$ a.s.

By Lemma 3.3.2, $I_2 \xrightarrow[k\to\infty]{} 0$ a.s., because $e > 0$ and $2\gamma > \frac{1}{2}$.

By (4.1.28) and (4.1.3) it follows that $\|w_j\| + \|u_j\| = O(c_j)$, and hence by i) and (4.2.18)

$$\begin{aligned}\|I_3\| &\le \sum_{j=N}^{k} \lambda_1 \exp(-\lambda \sum_{s=j+1}^{k} a_s) a_j^{\frac{1}{2}+\gamma} O(a_j^{\gamma}) \\ &\le \lambda_2 \sum_{j=N}^{k} \exp(-\lambda \sum_{s=j+1}^{k} a_s) a_j^{1+\frac{1}{2}-2\mu}, \end{aligned} \tag{4.2.20}$$

where λ_2 is a constant.

By Lemma 3.3.2 and $\mu < \frac{1}{4}$, the right-hand side of (4.2.20) tends to zero a.s. as $k \longrightarrow \infty$.

To estimate I_4 let us consider the following linear recursion

$$\begin{aligned}\eta_{k+1} &= \Big(I + a_k\big(F + \alpha\mu I + o(1)\big)\Big)\eta_k + a_k^{\frac{1}{2}+\gamma}(h_k + g_k), \quad k \ge N, \\ \eta_N &= a_{N+1}^{-\mu} \cdot a_N(h_N + g_N).\end{aligned} \tag{4.2.21}$$

By (4.2.17) it follows that

$$\eta_{k+1} = \sum_{j=N}^{k} \Phi_{k,j+1}(1 + O(a_j)) a_j^{\frac{1}{2}+\gamma}(h_j + g_j). \tag{4.2.22}$$

By (4.2.11), $\sum\limits_{k=1}^{\infty} a_k^{2(1-\gamma)} < \infty$. Since $\gamma > \frac{1}{4}$, $2 - 2\gamma < 1 + 2\gamma$ and $\sum\limits_{k=1}^{\infty} a_k^{1+2\gamma} < \infty$. Then by the convergence theorem for martingale difference sequences it follows that

$$\sum_{k=1}^{\infty} a_k^{\frac{1}{2}+\gamma} \sum_{j\ne i}^{l} h_{kj}^i \frac{1}{\Delta_k^i} < \infty \quad \text{a.s. } \forall i = 1,\dots,l,$$

i.e.,

$$\sum_{k=1}^{\infty} a_k^{\frac{1}{2}+\gamma} h_k < \infty \quad \text{a.s.}$$

Similarly,

$$\sum_{k=1}^{\infty} a_k^{\frac{1}{2}+\gamma} g_k < \infty \quad \text{a.s.}$$

Applying Lemma 3.1.1, we find that $\eta_k \xrightarrow[k\to\infty]{} 0$ a.s. From (4.2.22), it follows that $I_4 \xrightarrow[k\to\infty]{} 0$ a.s.

Since ξ_k is an MA process driven by a martingale difference sequence satisfying (4.2.6), $\sum_{j=1}^{\infty} O(a_j) a_j^{\frac{1}{2}} \xi_{j+1} < \infty$ a.s.

By the argument similar to that used for (4.2.21) and (4.2.22), from Lemma 3.1.1 it follows that

$$\sum_{j=N}^{k} \Phi_{k,j+1} O(a_j) a_j^{\frac{1}{2}} \xi_{j+1} \Delta_j^{-1} \xrightarrow[k\to\infty]{} 0 \quad \text{a.s.}$$

Therefore, putting all these convergence results into (4.2.19) yields

$$\lim_{k\to\infty} (z_{k+1} - \sum_{j=N}^{k} \Phi_{k,j+1} a_j^{\frac{1}{2}} \xi_{j+1} \Delta_j^{-1}) = 0 \quad \text{a.s.} \tag{4.2.23}$$

By (3.3.37),

$$\sum_{j=1}^{k} \Phi_{k,j+1} a_j^{\frac{1}{2}} \xi_{j+1} \Delta_j^{-1} \xrightarrow[k\to\infty]{d} N(0,S), \tag{4.2.24}$$

where S is given by (4.2.10). By (4.2.18), from (4.2.23) and (4.2.24) it follows that $z_k \xrightarrow[k\to\infty]{d} N(0,S)$, which together with the definition (4.2.14) for z_k proves the theorem. □

Example 4.2.1 The following example of $\{a_k\}$ and $\{c_k\}$ satisfies Conditions i) and iii) of Theorem 4.2.2:

$$a_k = \frac{1}{k}, \quad \alpha = 1, \quad c_k = \frac{1}{k^{\frac{1}{3}}}, \quad \gamma = \frac{1}{3}, \quad \beta = 1, \quad \delta = \frac{1}{5}.$$

In this example, $\mu = \frac{1}{6}$, $\frac{1}{5} = \delta \in (\frac{\gamma}{1+\beta}, \gamma) = (\frac{1}{6}, \frac{1}{3})$, and $\sum_{j=1}^{\infty} a_j^{2(1-\gamma-\delta)} = \sum_{j=1}^{\infty} \frac{1}{k^{\frac{7}{6}}} < \infty$.

Remark 4.2.2 Results in Sections 4.1 and 4.2 are proved for the case, where the two-sided randomized differences $y_{k+1} \triangleq \frac{y^+_{K+1}-y^-_{k+1}}{2c_k}\Delta_k^{-1}$ are used where y^+_{k+1} and y^-_{k+1} are given by (4.1.3) and (4.1.4), respectively. But, all results presented in Sections 4.1 and 4.2 are also valid for the case where the one-sided randomized differences

$$y_{k+1} \triangleq \frac{y^+_{k+1} - y^0_{k+1}}{2c_k}\Delta_k^{-1}$$

are used, where y^+_{k+1} and y^0_{k+1} are given by (4.1.3) and (4.1.6), respectively.

In this case, $\frac{1}{2}$ in (4.1.27), (4.1.28) and in the expression of h^i_{kj} should be replaced by 1, and (4.1.29)–(4.1.32) disappear. Accordingly, (4.1.36) changes to

$$\epsilon_{k+1} = w_k + h_k + \frac{\xi_{k+1}}{c_k}\Delta_k^{-1}, \quad \xi_{k+1} = \xi^+_{k+1} - \xi^0_{k+1}. \tag{4.2.25}$$

Theorems 4.1.1-4.1.4 and 4.2.1 remain unchanged. The conclusion of Theorem 4.2.2 remains valid too, if in Condition iv) $\xi_k = 2\sum_{i=0}^{r} b_i w_{k-i}$ changes to $\xi_k = \sum_{i=0}^{r} b_i w_{k-i}$.

4.3. Global Optimization

As pointed out at the beginning of the chapter, the KW algorithm may lead to a local minimizer of $L(\cdot)$. Before the 1980s, the random search or its combination with a local search method was the main stochastic approach to achieve the global minimum when the values of L can exactly be observed without noise. When the structural property of L is used for local search, a rather rapid convergence rate can be derived, but it is hard to escape a local attraction domain. The random search has a chance to fall into any attraction domain, but its convergence rate decreases exponentially as the dimension of the problem increases.

Simulating annealing is an attractive method for global optimization, but it provides only convergence in probability rather than path-wise convergence. Moreover, simulation shows that for functions with a few local minima, simulated annealing is not efficient. This motivates one to combine KW-type method with random search. However, a simple combination of SA and random search does not work: in order to reach the global minimum one has to reduce the noise effect as time goes on.

A hybrid algorithm composed of a search method and the KW algorithm is presented in the sequel with main effort devoted to design easily realizable switching rules and to provide an effective noise-reducing method.

We define a global optimization algorithm, which consists of three parts: search, selection, and optimization. To be fixed, let us discuss the global minimization problem. In the search part, we choose an initial value and make the local search by use of the KW algorithm with randomized differences and expanding truncations described in Section 4.1 to approach the bottom of the local attraction domain. At the same time, the average of the observations for L is used to serve as an estimate of the local minimum of L in this attraction domain. In the selection part, the estimates obtained for the local minima of L are compared with each other, and the smallest one among them together with the corresponding minimizer given by the KW algorithm are selected. Then, the optimization part takes place, where again the local search is carried out, i.e., the KW algorithm without any truncations is applied to improve the estimate for the minimizer. At the same time, the corresponding minimum of L is reestimated by averaging the noisy observations. After this, the algorithm goes back to the search part again.

For the local search, we use observations (4.1.3) and (4.1.4), or (4.1.5) and (4.1.6). To be fixed, let us use (4.1.5) and (4.1.6).

In the sequel, by KW algorithm with expanding truncations we mean the algorithm defined by (4.1.11) and (4.1.12) with

$$y_{k+1} = \frac{y^0_{K+1} - y^+_{k+1}}{2c_k}\Delta_k^{-1}, \tag{4.3.1}$$

where y^+_{k+1} and y^0_{k+1} are given by (4.1.5) and (4.1.6), respectively. Similar to (4.1.9) and (4.1.10) we have

$$y_{k+1} = \frac{L(x_k) - L(x_k + c_k\Delta_k)}{c_k}\Delta_k^{-1} + \frac{\xi_{k+1}}{c_k}\Delta_k^{-1}, \tag{4.3.2}$$

where

$$\xi_{k+1} = \xi^0_{k+1} - \xi^+_{k+1}. \tag{4.3.3}$$

By KW algorithm we mean

$$x_{k+1} = x_k + a_k y_{k+1} \tag{4.3.4}$$

with y_{k+1} defined by (4.3.2).

It is worth noting that unlike (4.1.8), $y^0_{k+1} - y^+_{k+1}$ is used in (4.3.1). Roughly speaking, this is because in the neighborhood of a miminizer of $L(\cdot)$, $f(x) \triangleq L_x(x)$ is increasing, and y_{k+1} in (4.1.11) should be an observation on $-f(\cdot)$.

In order to define switching rules, we have to introduce integer-valued and increasing functions $g(i)$ and $e(i)$ such that $g(i) > e(i)$, $\forall i = 1, 2, \ldots$, and $g(i+1) - g(i) \longrightarrow \infty$ as $i \longrightarrow \infty$.

Define

$$G(i) = g(i) - e(i), \quad i = 1, 2, \ldots \tag{4.3.5}$$

In the sequel, by the ith search period we mean the part of algorithm starting from the ith test of selecting the initial value $x_0^{(i)}$ up to the next selection of initial value. At the end of the $(i-1)$th search period, we are given $x_{g(i)}$ and $\Lambda_{g(i)}$ being the estimates for the global minimizer and the minimum of L, respectively. Variables such as x_k, ξ_k^0, ξ_k^+, δ_k, Δ_k and Δ_k^{-1} etc. in the ith search period are equipped by superscript (i), e.g., $x_k^{(i)}$, $\xi_k^{(i,0)}$ etc.

The global optimization algorithm is defined by the following five steps.

(GO1) Starting from $i = 1$ at the ith search period, the initial value $x_0^{(i)}$ is chosen according to a given rule (deterministic or random), and then $\{x_k^{(i)}\}$ is calculated by the KW algorithm with expanding truncations (4.1.11) and (4.1.12) with y_{k+1} defined by (4.3.1), for which x^*, step sizes $\{a_k\}$ and $\{c_k\}$, and $\{M_k\}$ used for truncation are defined as follows:

$$x^* = x_0^{(i)}, \quad M_k = M_k^{(i)} \triangleq \|x_0^{(i)}\| \vee M^{(i)} + \tilde{M}_k, \tag{4.3.6}$$

$$a_k = a/e(i) + k, \tag{4.3.7}$$

$$c_k = c/(e(i)+k)^{\mu} \quad \text{for } k \le G(i) - 1, \tag{4.3.8}$$

where $c > 0$ and $\mu \in (\frac{1}{4}, \frac{1}{2})$ are fixed constants, $\{M^{(i)}\}$ and $\{\tilde{M}_k\}$ are two sequences of positive real numbers increasingly diverging to infinity.

(GO2) Set the initial estimate $\Lambda_0^{(i)} = 0$ for $L(x_0^{(i)})$, and update the estimate for $L(x_k^{(i)})$ by

$$\Lambda_{k+1}^{(i)} = \Lambda_k^{(i)} + \frac{1}{k+1}(L(x_{k+1}^{(i)}) + \xi_{k+2}^{(i,0)} - \Lambda_k^{(i)}), \tag{4.3.9}$$

where $\xi_k^{(i,0)}$ is the noise when observing $L(x_k^{(i)})$, $0 \le k \le G(i) - 1$. After $G(i)$ steps, $\Lambda_{G(i)}^{(i)}$ is obtained.

(GO3) Let $\{\lambda(i)\}$ be a given sequence of real numbers such that $\lambda(i) > 0$ and $\lambda(i) \longrightarrow 0$ as $i \longrightarrow \infty$. Set $x_{g(1)} = x_{G(1)}^{(1)}$. For $i > 1$ if

$\Lambda^{(i)}_{G(t)} + \lambda(i) < \Lambda_{g(i)} - \lambda(i)$, then set $x_{g(i)} = x^{(i)}_{G(i)}$. Otherwise, keep $x_{g(i)}$ unchanged.

(GO4) Improve $x_{g(i)}$ to $x_{g(i+1)}$ by the KW algorithm with expanding truncations (4.1.11) and (4.1.12) with y_{k+1} defined by (4.3.1), for which

$$a_k = \frac{a}{k}, \quad c_k = c/k^{\mu}, \quad k \geq g(i), \quad \mu \in \left(\frac{1}{4}, \frac{1}{2}\right), \tag{4.3.10}$$

where in (4.1.11) and (4.1.12) $\{M_k\}$ may be an arbitrary sequence of numbers increasingly diverging to infinity, and

$$x^* = x_{g(i)}.$$

At the same time, update the estimate Λ_k for $L(x_k)$ by

$$\Lambda_{k+1} = \Lambda_k + \frac{1}{k+1-g(i)}(L(x_{k+1}) + \xi^0_{k+2} - \Lambda_k), \tag{4.3.11}$$
$$\Lambda_{g(i)} = 0, \qquad g(i) \leq k < g(i+1),$$

where ξ^0_{k+1} is the noise when observing $L(x_k)$. At the end of this step, $x_{g(i+1)}$ and $\Lambda_{g(i+1)}$ are derived.

(GO5) Go back to (GO1) for the $(i+1)$th search period.

We note that for the ith search period $e(i)$ is added to a_k and c_k (see (4.3.7) and (4.3.8)). The purpose of this is to diminish the effect of the observation noise as i increases. Therefore, a_k and c_k both tend to zero, not only as $k \longrightarrow \infty$ but also as $i \longrightarrow \infty$. The following example shows that adding an increasing $e(i)$ to the denominators of a_k and c_k is necessary.

Example 4.3.1 Let

$$L(x) = \begin{cases} (x+1)^2, & x \leq -0.75, \\ -3x^2 - 4x - 1.25, & -0.75 < x \leq -0.5, \\ x^2 - 0.25, & -0.5 < x \leq 0.5, \\ -3x^2 + 4x - 1.25, & 0.5 < x \leq 0.75, \\ (x-1)^2, & x > 0.75. \end{cases}$$

It is clear that the global minimizer is $J \triangleq \{0\}$ and $x = \pm 1$ are two local minima. Furthermore, $(-\infty, -0.75)$ and $(0.75, +\infty)$ are attraction domains for -1 and $+1$, respectively.

Since $L_x(x)$ is linear, for local search we apply the ordinary KW algorithm without truncation

$$x_{k+1} = x_k + a_k y_{k+1},$$
$$y_{k+1} = (y_{k+1}^+ - y_{k+1}^-)/2c_k,$$
$$y_{k+1}^- = L(x_k - c_k) + \xi_{k+1}^-, \qquad y_{k+1}^+ = L(x_k + c_k) + \xi_{k+1}^+.$$

Here, no randomized differences are introduced, because this is a one-dimentional problem.

Assume

$$\xi_{k+1}^+ = b(x_k)\epsilon_{k+1}^+, \qquad \xi_{k+1}^- = b(x_k)\epsilon_{k+1}^-, \tag{4.3.12}$$

where

$$b(x) = \begin{cases} 5, & -0.5 < x \le 0.5 \\ 0, & \text{otherwise}, \end{cases} \tag{4.3.13}$$

and $\{\epsilon_k^+\}$ and $\{\epsilon_k^-\}$ are mutually independent and both are sequences of iid random variables with

$$P(\epsilon_k^+ = 1) = P(\epsilon_k^- = -1) = P(\epsilon_k^- = 1) = P(\epsilon_k^1 = -1) = \frac{1}{2}.$$

Let us start from (GO1) and take

$$e(i) \equiv 1 \text{ (not tending to infinity)}, \quad a = c = \frac{1}{2}, \quad \mu = \frac{1}{3},$$
$$\text{i.e.,} \quad c_k = \frac{1}{2}(k+1)^{-\frac{1}{3}}, \quad a_k = \frac{1}{2}(k+1)^{-1}.$$

If $x_0 \in [-1, 1]$, then, by noticing $c_0 = a_0 = \frac{1}{2}$, one of $x_0 + c_0$ and $x_0 - c_0$ must belong to $[-\frac{1}{2}, \frac{1}{2}]$. Elementary calculation shows that

$$\left|L(x_0 + \frac{1}{2}) - L(x_0 - \frac{1}{2})\right| \le \left|L(x_0 + \frac{1}{2})\right| + \left|L(x_0 - \frac{1}{2})\right| \le \frac{1}{3}.$$

Paying attention to (4.3.13), we see

$$|y_1^- - y_1^+| \ge 5 - \frac{1}{3} = 4\frac{2}{3},$$

and

$$|x_1| = \left|x_0 + \frac{1}{2}(y_1^- - y_1^+)\right| \ge \left||x_0| - \frac{1}{2}|y_1^- - y_1^+|\right| \ge 1\frac{1}{3}.$$

This means that x_1 is located in one of the attraction domains $(-\infty, -0.75)$ and $(0.75, +\infty)$. Furthermore, by (4.3.12) and (4.3.13), the observations carried out at these domains are free of noise. Let us consider the further development of the algorithm, once x_k has fallen into the interval $(-\infty, -\frac{4}{3})$ or $(\frac{4}{3}, +\infty)$. To be fixed, let us assume $x_k \in (-\infty, -\frac{4}{3})$, $k \geq 1$. For $k \geq 1$, $x_k + c_k < -\frac{4}{3} + 0.256 < -1$, we have

$$\begin{aligned} x_{k+1} =& x_k + a_k\Big(\frac{1}{2}(x_k - c_k + 1)^2 - \frac{1}{2}(x_k + c_k + 1)^2\Big)\frac{1}{2c_k} \\ =& x_k - a_k(x_k + 1), \end{aligned} \tag{4.3.14}$$

or $(x_{k+1} + 1) = (1 - a_k)(x_k + 1)$, which implies $x_k \longrightarrow -1$.

If $x_0 \notin [-1, 1]$, say, $x_0 < -1$, then $x_0 + c_0 < -0.5$, since $c_0 = \frac{1}{2}$. It suffices to consider the case where $-0.75 < x_0 + c_0 < -0.5$, i.e., $-1.25 < x_0 < -1$, because for the case $x_0 + c_0 \leq -0.75$ we again have (4.3.14) and $x_k \longrightarrow -1$.

Simple computation shows that starting from $x_0 \in (-1.25, -1)$ the observations are free of noise, and the algorithm becomes

$$x_{k+1} = x_k + \frac{a_k}{2c_k}\Big[\frac{1}{2}(x_k - c_k + 1)^2 + \frac{3}{2}(x_k + c_k)^2 + 2(x_k + c_k) + 0.625\Big].$$

As a result of computation, we have

$$(x_k + c_k)|_{k=15} = -0.7545.$$

Then, starting from $k = 15$, the algorithm will be iterated according to (4.3.14), and hence $x_k \longrightarrow -1$.

For the case $x_0 > 1$, it can similarly be shown that $x_k \longrightarrow 1$.

Therefore, whatever the initial value is chosen, x_k will never converge to the global minimizer $\{0\}$, if $e(i)$ in (GO1) does not diverge to infinity.

Let us introduce conditions to be used.

Since we are seeking for global minima of $L(\cdot)$, Condition A4.1.2' should be modified.

A4.3.1 *$f(x) \triangleq L_x(x)$ is locally Lipschitz continuous, $\liminf_{\|x\| \longrightarrow \infty} L(x) = \infty$, and $L(J)$ is nowhere dense, where $J \triangleq \{x : f(x) = 0\}$, the set of extremes of L.*

Note that for seeking minima of $L(\cdot)$, the corresponding part in A4.1.2', should be modified as follows: x^* used in (4.1.11) is such that $L(x^*) < \inf_{\|x\|=c} L(x)$ for some c and $\|x^*\| < c$. But this is implied by assuming $\liminf_{\|x\| \longrightarrow \infty} L(x) = \infty$.

A4.3.2

$$\lim_{i\longrightarrow\infty}\frac{1}{G(i)}\sum_{k=0}^{G(i)}\xi_k^{(i,0)}=0 \quad a.s. \tag{4.3.15}$$

$$\lim_{i\longrightarrow\infty}\frac{1}{g(i+1)-g(i)}\sum_{k=g(i)+1}^{g(i+1)}\xi_k^0=0 \quad a.s. \tag{4.3.16}$$

A4.3.3 *For any convergent subsequence* $\{x_{n_k}^{(i_k)}\}$ *of* $\{x_k^{(i)}, k=1,\dots,G(i), i=1,2,\dots\}$

$$\lim_{T\longrightarrow 0}\limsup_{k\longrightarrow\infty}\frac{1}{T}\sup_{t\in[0,T]}\sum_{j=n_k}^{n_{i_k}(n_k,t)} a\xi_{j+1}^{(i_k)}\Delta_j^{-1(i_k)}/c\big(e(i_k)+j\big)^{1-\mu}=0 \ a.s., \tag{4.3.17}$$

where $\xi_{k+1}^{(i)}$ *denotes* ξ_{k+1} *given by (4.3.3) with* x_k *replaced by* $x_k^{(i)}$, $\Delta_k^{-1(i)}$ *denotes* Δ_k^{-1} *used for the ith search period, and*

$$n_i(k,T)=G(i)\wedge\max\{m:\sum_{j=k}^{m}\frac{a}{e(i)+j}<T\}. \tag{4.3.18}$$

A4.3.4 *For any convergent subsequence* $\{x_{n_k}\}$

$$\lim_{T\longrightarrow 0}\limsup_{k\longrightarrow\infty}\frac{1}{T}\sup_{t\in[0,T]}\sum_{j=n_k}^{m(n_k,t)} a\xi_{j+1}\Delta_j^{-1}/cj^{1-\mu}=0 \ a.s., \tag{4.3.19}$$

where $m(k,t)$ *is given by (1.3.2).*

It is worth emphasizing that each Δ_k^i in the sequence (Δ_k^i, $i=1,\dots,l$, $k=1,2,\dots$) is used only once when we form $\Delta_k^{(i)}$ and Δ_k.

We now give sufficient conditions for A4.1.2, A4.3.3, and A4.4.4. For this, we first need to define σ-algebras generated by estimates $x_j^{(i)}$ and x_j derived up-to current time. Precisely, for k running in the ith search period of Step (GO1) define

$$\mathcal{F}_k^{(i)}\triangleq\sigma\Big\{x_s^{(m)},x_t^{(i)},x_j,\forall m:1\le m<i,\forall s:0\le s\le G(m),$$
$$\forall t:0\le t\le k,\forall j:j\le f(i)\Big\}, \tag{4.3.20}$$

and for k running in Step (GO4) define

$$\mathcal{F}_k\triangleq\sigma\{x_s^{(i)},x_j,\forall i:g(i)<k,\forall s:0\le s\le G(i),\forall j:j\le k\}. \tag{4.3.21}$$

Remark 4.3.1 If both sequences

$$\{\xi_k^{(i,0)}\} = \{\xi_1^{(1,0)}, \xi_2^{(1,0)}, \dots, \xi_{G(1)}^{(1,0)}, \xi_1^{(2,0)}, \xi_2^{(2,0)}, \dots, \xi_{G(2)}^{(2,0)}, \dots, \xi_1^{(i,0)}, \dots, \xi_{G(i)}^{(i,0)}, \dots\}$$

and $\{\xi_k^0\}$ are martingale difference sequences with

$$\sup_{k,i} E\big((\xi_{k+1}^{(i,0)})^2|\mathcal{F}_k^{(i)}\big) < \infty, \quad \sup_k E\big((\xi_{k+1}^{(0)})^2|\mathcal{F}_k\big) < \infty, \tag{4.3.22}$$

and if

$$\liminf_{i\longrightarrow\infty} G(i)i^{-(1+\delta)} > 0, \quad \liminf_{i\longrightarrow\infty}[g(i+1) - g(i)]i^{-(1+\delta)} > 0 \tag{4.3.23}$$

for some $\delta > 0$, then A4.3.2 holds.

This is because

$$\eta_i \triangleq \frac{1}{\sqrt{G(i)}} \sum_{k=0}^{G(i)-1} \xi_{k+1}^{(i,0)}$$

is a maringale difference sequence with bounded second conditional moment, and hence

$$\sum_{i=1}^{\infty} i^{-(1+\delta)/2}\eta_i < \infty \qquad \text{a.s.},$$

which implies (4.3.15).

By using the second parts of conditions (4.3.22) and (4.3.23), (4.3.16) can be verified in a similar way.

Remark 4.3.2 If $\sup\limits_{k,i} E(\xi_k^{(i)})^2 < \infty$ and $\{\xi_{k+1}^{(i)}\}$ is independent of $\{\Delta_s^{(m)}, \Delta_j^{(i)}, \forall m : 1 \le m < i, \forall s : 0 \le s \le G(m), \forall j \le k\}$, and if $\delta > 0$ exists such that $\liminf\limits_{i\longrightarrow\infty} e(i)/i^{\left(\delta+1/(1-2\mu)\right)} > 0$, then by the uncorrelatedness of $\xi_k^{(i)}\Delta_k^{(i)}$ with $\xi_s^{(j)}\Delta_s^{(j)}$ for $i \ne j$ or $k \ne s$

$$\begin{aligned} &E\Big(\sum_{i=1}^{\infty}\sum_{k=1}^{G(i)}[a\xi_{k+1}^{(i)}\Delta_k^{(i)}/c(e(i)+k)^{1-\mu}]\Big)^2 \\ &\le M\sum_{i=1}^{\infty}\sum_{k=e(i)}^{g(i)}(1/k^{2-2\mu}) \le M(1-2\mu)^{-1}\sum_{i=1}^{\infty}[1/e(i)^{1-2\mu}] < \infty, \end{aligned}$$

where M is a constant. From this, it follows that

$$\sum_{k=1}^{G(i)} \left[a\xi_{k+1}^{(i)} \Delta_k^{(i)} / c\big(e(i)+k\big)^{1-\mu} \right] \xrightarrow[i\longrightarrow\infty]{} 0 \qquad \text{a.s.,}$$

and hence A4.3.3 holds.

Remark 4.3.3 If $\sup_k E(\xi_k)^2 < \infty$ and ξ_{k+1} is independent of $\{\Delta_j, j \leq k\}$, then by the martingale convergence theorem, A4.3.4 holds.

We now formulate the convergence theorem for the global optimization algorithm (GO1)–(GO5).

Theorem 4.3.1 *Assume A4.1.1, A4.3.2, A4.3.3, and A4.3.4 hold. Further, assume that* $\{x_0^{(i)}, i = 1, 2, \dots\}$ *selected in (GO1) is dense in an open set* U, $\liminf_{i\longrightarrow\infty} e(i) i^{-(1+\delta)} > 0$ *for some* $\delta > 0$, *and* $g(i)/e(i) \xrightarrow[i\longrightarrow\infty]{} \infty$. *If* $J^0 \subset U$, *then*

$$d(x_k, J^0) \xrightarrow[k\longrightarrow\infty]{} 0 \qquad a.s.,$$

where $\{x_k\}$ *is derived at Step (GO4) and* J^0 *is the set of global minimizers of* $L(\cdot)$:

$$J^0 \triangleq \{x : L(x) = \min_{y\in R^l} L(y)\}.$$

The proof of theorem is separated into lemmas.

We recall that the essence of the proof for the basic convergence Theorem 2.2.1 consists in showing the following property that $v(x_k)$ cannot cross a nonempty interval $[\delta_1, \delta_2]$ infinitely often if $d([\delta_1, \delta_2], v(J)) > 0$. We need to extend this property to a family of algorithms.

Assume for each fixed i the observation is

$$y_{k+1}^{(i)} = -L_x(x_k^{(i)}) + \epsilon_{k+1}^{(i)} \tag{4.3.24}$$

and the algorithm develops as follows

$$\begin{aligned} x_{k+1}^{(i)} =& (x_k^{(i)} + a_k y_{k+1}^{(i)}) I_{[\|x_k^{(i)} + a_k^{(i)} y_{k+1}^{(i)}\| \leq M_{\sigma_k^{(i)}}^{(i)}]} \\ &+ x_0^{(i)} I_{[\|x_k^{(i)} + a_k^{(i)} y_{k+1}^{(i)}\| > M_{\sigma_k^{(i)}}^{(i)}]}, \end{aligned} \tag{4.3.25}$$

$$\sigma_k^{(i)} = \sum_{j=0}^{k-1} I_{[\|x_j^{(i)} + a_j^{(i)} y_{j+1}^{(i)}\| > M_{\sigma_j^{(i)}}^{(i)}]}, \qquad \sigma_0^{(i)} = 0, \tag{4.3.26}$$

where

$$a_k^{(i)} > 0, \quad a_k^{(i)} \xrightarrow[k \longrightarrow \infty]{} 0, \quad \sum_{k=1}^{\infty} a_k^{(i)} = \infty, \quad \text{and} \quad M_k^{(i)} \xrightarrow[k \longrightarrow \infty]{} \infty.$$

Assume, further, for fixed k

$$a_k^{(i)} \xrightarrow[i \longrightarrow \infty]{} 0 \quad \text{and} \quad M_k^{(i)} \xrightarrow[i \longrightarrow \infty]{} \infty.$$

Lemma 4.3.1 *Assume $L(J)$ is nowhere dense, where $J \triangleq \{x : L_x(x) = 0\}$. Let $[a,b]$ be a nonempty interval such $[a,b] \cap L(J) = \emptyset$. If there are two sequences $\{x_{s_k}^{(i_k)}\}$ and $\{x_{t_k}^{(i_k)}\}$ such that*

$$\begin{aligned} &L(x_{s_k}^{(i_k)}) \le a, \qquad L(x_{t_k}^{(i_k)}) \ge b, \\ &a < L(x_j^{(i_k)}) < b, \quad \forall j : s_k < j < t_k, \quad k = 1, 2, \ldots, \end{aligned} \tag{4.3.27}$$

and $\{x_{s_k}^{(i_k)}\}$ is bounded, then it is impossible to have

$$\lim_{T \longrightarrow 0} \limsup_{k \longrightarrow \infty} \frac{1}{T} \Big\| \sum_{j=s_k}^{m_{i_k}(s_k,t)} a_j^{(i_k)} \epsilon_{j+1}^{(i_k)} \Big\| = 0, \quad \forall t \in [0,T], \tag{4.3.28}$$

where

$$m_i(k,T) = \max\Big\{m : \sum_{j=k}^{m} a_j^{(i)} \le T\Big\}.$$

Proof. Without loss of generality we may assume $\{x_{s_k}^{(i_k)}\}$ converges as $k \longrightarrow \infty$; otherwise, it suffices to select a subsequence.

Assume the converse: i.e., (4.3.28) holds. Along the lines of the proof for Theorem 2.2.1 we can show that

$$\Big\| \sum_{j=s_k}^{m+1} a_j^{(i_k)} y_{j+1}^{(i_k)} \Big\| \le M, \quad \forall m : s_k - 1 \le m \le m_{i_k}(s_k, T) \tag{4.3.29}$$

for some constant M if k is sufficiently large. As a matter of fact, this is an analogue of (2.2.3). From (4.3.29) the following analogue of (2.2.15) takes place:

$$\|x_{m+1}^{(i_k)} - x_{s_k}^{(i_k)}\| \le \alpha_1 T, \quad \forall m : s_k \le m \le m_{i_k}(s_k, T), \tag{4.3.30}$$

and the algorithm for $\{x_m^{(i_k)}\}$ has no truncation for $m : s_k \le m \le m_{i_k}(s_k, T)$ if k is large enough, where $\alpha_1 > 0$ is a constant. Similar to (2.2.27), we then have

$$L(x_{m_{i_k}(s_k,T)+1}^{(i_k)}) - L(x_{s_k}^{(i_k)}) \le -\alpha T, \quad \alpha > 0 \tag{4.3.31}$$

for some small $T > 0$ and all sufficiently large k.

From this, by (4.3.27) and convergence of $\{x_{s_k}^{(i_k)}\}$ it follows that

$$\limsup_{k\longrightarrow\infty} L\big(x^{(i_k)}_{m_{i_k}(s_k,T)+1}\big) \le a - \alpha T. \tag{4.3.32}$$

By continuity of $L(\cdot)$ and (4.3.30) we have

$$\lim_{T\longrightarrow 0} \max_{s_k\le m\le m_{i_k}(s_k,T)} |L(x^{(i_k)}_{m+1}) - L(x^{(i_k)}_{s_k})| = 0,$$

which implies that $m_{i_k}(s_k, T) + 1 < t_k$ for small enough T.

Then by definition,

$$L(x^{(i_k)}_{m_{i_k}(s_k,T)+1}) \in [a, b), \tag{4.3.33}$$

which contradicts (4.3.32). The obtained contradiction shows the impossibility of (4.3.28). □

Introduce

$$J_h \stackrel{\Delta}{=} \{x \in \mathbb{R}^l, \quad L(x) < h\}, \quad h > L_{\min}(\stackrel{\Delta}{=} \min_x L(x)), \tag{4.3.34}$$

$$S^{(i)}_{[\delta_1,\delta_2]} \stackrel{\Delta}{=} \{\omega, \exists s \text{ and } t : 0 < s < t < G(i) \\ \text{such that } x_s^{(i)} \in J_{\delta_1}, x_t^{(i)} \in J^c_{\delta_2}\}. \tag{4.3.35}$$

Lemma 4.3.2 *Let $\{x_k^{(i)}, k = 1, \dots, G(i)\}$ be given by (GO1). Assume A4.3.1 and A4.3.3 hold and $\liminf_{i\longrightarrow\infty} e(i)i^{-(1+\delta)} > 0$ for some $\delta > 0$. Then for any $[\delta_1, \delta_2] \subset (L_{\min}, \infty)$, $S^{(i)}_{[\delta_1,\delta_2]}$ may occur infinitely many often with probability 0, i.e.,*

$$P\Big\{\bigcap_{k=1}^{\infty}\bigcup_{i=k}^{\infty} S^{(i)}_{[\delta_1,\delta_2]}\Big\} = 0. \tag{4.3.36}$$

Proof. Since $L(J)$ is nowhere dense, for any ω belonging to infinitely many of $S^{(i)}_{[\delta_1,\delta_2]}$, there are subsequences $\{i_k\}$, $\{s_k\}$, $\{t_k\}$ such that

$$i_k \xrightarrow[k\longrightarrow\infty]{} \infty, \quad x^{(i_k)}_{s_k} \in J_{\delta_\alpha}, \quad x^{(i_k)}_{t_k} \in J^c_{\delta_\beta}, \quad t_k < G(i_k),$$

and

$$x_j^{(i_k)} \in J_{\delta_\beta} \cap J^c_{\delta_\alpha}, \quad \forall j : s_k < j < t_k,$$

where

$$\delta_1 < \delta_\alpha < \delta_\beta < \delta_2 \text{ and } [\delta_\alpha, \delta_\beta] \cap L(J) = \emptyset.$$

By assumption $L(x) \longrightarrow \infty$ as $\|x\| \longrightarrow \infty$, J_{δ_α} must be bounded. Hence, $\{x_{s_k}^{(i_k)}\}$ is bounded. Without loss of generality we may assume that $\{x_{s_k}^{(i_k)}\}$ is convergent.

Notice that at Step (GO1), $\{x_k^{(i)}\}$ is calculated according to (4.1.11) and (4.1.12) with y_{k+1} given by (4.3.2) and (4.3.3), i.e.,

$$\begin{aligned} x_{k+1}^{(i)} =&(x_k^{(i)} + a_k^{(i)} y_{k+1}^{(i)}) I_{[\|x_k^{(i)}+a_k^{(i)}y_{k+1}^{(i)}\| \le M_{\sigma_k^{(i)}}^{(i)}]} \\ &+ x_0^{(i)} I_{[\|x_k^{(i)}+a_k^{(i)}y_{k+1}^{(i)}\| > M_{\sigma_k^{(i)}}^{(i)}]}, \end{aligned} \tag{4.3.37}$$

$$\sigma_k^{(i)} = \sum_{j=0}^{k-1} I_{[\|x_j^{(i)}+a_j^{(i)}y_{j+1}^{(i)}\| > M_{\sigma_j^{(i)}}^{(i)}]}, \quad \sigma_0^{(i)} = 0 \tag{4.3.38}$$

$$y_{k+1}^{(i)} = \frac{L(x_k^{(i)}) - L(x_k^{(i)} + c_k^{(i)}\Delta_k^{(i)})}{c_k^{(i)}} \Delta_k^{-1(i)} + \frac{\xi_{k+1}^{(i)}}{c_k^{(i)}} \Delta_k^{-1(i)}, \tag{4.3.39}$$

which differ from (4.1.11) (4.1.12), (4.3.2), and (4.3.3) by superscript (i), which means the calculation is carried out in the ith search period.

By (4.1.27) with notations (4.1.33) and (4.1.34), equipped by superscript (i) we have

$$\begin{aligned} y_{k+1}^{(i)} =& -L_x(x_k^{(i)}) + \frac{L(x_k^{(i)}) - L(x_k^{(i)} + c_k^{(i)}\Delta_k^{(i)})}{c_k^{(i)}} \Delta_k^{-1(i)} \\ &+ L_x(x_k^{(i)}) + \frac{\xi_{k+1}^{(i)}}{c_k^{(i)}} \Delta_k^{-1(i)} \\ =& -L_x(x_k^{(i)}) - 2w_k^{(i)} - 2h_k^{(i)} + \frac{\xi_{k+1}^{(i)}}{c_k^{(i)}} \Delta_k^{-1(i)} \\ =& -L_x(x_k^{(i)}) + \epsilon_{k+1}^{(i)}, \end{aligned} \tag{4.3.40}$$

where

$$\epsilon_{k+1}^{(i)} = -2w_k^{(i)} - 2h_k^{(i)} + \frac{(e(i)+k)^\mu}{c} \xi_{k+1}^{(i)} \Delta_k^{-(i)} \tag{4.3.41}$$

If we can show that $m_{i_k}(s_k, T) \le G(i_k)$ and

$$\lim_{T\longrightarrow 0} \limsup_{k\longrightarrow\infty} \left\| \frac{1}{T} \sum_{j=s_k}^{m_{i_k}(s_k,t)} \frac{a}{e(i_k)+j} \epsilon_{j+1}^{(i_k)} \right\| = 0, \quad \forall t \in [0,T], \tag{4.3.42}$$

where

$$m_i(k,T) = \max\Big\{m : \sum_{j=k}^{m} \frac{a}{e(i)+j} \le T\Big\},$$

then by Lemma 4.3.1, (4.3.42) contradicts with that all sequences $\{x_{s_k}^{(i_k)}, \ldots, x_{t_k}^{(i_k)}\}$, $k = 1, 2, \ldots$, cross the interval $[\delta_\alpha, \delta_\beta]$, which is disjoint with $L(J)$.

This then proves (4.3.36).

We now show $m_k(s_k, t) \le G(i_k)$ for all sufficiently large k if T is small enough.

Since $\sup_{x \in J_{\delta_2}} \|x\| < \infty$, p_1 and p_2 are finite, where

$$p_1 \triangleq \sup_{x \in J_{\delta_2}} \|x\|, \qquad p_2 \triangleq \sup_{\|x\| \le 2p_1} \|L_x(x)\|. \tag{4.3.43}$$

We now show that on the ω-set $\{\omega : \|x_j^{(i)}\| \le 2p_1, j = 1, \ldots, G(i), i = 1, 2, \ldots\}$, $m_{i_k}(s_k, T) \le G(i_k)$ if k is sufficiently large and T is small enough.

Suppose the converse: for any fixed $T > 0$, there always exists $k \ge k_1$ whatever large k_1 is taken such that $m_{i_k}(s_k, T) > G(i_k)$.

Since $x_{s_k}^{(i_k)} \in J_{\delta_\alpha}$, $x_{t_k}^{(i_k)} \in J_{\delta_\beta}^c$, by continuity of $L(\cdot)$ there is a constant $q > 0$ such that $\|x_{s_k}^{(i_k)} - x_{t_k}^{(i_k)}\| \ge q$, $\forall k$.

For any m: $s_k \le m \le G(i_k)$ let us estimate $\|\sum_{j=s_k}^{m} a_j^{(i)} y_{j+1}^{(i_k)}\|$. By $\|x_j^{(i)}\| \le 2p$ and the local Lipschitz continuity of $L(\cdot)$, it is seen that

$$\left\{ \frac{L(x_j^{(i_k)}) - L(x_j^{(i_k)} + c_j^{(i_k)} \Delta_j^{(i_k)}) \Delta_j^{-1(i_k)}}{c_j^{(i_k)}} \right\}$$

is uniformly bounded with respect to $j = 1, 2, \ldots, G(i_k)$ and all $k \ge 1$. Then by A4.3.3, it follows that there is a constant q_1 such that

$$\|\sum_{j=s_k}^{m} a_j^{(i_k)} y_{j+1}^{(i_k)}\| \le q_1 T, \qquad \forall m : s_k \le m \le G(i_k).$$

From this it follows that there is no truncation for m: $s_k \le m \le G(i_k)$, and

$$\|x_{m+1}^{(i_k)} - x_{s_k}^{(i_k)}\| \le q_1 T, \qquad \forall m : s_k \le m \le G(i_k).$$

Let T be so small that $q_1 T < q$.

On the other hand, however, we have $t_k < G(i_k)$ and $\|x^{(i_k)}_{t_k} - x^{(i_k)}_{s_k}\| \geq q$. The obtained contradiction shows $m_{i_k}(s_k, T) \leq G(i)$ for all sufficiently large k if T is small enough.

We now prove (4.3.42). Let us order $h^{(i)}_j$ in the following way $\{h^{(1)}_1, h^{(1)}_2, \ldots, h^{(1)}_{G(1)}, h^{(2)}_1, \ldots, h^{(2)}_{G(2)}, \ldots, h^{(i)}_1, \ldots, h^{(i)}_{G(i)}, \ldots\} \triangleq \{h^{(i)}_j\}$.

From (4.1.34) and by the fact that $\{\Delta_k\}$ is an iid sequence and is independent of sums appearing in (4.1.34), it is easy to be convinced that $\{h^{(i)}_j\}$ is a martingale difference sequence.

By the condition $\liminf_{i \longrightarrow \infty} e(i) i^{-(1+\delta)} > 0$ for some $\delta > 0$, it is clear that $e(i) > e i^{1+\delta}$ for $i \geq i_0$ with e being a constant. Then we have

$$\sum_{i=1}^{\infty} \sum_{j=1}^{G(i)} \frac{1}{(e(i)+j)^2} \leq \sum_{i=1}^{i_0} \sum_{j=1}^{G(i)} \frac{1}{(e(i)+j)^2} + \sum_{i=i_0+1}^{\infty} \sum_{j=1}^{G(i)} \frac{1}{(e i^{1+\delta}+j)^2} < \infty,$$

and by the convergence theorem for martingale difference sequences

$$\sum_{i=1}^{\infty} \sum_{j=1}^{G(i)} \frac{a}{e(i)+j} h^{(i)}_j I_{[\|x^{(i)}_j\| \leq 2p_1]} < \infty \quad \text{a.s.} \tag{4.3.44}$$

By (4.1.28) and (4.3.8), we have

$$\|2 w^{(i_k)}_j I_{[\|x^{(i_k)}_j\| \leq 2p_1]}\| \leq \lambda (e(i_k))^{-\mu}, \tag{4.3.45}$$

where λ is a constant. Noticing that $m_{i_k}(s_k, T) \leq G(i_k)$ for large k and small T, by (4.3.44),(4.3.45), and A4.3.3 we may assume k_0 sufficiently large and T small enough such that

$$\Big\| \sum_{j=s_k}^{m_{i_k}(s_k,t)} \frac{a}{e(i_k)+j} \Big[(2 w^{(i_k)}_j + 2 h^{(i_k)}_j) I_{[\|x^{(i_k)}_j\| \leq 2p_1]} - \frac{(e(i_k)+j)^{\mu}}{c} \xi^{(i_k)}_{j+1} \Delta^{-1(i_k)}_j \Big] \Big\| \leq o(T), \quad \forall k \geq k_0, \quad \forall t \in [0, T]. \tag{4.3.46}$$

This will imply (4.3.42) if we can show

$$\|x^{(i_k)}_j\| < 2p_1, \quad \forall j : s_k \leq j \leq m_{i_k}(s_k, T), \quad \forall k \geq k_0. \tag{4.3.47}$$

We prove (4.3.47) by induction.

We have $\|x^{(i_k)}_{s_k}\| \leq p_1$ by definition of s_k. Assume that

$$\|x^{(i_k)}_j\| < 2p_1, \quad \forall j : s_k \leq j \leq \nu < m_{i_k}(s_k, T), \tag{4.3.48}$$

and $M_0^{(i_{k0})} > 3p_1$. Then there is no truncation at time $j : s_k \le j \le \nu$, since by (4.3.46) (with t chosen such that $m_{i_k}(s_k, t) = \nu$)

$$\Big\| \sum_{j=s_k}^{\nu} \frac{a}{e(i_k)+j} \big[(2w_j^{(i_k)} + 2h_j^{(i_k)}) - \frac{(e(i_k)+j)^\mu}{c} \xi_{j+1}^{(i_k)} \Delta_j^{-1(i_k)}\big] \Big\| < p_1$$

if T in (4.3.46) is sufficiently small.

Then by (4.3.40), we have

$$\begin{aligned} x_{\nu+1}^{(i_k)} &= x_{s_k}^{(i_k)} + \sum_{j=s_k}^{\nu} a_j^{(i_k)} y_{j+1}^{(i_k)} \\ &= x_{s_k}^{(i_k)} - \sum_{j=s_k}^{\nu} \frac{a}{e(i_k)+j} \Big(L_x(x_j^{(i_k)}) + 2w_j^{(i_k)} + 2h_j^{(i_k)} \\ &\quad - \frac{(e(i_k)+j)^\mu}{c} \xi_{j+1}^{(i_k)} \Delta_j^{-1(i_k)} \Big), \end{aligned}$$

and by (4.3.43) and (4.3.46)

$$\|x_{\nu+1}^{(i_k)}\| \le p_1 + p_2 T + o(T) < 2p_1$$

for small T. This completes induction, and (4.3.42) is proved, which, in turn, concludes the lemma. □

Lemma 4.3.3 *Assume A4.3.1–A4.3.3 hold. Further, assume that* $\liminf_{i\to\infty} e(i) i^{-(1+\delta)} > 0$ *for some* $\delta > 0$ *and* $g(i)/e(i) \xrightarrow[i\to\infty]{} \infty$. *If there exists a subsequence* $\{i_k\}$ *such that* $x_0^{(i_k)} \in J_k$, *then*

$$\limsup_{k\to\infty} \Lambda_{G(i_k)}^{(i_k)} \le h. \tag{4.3.49}$$

Proof. For any $\epsilon > 0$, by Lemma 4.3.2 there exists k_0 such that for any $k > k_0$, $x_j^{(i_k)} \in J_{h+\epsilon}$, $\forall j : 0 \le j \le G(i_k)$, if $x_0^{(i_k)} \in J_k$. By (GO2), we have

$$\Lambda_{G(i_k)}^{(i_k)} = \frac{1}{G(i_k)} \sum_{j=0}^{G(i_k)-1} L(x_j^{(i_k)}) + \frac{1}{G(i_k)} \sum_{j=0}^{G(i_k)-1} \xi_j^{(i_k,0)}.$$

Then by A4.3.2, there exists $k_1 > k_0$ such that, for any $j > k_1$, $\Lambda_{G(i_k)}^{(i_k)} \le h + 2\epsilon$.

This implies the conclusion of the lemma by the arbitrariness of ϵ. □

Lemma 4.3.4 *Assume A4.3.1–A4.3.3 hold,* $\liminf_{i\longrightarrow\infty} e(i)i^{-(1+\delta)} > 0$ *for some* $\delta > 0$ *and* $g(i)/e(i) \xrightarrow[i\longrightarrow\infty]{} \infty$. *If subsequence* $\{i_k\}$ *is such that*

$$\infty > \liminf_{k\longrightarrow\infty} L(x^{(i_k)}_{G(i_k)}) = h \notin \overline{L}(J), \tag{4.3.50}$$

then

$$\liminf_{k\longrightarrow\infty} \Lambda^{(i_k)}_{G(i_k)} \triangleq h' > h, \tag{4.3.51}$$

where $\overline{L}(J)$ *denotes the closure of* $L(J)$, *and* $\{x^{(i_k)}_{G(i_k)}\}$ *and* $\{\Lambda^{(i_k)}_{G(i_k)}\}$ *are given by (GO1) and (GO2) for the* i_k*th search period.*

Proof. Since $L(x) \longrightarrow \infty$ as $\|x\| \longrightarrow \infty$ by A4.3.1, for (4.3.50) it is seen that $\{x^{(i_k)}_{G(i_k)}\}$ contains a bounded infinite subsequence, and hence, a convergent subsequence (for simplicity of notation, assume $x^{(i_k)}_{G(i_k)} \xrightarrow[k\longrightarrow\infty]{} \overline{x}$), such that

$$\lim_{k\longrightarrow\infty} L(x^{(i_k)}_{G(i_k)}) = L(\overline{x}) = h. \tag{4.3.52}$$

Since $h \notin \overline{L}(J)$, there exists a $\delta > 0$ such that $[h-\delta, h+\delta] \cap \overline{L}(J) = \emptyset$, and hence $d(\overline{x}, J) > 0$.

Define

$$n(i,T) \triangleq \max\left\{k : \sum_{j=k}^{G(i)} \frac{a}{e(i)+j} > T\right\}.$$

It is worth noting that for any $T > 0$, $n(i,T)$ is well defined for all sufficiently large i, because $\frac{g(i)}{e(i)} \xrightarrow[i\longrightarrow\infty]{} \infty$ and hence $\sum_{j=1}^{G(i)} \frac{a}{e(i)+j} \xrightarrow[i\longrightarrow\infty]{} \infty$.

We now show that

$$\liminf_{k\longrightarrow\infty} L\big(x^{(i_k)}_{n(i_k,T)}\big) \triangleq h_1 > h. \tag{4.3.53}$$

By the same argument as that just used before, without loss of generality, we may assume $x^{(i_k)}_{n(i_k,T)}$ is convergent (otherwise, a convergent subsequence should be extracted) and thus

$$\lim_{k\longrightarrow\infty} L\Big(x^{(i_k)}_{n(i_k,T)}\Big) = h_1. \tag{4.3.54}$$

We have to show $h_1 > h$.

By the same argument as that used for deriving (2.2.27), it follows that there is $\alpha > 0$ such that

$$\lim_{k\longrightarrow\infty} L\big(x^{(i_k)}_{G(i_k)}\big) - L\big(x^{(i_k)}_{n(i_k,T)}\big) \leq -\alpha T,$$

which implies the correctness of (4.3.53).

From (4.3.53) it follows that

$$\liminf_{k\longrightarrow\infty} \min_{n(i_k,T)\leq j\leq G(i_k)} L(x_j^{(i_k)}) \geq h, \tag{4.3.55}$$

because, otherwise, we would have a subsequence $\{x^{(i_k)}_{j_k}\}$ with $n(i_k,T) \leq j_k \leq G(i_k)$ such that $L(x^{(i_k)}_{j_k}) \xrightarrow[k\longrightarrow\infty]{} h_3 < h$, and by (4.3.54)

$$|L(x^{(i_k)}_{j_k}) - L(x^{(i_k)}_{n(i_k,T)})| > \frac{h_1 - h_3}{2} \tag{4.3.56}$$

for large k. However, by (2.2.15), $\|x^{(i_k)}_{j_k} - x^{(i_k)}_{n(i_k,T)}\| = O(T)$, so for small enough $T > 0$, (4.3.56) is impossible. This verifies (4.3.55).

We now show

$$\liminf_{k\longrightarrow\infty} \inf_{0\leq j\leq n(i_k,T)} L\left(x_j^{(i_k)}\right) = h_1. \tag{4.3.57}$$

Assume the converse, i.e.,

$$\liminf_{i\longrightarrow\infty} \inf_{0\leq j\leq n(i_k,T)} L\left(x_j^{(i_k)}\right) = h_2 < h_1. \tag{4.3.58}$$

From (4.3.54) and (4.3.58) it is seen that for all sufficiently large k, the sequence

$$L(x_0^{(i_k)}), L(x_1^{(i_k)}), \ldots, L(x^{(i_k)}_{n(i_k,T)})$$

contains at least a crossing the interval $[h_2+\epsilon, h_1-\epsilon]$ with $\epsilon \in (0, \frac{h_1-h_2}{3})$. In other words, we are dealing with a sample path ω on which both (4.3.54) and (4.3.58) are satisfied. Thus, ω belongs to $S^{(i_k)}_{[h_2+\epsilon,h_1-\epsilon]}$. By Lemma 4.3.2, the set composed of such ω is with zero probability. This verifies (4.3.57).

From (4.3.57) it follows that

$$L(x_j^{(i_k)}) > \frac{2}{3}h_1 + \frac{1}{3}h, \quad \forall j : 0 \leq j \leq n(i_k,T) \tag{4.3.59}$$

for all sufficiently large k.

Notice that from the following elementary inequalities

$$\int_{n(i,T)+1}^{G(i)+1} \frac{ads}{e(i)+s} < \sum_{j=n(i,T)+1}^{G(i)} \frac{a}{e(i)+j} \le T$$
$$< \sum_{j=n(i,T)}^{G(i)} \frac{a}{e(i)+j} < \int_{n(i,T)-1}^{G(i)} \frac{ads}{e(i)+s}, \tag{4.3.60}$$

by (4.3.5) it follows that

$$e^{-\frac{T}{a}}(g(i)+1) - e(i) - 1 < n(i,T) < e^{-\frac{T}{a}} g(i) - e(i) + 1. \tag{4.3.61}$$

By definition of $\Lambda_{G(i_k)}^{(i_k)}$, we write

$$\Lambda_{G(i_k)}^{(i_k)} = \frac{1}{G(i_k)} \Big[\sum_{j=0}^{n(i_k,T)} L\big(x_j^{(i_k)}\big) + \sum_{j=n(i_k,T)+1}^{G(i_k)} L\big(x_j^{(i_k)}\big) + \sum_{j=0}^{G(i_k)} \xi_j^{(i_k,0)} \Big]. \tag{4.3.62}$$

By (4.3.59) and (4.3.61), noticing $G(i) = g(i) - e(i)$, we have

$$\liminf_{k\longrightarrow\infty} \frac{1}{G(i_k)} \sum_{j=0}^{n(i_k,T)} L(x_j^{(i_k)}) > (\frac{2}{3}h_1 + \frac{1}{3}h) \liminf_{k\longrightarrow\infty} \frac{e^{-\frac{T}{a}}(g(i_k)+1) - e(i_k)}{G(i_k)}$$
$$= e^{-\frac{T}{a}} (\frac{2h_1}{3} + \frac{h}{3}), \tag{4.3.63}$$

because $e(i)/g(i) \xrightarrow[i\longrightarrow\infty]{} 0$.

By (4.3.55) and (4.3.61) we have

$$\liminf_{k\longrightarrow\infty} \frac{1}{G(i_k)} \sum_{j=n(i_k,T)+1}^{G(i_k)} L(x_j^{(i_k)})$$
$$\ge \liminf_{k\longrightarrow\infty} \frac{G(i_k) - n(i_k,T) - 1}{G(i_k)} \min_{n(i_k,T)\le j \le G(i_k)} L(x_j^{(i_k)})$$
$$\ge \lim_{k\longrightarrow\infty} (1 - \frac{e^{-\frac{T}{a}} g(i_k) - e(i_k) + 2}{G(i_k)}) h = (1 - e^{-\frac{T}{a}}) h. \tag{4.3.64}$$

Since $\frac{1}{G(i_k)} \sum_{j=0}^{G(i_k)} \xi_j^{(i_k,0)} \xrightarrow[k\longrightarrow\infty]{} 0$ by (4.3.15), combining (4.3.62)–(4.3.64) leads to

$$\liminf_{k\longrightarrow\infty} \Lambda_{G(i_k)}^{(i_k)} \ge h + \frac{2}{3} e^{-\frac{T}{a}} (h_1 - h) > h,$$

which completes the proof of the lemma. □

Lemma 4.3.5 *Let $\{x_k\}$ be given by (GO1)–(GO5). Assume that A4.3.1–A4.3.4 hold, initial values $\{x_0^{(i)}\}$ selected in (GO1) are dense in an open set U containing the set J^0 of global minima of $L(\cdot)$,* $\liminf_{i\longrightarrow\infty} e(i) \cdot i^{-(1+\delta)} > 0$ *for some $\delta > 0$, and $g(i)/e(i) \xrightarrow[i\longrightarrow\infty]{} \infty$. Then for any $[h_1, h_2] \subset (L_{\min}, \infty)$,*

$$P\{\omega : L(x_k(\omega)) \text{ crosses } [h_1, h_2] \text{ infinitely many times }\} = 0.$$

Proof. Among the first n search periods denote by τ_n the number of those search periods for which $x_{g(i)}$ are reset to be $x_{G(i)}^{(i)}$, i.e.,

$$\tau_n = \sum_{i=1}^{n} I_{[x_{g(i)}=x_{G(i)}^{(i)}]}.$$

Since $L(J)$ is not dense in any interval, there exists an interval $[h_\alpha, h_\beta] \subset [h_1, h_2]$ such that $[h_\alpha, h_\beta] \cap L(J) = \emptyset$. So, for lemma it suffices to prove that $L(x_k(\omega))$ cannot cross $[h_\alpha, h_\beta]$ infinitely many times a.s.

If $\lim_{n\longrightarrow\infty} \tau_n < \infty$, then after a finite number of steps, $\{x_k\}$ is generated by (GO4). By Lemma 4.3.1 the assertion of the lemma follows immediately. Therefore, we need only to consider the case where $\tau_k \xrightarrow[k\longrightarrow\infty]{} \infty$.

Denote by i_s the search period for which a resetting happens, i.e., $x_{g(i_s)} = x_{G(i_s)}^{(i_s)}$. It is clear that $i_s \xrightarrow[s\longrightarrow\infty]{} \infty$ by $\tau_k \xrightarrow[k\longrightarrow\infty]{} \infty$.

In the case $i_s \xrightarrow[s\longrightarrow\infty]{} \infty$ by (GO4) the algorithm generates a family of consecutive sequences:

$$\{x_{g(i_s)}, x_{g(i_s)+1}, \ldots, x_{g(i_{s+1})}\} \text{ with } x_{g(i_s)} = x_{G(i_s)}^{(i_s)}.$$

Let us denote the sequence by

$$X_{(i_s)} \triangleq \{x_{g(i_s)}, x_{g(i_s)+1}, \ldots, x_{g(i_{s+1})}\}, \quad s = 1, 2, \ldots. \tag{4.3.65}$$

and the corresponding sequence of the values of $L(\cdot)$ by

$$L(X_{(i_s)}) \triangleq \{L(x_{g(i_s)}), L(x_{g(i_s)+1}), \ldots, L(x_{g(i_{s+1})})\}, \quad s = 1, 2, \ldots. \tag{4.3.66}$$

Let $\epsilon > 0$ be sufficiently small such that

$$0 < \epsilon < (h_\alpha - L_{\min})/6, \quad J_{L_{\min}+\epsilon} \subset U \tag{4.3.67}$$

and $[L_{\min} + \frac{10\epsilon}{3}, L_{\min} + 6\epsilon] \cap \overline{L(J)} = \emptyset$, which is possible because $L(J)$ is nowhere dense.

Since $x_0^{(i)}$ is dense in U, $x_0^{(i)}$ visits $J_{L_{\min}+\frac{\epsilon}{2}}$ infinitely often. Assume $x_0^{(i(k))} \in J_{L_{\min}+\frac{\epsilon}{2}}$, $k = 1, 2, \ldots$.

By Lemma 4.3.2

$$x_j^{(i(k))} \in J_{L_{\min}+\epsilon}, \quad \forall j : 0 \le j \le G(i(k)), \quad \forall k \ge k_0, \tag{4.3.68}$$

if k_0 is large enough.

Define

$$i_{s(k)} = \inf\{j : j \ge i(k), \quad \tau_j > \tau_{j-1}\}.$$

This means that the first resetting in or after the $i(k)$th search period occurs in the $i_{s(k)}$th search period.

We now show that there is a large enough $S \ge s(k_0)$ such that the following requirements are simultaneously satisfied:

i) $\Lambda_{G(i_s)}^{(i_s)} \le L_{\min} + 6\epsilon$ implies

$$L(x_{g(i_s)}) = L(x_{G(i_s)}^{(i_s)}) < L_{\min} + 6\epsilon - 2\delta, \quad \forall s \ge S, \tag{4.3.69}$$

where $\delta \in (0, \epsilon)$ is fixed;

ii) $L(x_{(i_s)})$ does not cross intervals $[L_{\min} + \frac{10\epsilon}{3}, L_{\min} + \frac{11\epsilon}{3}]$,

$$[L_{\min} + 4\epsilon, L_{\min} + 5\epsilon], \quad \text{and} \quad [L_{\min} + 6\epsilon - 2\delta, L_{\min} + 6\epsilon - \delta], \quad \forall s \ge S;$$

iii)

$$\left\| \frac{1}{g(i) - g(i-1)} \sum_{j=g(i-1)+1}^{g(i)} \xi_{j+1}^0 \right\| < \frac{\delta}{3}, \quad \forall i \ge i_S; \tag{4.3.70}$$

vi)

$$\left\| \frac{1}{G(i) + 1} \sum_{j=0}^{G(i)} \xi_{j+1}^{(i,0)} \right\| \le \epsilon, \qquad \forall i \ge i_S; \tag{4.3.71}$$

v)

$$\lambda(i) \le \delta/2, \quad \forall i \ge i_S. \tag{4.3.72}$$

We first show ii)-v).

Since all three intervals indicated in ii) have an empty intersection with $L(J)$, by Lemma 4.3.1 ii) is true if S is large enough. It is clear that iii) and vi) are correct for fixed $\delta > 0$ and $\epsilon > 0$ if S is large enough, while v) is true because $\lambda(i) \xrightarrow[i\to\infty]{} 0$.

For i) we first show that there are infinitely many s for which

$$\Lambda_{G(i_s)}^{(i(k))} \le L_{\min} + 6\epsilon. \tag{4.3.73}$$

By (4.3.68) and (4.3.71) we have

$$\Lambda_{G(i(k))}^{(i(k))} \le L_{\min} + 2\epsilon, \quad \forall i(k) \ge i_S. \tag{4.3.74}$$

Consider two cases.

1) There is no resetting in the $i(k)$th search period. Then

$$\Lambda_{G(i(k))}^{(i(k))} + \lambda(i(k)) \ge \Lambda_{g(i(k))} - \lambda(i(k)),$$

and by (4.3.72) and (4.3.74) it follows that

$$\Lambda_{g(i(k))} < L_{\min} + 3\epsilon, \quad \forall i(k) \ge i_S. \tag{4.3.75}$$

By (4.3.70) and the definition of $\Lambda_{g(i(k))}$, there exists at least an integer j among $\{g(i(k)-1), g(i(k)-1)+1, \dots, g(i(k))\}$ such that

$$L(x_j) \le L_{\min} + \frac{10}{3}\epsilon,$$

because, otherwise, we would have $\Lambda_{g(i(k))} \ge L_{\min} + 3\epsilon$ which contradicts (4.3.74).

By ii) we conclude that

$$\sup_{j \le \lambda \le g(i(k))} L(x_\lambda) \le L_{\min} + \frac{11\epsilon}{3}, \quad \forall i(k) \ge i_S, \tag{4.3.76}$$

and, in particular, $L(x_{g(i(k))}) < L_{\min} + 4\epsilon$.

2) If there is a resetting in the $i(k)$th search period, then

$$x_{g(i(k))} = x_{G(i(k))}^{(i(k))}$$

and by (4.3.68) we also have (4.3.76).

From (4.3.76), by ii) $L(X(i_s))$ does not cross $[L_{\min} + 4\epsilon, L_{\min} + 5\epsilon]$ for $i_s \ge i_S$. Consequently,

$$L(x_j) < L_{\min} + 5\epsilon, \quad \forall j : g(i(k)) \le j \le g(i_{s(k)}).$$

This together with (4.3.70) implies that

$$\Lambda_{g(i_{s(k)})} = \frac{1}{g(i_{s(k)}) - g(i_{s(k)} - 1)} \sum_{j=g(i_{s(k)}-1)+1}^{g(i_{s(k)})} (L(x_j) + \xi_{j+1}^0) < L_{\min} + 6\epsilon.$$

By (GO3) we then have

$$\Lambda^{(i_{s(k)})}_{G(i_{s(k)})} < L_{\min} + 6\epsilon, \quad \forall i_{s(k)} > i_S. \tag{4.3.77}$$

Noticing $s(k) \longrightarrow \infty$ as $k \longrightarrow \infty$, we conclude that there are infinitely many s for which (4.3.73) holds.

We now show that there is a $\delta > 0$ such that

$$\limsup_{s \longrightarrow \infty} L\big(x^{(i_s)}_{G(i_s)}\big) < L_{\min} + 6\epsilon - 2\delta, \tag{4.3.78}$$

where lim sup is taken along those s for which (4.3.73) holds.

Assume the converse: there is a subsequence of s_k such that

$$\lim_{k \longrightarrow \infty} L\big(x^{(i_{s_k})}_{G(i_{s_k})}\big) \geq L_{\min} + 6\epsilon.$$

Then by Lemma 4.3.4,

$$\liminf_{k \longrightarrow \infty} \Lambda\big(x^{(i_{s_k})}_{G(i_{s_k})}\big) > L_{\min} + 6\epsilon,$$

which contradicts (4.3.73). This proves (4.3.78), and also i). As a matter of fact, we have proved more than i): Precisely, we have shown that there are infinitely many s for which (4.3.73) holds, and for $s \geq S$ (4.3.73) implies the following inequality:

$$L(x^{(i_{s_k})}_{G(i_{s_k})}) < L_{\min} + 6\epsilon - 2\delta. \tag{4.3.79}$$

Let us denote by T_ϵ the totality of those $L(x_{(i_s)})$ for which (4.3.73) holds and $s \geq S$. What we have just proved is that T_ϵ contains infinitely many $L(x_{(i_s)})$, if $i_s \longrightarrow \infty$.

Consider a sequence $L(x_{(i_{s'})}) \in T_\epsilon$. By ii) it cannot cross the interval $[L_{\min} + 6\epsilon - 2\delta, L_{\min} + 6\epsilon - \delta]$. This means that

$$\sup_{g(i_{s'}) \leq j \leq g(i_{s'+1})} L(x_j) < L_{\min} + 6\epsilon - \delta.$$

Then by (4.3.70)

$$\Lambda_{g(s_{i'+1})} = \frac{1}{g(i_{s'+1}) - g(i_{s'+1} - 1)} \sum_{j=g(i_{s'+1}-1)+1}^{g(i_{s'+1})} \big(L(x_j) + \xi^0_{j+1}\big) < L_{\min} + 6\epsilon,$$

and by (GO3)

$$\Lambda^{(i_{s'+1})}_{G(i_{s'+1})} < L_{\min} + 6\epsilon,$$

since $i_{s'+1}$ is a search period with resetting.

Thus, we have shown that if $L(x_{(i_{s'})}) \in T_\epsilon$, then $L(x_{(i_{s'+1})})$ also belongs to T_ϵ. Therefore, $L(x_{(i_{s'+j})}) \in T_\epsilon$, $\forall j \geq 1$, and

$$L\left(x^{(i_{i'+j})}_{G(i_{i'+j})}\right) < L_{\min} + 6\epsilon - 2\delta, \quad \forall j = 0, 1, 2, \ldots.$$

From here and (4.3.67) it follows that

$$\limsup_{s \longrightarrow \infty} L\left(x^{(i_s)}_{G(i_s)}\right) \leq L_{\min} + 6\epsilon - 2\delta < h_\alpha.$$

Since $[h_\alpha, h_\beta] \cap L(J)$, $L(x_{(i_s)})$ may cross the interval $[h_\alpha, h_\beta]$ only for finite number of times by Lemma 4.3.1. This completes the proof of the lemma. □

Proof of Theorem 4.3.1.

By Lemma 4.3.5 the limit $\lim_{k \longrightarrow \infty} L(x_k)$ exists. By arbitrariness of ϵ, from (4.3.69) it follows that

$$\lim_{k \longrightarrow \infty} L(x_k) = L_{\min}, \quad \text{a.s.}$$

By continuity of $L(\cdot)$, we conclude that

$$d(x_k, J^0) \xrightarrow[k \longrightarrow \infty]{} 0 \quad \text{a.s.}$$

□

4.4. Asymptotic Behavior of Global Optimization Algorithm

In last section a global optimization algorithm combining the KW algorithm with search method was proposed, and it was proved that the algorithm converges to the set J^0 of global minimizers, i.e., $d(x_k, J^0) \xrightarrow[k \longrightarrow \infty]{} 0$ a.s. However, in the algorithm defined by (GO1)–(GO5), resettings are involved. The convergence $d(x_k, J^0) \xrightarrow[k \longrightarrow \infty]{} 0$ a.s. by no means excludes the algorithm from resettings asymptotically. In other words, although $d(x_k, J^0) \xrightarrow[k \longrightarrow \infty]{} 0$ a.s., it may still happen that $\tau_k \xrightarrow[k \longrightarrow \infty]{} \infty$, where τ_k is defined in Lemma 4.3.5, i.e., it may still be possible to have infinitely many resettings.

In what follows we will give conditions under which $\lim_{k \longrightarrow \infty} \tau_k < \infty$ a.s. In this case, the global optimization algorithm (GO1)–(GO5) asymptotically behaves like a KW algorithm with expanding truncations and randomized differences, because for large k, $\{x_k\}$ is purely generated by (GO4) without resetting.

A4.4.1 *$J^0 = \{x^0\}$ is a singleton, $L(\cdot)$ is twice continuously differentiable in $\mathcal{B}(x^0, \delta)$, the ball centered at x^0 with radius δ, for some $\delta > 0$, and $H \triangleq$ Hessian of $L(x^0)$ is positive definite.*

A4.4.2 *$\{\xi_k^0, \mathcal{F}_k\}$ and $\{\xi_k^{(i,0)}, \mathcal{F}_k^{(i)}\}$ ordered as in (4.3.20) (4.3.21) and Remark 4.3.1 are martingale difference sequences with*

$$\sup_{k,i} E\{(\xi_{k+1}^{(i,0)})^2 | \mathcal{F}_k^{(i)}\} < \infty,$$

$$\sup_{k} E\{(\xi_{k+1}^{0})^2 | \mathcal{F}_k\} < \infty.$$

A4.4.3 *$\xi_{k+1}^{(i)}$ is independent of*

$$\{\Delta_j^{(s)}, j \le G(s) \quad \text{for } s < i, \text{ and } \ j \le k \ \text{ for } s = i\}$$

and $\sup_{k,i} E(\xi_k^{(i)})^2 < \infty.$

We recall that $\xi_{k+1}^{(i)}$ is the observation noise $\xi_{k+1} = \xi_{k+1}^- - \xi_{k+1}^+$ in the ith search period.

A4.4.4 *ξ_{k+1} is independent of $\{\Delta_j, j \le k\}$ and $\sup_k E(\xi_k)^2 < \infty$, where ξ_{k+1} denotes the observation noise $\xi_{k+1}^- - \xi_{k+1}^+$ when $\{x_k\}$ is calculated in (GO4).*

Lemma 4.4.1 *Assume A4.4.2 holds and, in addition,*

$$\frac{g(i+1) - g(i)}{i^{(1+\delta)}} \xrightarrow[i \to \infty]{} \infty \quad \text{and} \quad \frac{G(i)}{i^{(1+\delta)}} \xrightarrow[i \to \infty]{} \infty. \tag{4.4.1}$$

Then, there exists an i_0 (maybe depending on ω) such that for any $i > i_0$

$$\frac{1}{G(i)} \Big| \sum_{k=1}^{G(i)} \xi_{k+1}^{(i,0)} \Big| \le \frac{i^{\frac{1+\delta}{2}}}{\sqrt{G(i)}} \tag{4.4.2}$$

and

$$\frac{1}{g(i+1) - g(i)} \Big| \sum_{k=g(i)+1}^{g(i+1)} \xi_k^0 \Big| \le \frac{i^{\frac{1+\delta}{2}}}{\sqrt{g(i+1) - g(i)}}. \tag{4.4.3}$$

Proof. Notice that by A4.4.2 $\{\frac{1}{\sqrt{G(i)}}\sum_{k=1}^{G(i)}\xi_{k+1}^{(i,0)}\mathcal{F}_{G(i)+1}^{(i)}\}$ is a martingale difference sequence with bounded conditional variance. By the convergence theorem for martingale difference sequences

$$\sum_{i=1}^{\infty}\frac{1}{i^{\frac{(1+\delta)}{2}}}\frac{1}{\sqrt{G(i)}}\sum_{k=1}^{G(i)}\xi_{k+1}^{(i,0)} < \infty \quad \text{a.s.}$$

which implies (4.4.2).

Estimate (4.4.3) can be proved by a similar way. □

Lemma 4.4.2 *Assume A4.4.3 and A4.4.4 hold. If* $e(i)i^{-(\frac{1}{1-2\mu-2\gamma}+\delta)} \xrightarrow[i\to\infty]{} \infty$ *for some* $\gamma \in (0, \frac{1}{2}-\mu)$, *then*

$$\sum_{i=1}^{\infty}\sum_{k=1}^{G(i)}\frac{a}{(e(i)+k)^{1-\gamma}}\Big(-2h_k^{(i)}I_{[x_k^{(i)}\in J_h]} + \frac{(e(i)+k)^{\mu}}{c}\xi_{k+1}^{(i)}\Delta_k^{-1(i)}\Big) < \infty \text{ a.s.,} \tag{4.4.4}$$

and

$$\sum_{k=1}^{\infty}\frac{a}{k^{1-\gamma}}(-2h_kI_{[x_k\in J_h]} + \frac{k^{\mu}}{c}\xi_{k+1}\Delta_k^{-1}) < \infty \tag{4.4.5}$$

for $\forall h > L_{\min}$, *where* $h_k^{(i)}$ *and* h_k *are given in (4.1.34), where superscript* (i) *denotes the corresponding values in the ith search period.*

Proof. Let us prove

$$\sum_{i=1}^{\infty}\sum_{k=1}^{G(i)}\frac{a}{c(e(i)+k)^{1-\gamma-\mu}}\xi_{k+1}^{(i)}\Delta_k^{-1(i)} < \infty \quad \text{a.s.} \tag{4.4.6}$$

Note that

$$\{\xi_1^{(1)}\Delta_0^{-1(1)},\dots,\xi_{G(1)+1}^{(1)}\Delta_{G(1)}^{-1(1)},\xi_1^{(2)}\Delta_0^{-1(2)},\dots,\xi_{G(2)+1}^{(2)}\Delta_{G(2)}^{-1(2)},\dots\}$$

is a martingale difference sequence with bounded conditional second moment. So, by the convergence theorem for martingale difference sequences for (4.4.6) it suffices to show

$$\sum_{i=1}^{\infty}\sum_{k=1}^{G(i)}\frac{1}{(e(i)+k)^{2-2\gamma-2\mu}} < \infty.$$

By assumption of the lemma $\gamma < \frac{1}{2} - \mu$, or $2\gamma + 2\mu < 1$ and $e(i) \cdot i^{-(\frac{1}{1-2\mu-2\gamma}+\delta)} > 1$ for large i. The last inequality yields

$$e(i)^{1-2\mu-2\gamma} > i^{(1+\delta(1-2\gamma-2\mu))} \quad \text{and hence} \quad \sum_{i=1}^{\infty} e(i)^{-(1-2\mu-2\gamma)} < \infty.$$

Therefore,

$$\sum_{i=1}^{\infty}\sum_{k=1}^{G(i)} \frac{1}{(e(i)+k)^{2-2\gamma-2\mu}} < \sum_{i=1}^{\infty}\int_0^{G(i)} \frac{dx}{(e(i)+x)^{2-2\gamma-2\mu}}$$
$$= \frac{1}{1-2\gamma-2\mu}\sum_{i=1}^{\infty}\Big[\frac{1}{e(i)^{1-2\gamma-2\mu}} - \frac{1}{(e(i)+G(i))^{1-2\gamma-2\mu}}\Big] < \infty.$$

Thus, (4.4.6) is correct. As noted in the proof of Lemma 4.3.2, $\{h_k^{(i)}\}$ is a martingale difference sequence. So, (4.4.4) is true.

Similarly, (4.4.5) is also verified by using the convergence theorem for martingale difference sequences. □

Lemma 4.4.3 *In addition to the conditions of Theorem 4.3.1, suppose that A4.4.1 and A4.4.3 hold, $H - \frac{\gamma}{a}I$ is positive definite, and*

$$e(i)i^{-\frac{1}{1-2\mu-2\gamma}-\delta} \xrightarrow[i\to\infty]{} \infty \tag{4.4.7}$$

for some $\gamma \in (0, \frac{1}{2} - \mu)$. Then there exists a sufficiently large i_0 such that, for $i \ge i_0$ if the inequality

$$2(e(i)+k)^{2\gamma}[L(x_k^{(i)}) - L_{\min}] \le q \tag{4.4.8}$$

holds for some $k > 0$ with $q > 1$, then the following inequality holds

$$2(e(i)+j)^{2\gamma}[L(x_j^{(i)}) - L_{\min}] \le (1+\gamma)q, \quad \forall j : k \le j \le G(i). \tag{4.4.9}$$

Proof. . By A4.4.1 and the Taylor's expansion, we have

$$L(x) = L_{\min} + \frac{1}{2}(x-x^0)^T H(x-x^0) + o(\|x-x^0\|^2),$$

i.e.,

$$2(L(x) - L_{\min}) = \Big(1+o(1)\Big)(x-x^0)^T H(x-x^0), \tag{4.4.10}$$

$$L_x(x) = \Big(H + F(x)\Big)(x-x^0), \tag{4.4.11}$$

where

$$\lim_{x \longrightarrow x^0} \|F(x)\| = 0.$$

Therefore, for any $\epsilon \in (0, \frac{1}{2})$ there is a $\delta_1 > 0$ such that for any x : $\|x - x^0\| < \delta_1$,

$$|o(1)| < \epsilon, \tag{4.4.12}$$

$$\Big(\frac{1}{2} - \epsilon\Big)\lambda_{\min}\|x - x^0\| \leq L(x) - L_{\min} \leq \Big(\frac{1}{2} + \epsilon\Big)\lambda_{\max}\|x - x^0\|, \tag{4.4.13}$$

and

$$\|F(x)\|I < \frac{1}{2a}(aH - \gamma I), \tag{4.4.14}$$

where $\lambda_{\min}$ and $\lambda_{\max}$ denote the minimum and maximum eigenvalue of H, respectively, and $o(1)$ is the one given in (4.4.10).

Since x^0 is the unique minimizer of $L(\cdot)$ and $L(\cdot)$ is continuous, there is ϵ^1 such that $\|x - x^0\| < \delta_1$, if $L(x) - L_{\min} \leq \epsilon^1$. We always assume that i is large enough such that

$$\frac{q(1+\gamma)}{2(e(i))^{2\gamma}} < \epsilon^1 \quad \text{and} \quad M^{(i)} > \|x^0\| + \delta_1, \tag{4.4.15}$$

where $M^{(i)}$ is used in (GO1). From (4.4.8) it then follows that $\|x_k^{(i)} - x^0\| < \delta_1$ and there is no truncation at time k.

Denote

$$\zeta_k^{(i)} = -2h_k^{(i)} + \frac{(e(i)+k)^\mu}{c}\xi_{k+1}^{(i)}\Delta_k^{-1(i)}, \tag{4.4.16}$$

$$\theta_k^{(i)} = (e(i)+k)^\gamma(x_k^{(i)} - x^0). \tag{4.4.17}$$

For $x_k^{(i)}$ satisfying (4.4.8) and $j \geq k$ we have

$$\begin{aligned}
\theta_{k+1}^{(i)} &= (e(i)+k+1)^\gamma(x_{k+1}^{(i)} - x^0) \\
&= (e(i)+k)^\gamma\Big(1 + \frac{1}{e(i)+k}\Big)^\gamma(x_{k+1}^{(i)} - x_k^{(i)} + x_k^{(i)} - x^0) \\
&= \Big(1 + \frac{1}{e(i)+k}\Big)^\gamma\Big[\theta_k^{(i)} + \frac{a(e(i)+k)^\gamma}{e(i)+k}(-L_x(x_k^{(i)}) + \epsilon_{k+1}^{(i)})\Big],
\end{aligned}$$

where $\epsilon_{k+1}^{(i)}$ is given by (4.3.41).

By (4.4.11) it then follows that

$$\begin{aligned}\theta_{k+1}^{(i)} &= \Big(1+\frac{1}{e(i)+k}\Big)^{\gamma}\Big[\theta_k^{(i)} - \frac{a}{e(i)+k}(H+F(x_k^{(i)}))\theta_k^{(i)} \\ &\quad + \frac{a}{(e(i)+k)^{1-\gamma}}\epsilon_{k+1}^{(i)}\Big] \\ &= \theta_k^{(i)} + \frac{a}{e(i)+k}\Big(\frac{\gamma}{a}I - F(x_k^{(i)}) - H + o(1)\Big)\theta_k^{(i)} \\ &\quad + \frac{a(e(i)+k+1)^{\gamma}}{e(i)+k}\epsilon_{k+1}^{(i)} \\ &= \theta_k^{(i)} + \frac{1}{e(i)+k}H_k^{(i)}\theta_k^{(i)} + \frac{a(e(i)+k+1)^{\gamma}}{e(i)+k}\big(-2w_k^{(i)} + \zeta_k^{(i)}\big),\end{aligned} \tag{4.4.18}$$

where $w_k^{(i)}$ is given by (4.1.33) with superscript (i) denoting the ith search period and

$$H_k^{(i)} \triangleq \frac{\gamma}{a}I - F(x_k^{(i)}) - H + o(1).$$

By (4.4.14) it is clear that

$$\frac{3}{2}(\gamma I - aH) < aH_k^{(i)} < \frac{1}{2}(\gamma I - aH). \tag{4.4.19}$$

Let

$$l_0 + 1 \triangleq \inf\{j > k : 2(e(i)+j)^{2\gamma}(L(x_j^{(i)}) - L_{\min}) > (1+\gamma)q\}. \tag{4.4.20}$$

For (4.4.9) it suffices to show that $l_0 \geq G(i)$.

Assume the converse: $l_0 < G(i)$.

Let

$$j_0 \triangleq \sup\{j < l_0 : 2(e(i)+j)^{2\gamma}\big(L(x_j^{(i)}) - L_{\min}\big) < q\}. \tag{4.4.21}$$

By (4.4.20), for all $j : j_0 \leq j \leq l_0$

$$L(x_j^{(i)}) - L_{\min} \leq \frac{(1+\gamma)c}{2(e(i)+j)^{2\gamma}} < \epsilon^1,$$

and hence,

$$\|x_j^{(i)} - x^0\| < \delta_1.$$

Thus, (4.4.12)-(4.4.14) are applicable.

By (4.4.17) and the second inequality of (4.4.13), we have for $j : j_0 < j \le l_0$

$$\|\theta_j^{(i)}\|^2 \ge \frac{2(e(i)+j)^{2\gamma}}{(1+2\epsilon)\lambda_{max}}\Big(L(x_j^{(i)}) - L_{\min}\Big)$$

which incorporating with (4.4.21) yields

$$\|\theta_j^{(i)}\|^2 \ge \frac{q}{(1+2\epsilon)\lambda_{max}} \quad \forall j : j_0 < j \le l_0. \tag{4.4.22}$$

Applying the first inequality of (4.4.13) and then (4.4.20) leads to

$$\begin{aligned}\|\theta_j^{(i)}\|^2 \le & \frac{2(e(i)+j)^{2\gamma}}{(1-2\epsilon)\lambda_{min}}(L(x_j^{(i)}) - L_{\min}) \\ \le & \frac{(1+\gamma)q}{(1-2\epsilon)\lambda_{min}} \quad \forall j : j_0 < j \le l_0. \end{aligned} \tag{4.4.23}$$

Since for $j : j_0 \le j \le l_0$, $\|x_j^{(i)} - x^0\| < \delta_1$, there is no truncation for $x_j^{(i)}$. Using (4.4.18) we have

$$\begin{aligned}\theta_{l_0+1}^{(i)}{}^T H \theta_{l_0+1}^{(i)} =& (\theta_{l_0+1}^{(i)} - \theta_{l_0}^{(i)} + \theta_{l_0}^{(i)})^T H (\theta_{l_0+1}^{(i)} - \theta_{l_0}^{(i)} + \theta_{l_0}^{(i)}) \\ =& \Big[\theta_{l_0}^{(i)} + \frac{a(e(i)+l_0+1)^\gamma}{e(i)+l_0}(-2w_{l_0}^{(i)} + \varsigma_{l_0}^{(i)})\Big]^T H \\ & \Big[\theta_{l_0}^{(i)} + \frac{a(e(i)+l_0+1)^\gamma}{e(i)+l_0}(-2w_{l_0}^{(i)} + \varsigma_{l_0}^{(i)})\Big] + Q_i(1), \end{aligned} \tag{4.4.24}$$

where

$$\begin{aligned}Q_i(1) =& \frac{2}{e(i)+l_0}\theta_{l_0}^{(i)T} H H_{l_0}^{(i)} \theta_{l_0}^{(i)} + \frac{2a(e(i)+l_0+1)^\gamma}{(e(i)+l_0)^2}\Big(-2w_{l_0}^{(i)} + \varsigma_{l_0}^{(i)}\Big)^T \\ & \cdot H H_{l_0}^{(i)} \theta_{l_0}^{(i)} + \frac{1}{(e(i)+l_0)^2}\theta_{l_0}^{(i)T} H_{l_0}^{(i)} H H_{l_0}^{(i)} \theta_{l_0}^{(i)} \\ & - \frac{4a(e(i)+l_0+1)^\gamma}{e(i)+l_0} w_{l_0}^{(i)T} H[\theta_{l_0}^{(i)} + \frac{a(e(i)+l_0+1)^\gamma}{e(i)+l_0} \\ & \cdot (-2w_{l_0}^{(i)} + \varsigma_{l_0}^{(i)})]. \end{aligned} \tag{4.4.25}$$

We now show that $Q_i(1)$ is negative for all sufficiently large i.

Let us consider terms in $(e(i)+l_0)Q_i(1)$. By assumption, $H - \frac{\gamma}{a}I > 0$, from (4.4.19) and (4.4.22) it follows that

$$2\theta_{l_0}^{(i)T} H H_0^{(i)} \theta_0^{(i)} < -\rho I, \tag{4.4.26}$$

where $\rho = \dfrac{q\|(aH-\gamma I)H\|}{(1+2\epsilon)\lambda_{\max}} > 0$.

We now estimate the second term on the right-hand side of (4.4.25) after multiplying it by $(e(i)+l_0)$.

From (4.4.4) and (4.4.16) it follows that

$$S_{mn}(i) \triangleq \sum_{k=m}^{n} \frac{a(e(i)+k+1)^{\gamma}}{e(i)+k} \zeta_k^{(i)} I_{[\|x_k^{(i)}-x^0\| \le \delta_1]} \xrightarrow[i\to\infty]{} 0 \tag{4.4.27}$$

uniformly with respect to n and m with $0 < m < n \le G(i)$.

Noticing that $\|w_{l_0}^{(i)}\| \le \lambda (e(i))^{-\mu}$ with λ being a constant, $\mu \in (\frac{1}{4}, \frac{1}{2})$, and that $\mu + \gamma < \frac{1}{2}$, which implies $\mu > \gamma$, we find

$$\varphi_i \triangleq \left\| 2a(e(i)+l_0+1)^{\gamma} w_{l_0}^{(i)} \right\| \xrightarrow[i\to\infty]{} 0.$$

Then, noticing that $\|2HH_{l_0}^{(i)}\theta_{l_0}^{(i)}\|$ is bounded by some constant $\rho_1 > 0$, we have

$$\begin{aligned} &\left\| \frac{2a(e(i)+l_0+1)^{\gamma}}{e(i)+l_0} (-2w_{l_0}^{(i)} + \zeta_{l_0}^{(i)}) HH_{l_0}^{(i)}\theta_{l_0}^{(i)} \right\| \\ &\quad \le \rho_1 \Big(\frac{\varphi_i}{e(i)+l_0} + \sup_{0<m\le n\le G(i)} \|S_{mn}(i)\| \Big) \xrightarrow[i\to\infty]{} 0. \end{aligned} \tag{4.4.28}$$

For the third term on the right-hand side of (4.4.25), multiplying it by $(e(i)+l_0)$ we have

$$\frac{1}{e(i)+l_0} \|\theta_{l_0}^{(i)T} H_{l_0}^{(i)} HH_{l_0}^{(i)} \theta_{l_0}^{(i)}\| \le \frac{\rho_2}{e(i)+l_0} \xrightarrow[i\to\infty]{} 0, \tag{4.4.29}$$

where $\rho_2 > 0$ is a constant.

Finally, for the last term of (4.4.25) we have the following estimate

$$\begin{aligned} &|4a(e(i)+l_0+1)^{\gamma} w_{l_0}^{(i)T} H[\theta_{l_0}^{(i)} + \frac{a(e(i)+l_0+1)^{\gamma}}{e(i)+l_0}(-2w_{l_0}^{(i)} + \xi_{l_0}^{(i)})]| \\ &\quad \le 2\varphi_i(\|H\theta_{l_0}^{(i)}\| + \|H\| \sup_{0<m<n\le G(i)} \|S_{mn}(i)\| + \frac{\varphi_i}{e(i)+l_0}) \xrightarrow[i\to\infty]{} 0. \end{aligned} \tag{4.4.30}$$

Combining (4.4.26)–(4.4.30) we find that

$$Q_i(1) \xrightarrow[i\to\infty]{} 0 \quad \text{and} \quad Q_i(1) < 0 \quad \text{for large} \quad i.$$

Consequently, from (4.4.25) it follows that

$$\theta_{l_0+1}^{(i)}{}^T H\theta_{l_0+1}^{(i)} < \Big[\theta_{l_0}^{(i)} + \frac{a(e(i)+l_0+1)^\gamma}{e(i)+l_0}(-2w_{l_0}^{(i)} + \zeta_{l_0}^{(i)})\Big]^T H \cdot \Big[\theta_{l_0}^{(i)} + \frac{a(e(i)+l_0+1)^\gamma}{e(i)+l_0}(-2w_{l_0}^{(i)} + \zeta_{l_0}^{(i)})\Big]. \quad (4.4.31)$$

We now show that

$$\theta_{l_0+1}^{(i)}{}^T H\theta_{l_0+1}^{(i)} \le \Big[\theta_{j_0+1}^{(i)} + \sum_{j=j_0+1}^{l_0} (\frac{a(e(i)+j+1)^\gamma}{e(i)+j}\zeta_j^{(i)})\Big]^T H \cdot \Big[\theta_{j_0+1}^{(i)} + \sum_{j=j_0+1}^{l_0} (\frac{a(e(i)+j+1)^\gamma}{e(i)+j}\zeta_j^{(i)})\Big] \quad (4.4.32)$$

by induction.

Assume it holds for $k+1$, i.e.,

$$\theta_{l_0+1}^{(i)}{}^T H\theta_{l_0+1}^{(i)} \le [\theta_{k+1}^{(i)} + \sum_{j=k+1}^{l_0} (\frac{a(e(i)+j+1)^\gamma}{e(i)+j}\zeta_j^{(i)})\Big]^T H \cdot \Big[\theta_{k+1}^{(i)} + \sum_{j=k+1}^{l_0} (\frac{a(e(i)+j+1)^\gamma}{e(i)+j}\zeta_j^{(i)})\Big], \quad (4.4.33)$$

which has been verified for $k+1 \le l_0$. We have to show it is true for $k, k \ge j_0+1$.

By (4.4.18) we have

$$\theta_{k+1}^{(i)} + \sum_{j=k+1}^{l_0} \Big(\frac{a(e(i)+j+1)^\gamma}{e(i)+j}\zeta_j^{(i)}\Big) = \theta_k^{(i)} + \sum_{j=k}^{l_0} \frac{a(e(i)+j+1)^\gamma}{e(i)+j}\zeta_j^{(i)} - \frac{2a(e(i)+k+1)^\gamma}{e(i)+k}w_k^{(i)} + \frac{1}{e(i)+k}H_k^{(i)}\theta_k^{(i)}$$

and

$$\Big[\theta_{k+1}^{(i)} + \sum_{j=k+1}^{l_0} (\frac{a(e(i)+j+1)^\gamma}{e(i)+j}\zeta_j^{(i)})\Big]^T H\Big[\theta_{k+1}^{(i)}$$

$$+\sum_{j=k+1}^{l_0}\Big(\frac{a(e(i)+j+1)^{\gamma}}{e(i)+j}\zeta_j^{(i)}\Big)\Big]$$

$$=\Big[\theta_k^{(i)}+\sum_{j=k}^{l_0}\Big(\frac{a(e(i)+j+1)^{\gamma}}{e(i)+j}\zeta_j^{(i)}\Big)\Big]^T H\Big[\theta_k^{(i)}$$

$$+\sum_{j=k}^{l_0}\Big(\frac{a(e(i)+j+1)^{\gamma}}{e(i)+j}\zeta_j^{(i)}\Big)\Big]+Q_i(l_0-k+1), \tag{4.4.34}$$

where

$$Q_i(l_0-k+1)=\frac{2}{e(i)+k}\theta_k^{(i)T}HH_k^{(i)}\theta_k^{(i)}$$

$$+2\Big(\sum_{j=k}^{l_0}\frac{a(e(i)+j+1)^{\gamma}}{e(i)+j}\zeta_j^{(i)}-\frac{2a(e(i)+k+1)^{\gamma}}{e(i)+k}w_k^{(i)}\Big)^T\frac{HH_k^{(i)}\theta_k^{(i)}}{e(i)+k}$$

$$+\frac{1}{(e(i)+k)^2}\theta_k^{(i)}H_k^{(i)T}HH_k^{(i)}\theta_k^{(i)}-\frac{4a(e(i)+k+1)^{\gamma}}{e(i)+k}w_k^{(i)T}H\Big[\theta_k^{(i)}$$

$$+\sum_{j=k}^{l_0}\frac{a(e(i)+j+1)^{\gamma}}{e(i)+j}\zeta_j^{(i)}-\frac{2a(e(i)+k+1)^{\gamma}}{e(i)+k}w_k^{(i)}\Big]. \tag{4.4.35}$$

Comparing (4.4.35) with (4.4.25), we find that in lieu of l_0 and $\frac{2a(e(i)+l_0+1)^{\gamma}}{e(i)+l_0}$ we now have k and $2a\sum_{j=k}^{l_0}\frac{(e(i)+j+1)^{\gamma}}{e(i)+j}\zeta_j^{(i)}$, respectively. But, for both cases we use the same estimate (4.4.27). Therefore, completely by the same argument as (4.4.26)–(4.4.30), we can prove that

$$Q_i(l_0-k+1)\xrightarrow[i\longrightarrow\infty]{}0 \quad \text{and} \quad Q_i(l_0-k+1)<0 \quad \text{for large} \quad i.$$

Thus, we have proved (4.4.32).

By the elementary inequality $(1+\delta)A^THA+(1+\frac{1}{\delta})B^THB\geq(A+B)^TH(A+B)$ for $\delta>0$, which is derived from $(\sqrt{\delta}A-\frac{1}{\sqrt{\delta}}B)^TH(\sqrt{\delta}A-\frac{1}{\sqrt{\delta}}B)\geq 0$ for any matrices A and B of compatible dimensions, we derive from (4.4.32)

$$\theta_{l_0+1}^{(i)}{}^T H\theta_{l_0+1}^{(i)} \leq \Big(1+\frac{\gamma}{2}\Big)\theta_{j_0+1}^{(i)T} H\theta_{j_0+1}^{(i)} + \Big(1+\frac{2}{\gamma}\Big)\lambda_{\max}\Big\| \sum_{j=j_0+1}^{l_0} \frac{a(e(i)+j+1)^\gamma}{e(i)+j}\zeta_{j_0}^{(i)}\Big\|^2. \tag{4.4.36}$$

As mentioned before, for $j: j_0 \leq j \leq l_0$, $\|x_j^{(i)} - x^0\| < \delta_1$ and there is no truncation. Then by (4.4.18)

$$\theta_{j_0+1}^{(i)} = \theta_{j_0}^{(i)} + d_{j_0}^{(i)}, \tag{4.4.37}$$

where

$$d_{j_0}^{(i)} \triangleq \frac{1}{e(i)+j_0} H_{j_0}^{(i)}\theta_{j_0}^{(i)} + \frac{a(e(i)+j_0+1)^\gamma}{e(i)+j_0}(-2w_{j_0}^{(i)} + \zeta_{j_0}^{(i)}) \xrightarrow[i\to\infty]{} 0. \tag{4.4.38}$$

Then from (4.4.36) and (4.4.27) it follows that

$$\begin{aligned}\theta_{l_0+1}^{(i)}{}^T H\theta_{l_0+1}^{(i)} &\leq (1+\frac{\gamma}{2})\theta_{j_0+1}^{(i)T} H\theta_{j_0+1}^{(i)} + (1+\frac{2}{\gamma})\lambda_{\max}\|S_{j_0+1}(i)\| \\ &= (1+\frac{\gamma}{2})\theta_{j_0}^{(i)T} H\theta_{j_0}^{(i)} + R(j_0, l_0, i),\end{aligned}$$

where

$$R(j_0,l_0,i) = \Big(1+\frac{\gamma}{2}\Big)\Big(2\theta_{j_0}^{(i)T} Hd_{j_0}^{(i)} + d_{j_0}^{(i)T} Hd_{j_0}^{(i)}\Big) + (1+\frac{2}{\gamma})\lambda_{\max}\|S_{j_0+1,l_0}(i)\|,$$

which tends to zero as $i \longrightarrow \infty$ by (4.4.27) and (4.4.38).

Then

$$\begin{aligned}\theta_{l_0+1}^{(i)T} H\theta_{l_0+1}^{(i)} &\leq (1+\frac{\gamma}{2})(e(i)+j_0)^{2\gamma}(x_{j_0}^{(i)} - x^0)^T H(x_{j_0}^{(i)} - x^0) + R(j_0,l_0,i) \\ &= 2(1+\frac{\gamma}{2})(e(i)+j_0)^{2\gamma}(1+o(1))(L(x_{j_0}^{(i)}) - L_{\min}) + R(j_0,l_0,i),\end{aligned} \tag{4.4.39}$$

where for the last equality (4.4.10) is used.

Finally, by (4.4.21), for large i from (4.4.39) it follows that

$$\theta_{l_0+1}^{(i)T} H\theta_{l_0+1}^{(i)} \leq (1+\frac{3\gamma}{4})q,$$

which incorporating with (4.4.10) yields

$$2\big(e(i)+l_0+1\big)^{2\gamma}(L(x^{(i)}_{l_0+1})-L_{\min}) = (1+o(1))\theta^{(i)T}_{l_0+1}H\theta^{(i)}_{l_0+1} \le (1+\gamma)q.$$

This contradicts (4.4.20), the definition of l_0+1. The contradiction shows $l_0 \ge G(i)$. □

Theorem 4.4.1 *Assume that A4.3.1, A4.4.1–A4.4.4 hold, and $H-\frac{\gamma}{a}I$ is positive definite for some $\gamma \in (0, \frac{1}{2}-\mu)$.*

Further, assume that

$$\lambda(i)i^{(\frac{4\mu+4\gamma-1}{2-4\mu-4\gamma})\wedge 2\gamma} \xrightarrow[i\to\infty]{} \infty, \tag{4.4.40}$$

$$e(i)i^{-\frac{1}{1-2\mu-2\gamma}-\delta} \xrightarrow[i\to\infty]{} \infty, \tag{4.4.41}$$

$$g(i)/e(i) \xrightarrow[i\to\infty]{} \infty, \tag{4.4.42}$$

and for some constants $q_2 > q_1 > 0$

$$1+\frac{q_1}{i} < \frac{g(i+1)}{g(i)} < 1+\frac{q_2}{i}. \tag{4.4.43}$$

Then the number of resettings is finite, i.e.,

$$\lim_{k\to\infty} \tau_k < \infty \quad \text{a.s.}, \tag{4.4.44}$$

where τ_k is the number of resettings among the first k search periods (GO1), and $\lambda(i)$ is given in (GO3).

Proof. If (4.4.44) were not true, then there would be an S with positive probability such that, for any $\omega \in S$, there exists a subsequence $\{i_m\}$ such that at the i_mth search period a resetting occurs, i.e.,

$$\Lambda^{(i_m)}_{G(i_m)} < \Lambda_{g(i_m)} - 2\lambda(i_m). \tag{4.4.45}$$

Notice that

$$\frac{g(i+1)-g(i)}{i^{(1+\delta)}} = \frac{g(i)(\frac{g(i+1)}{g(i)}-1)}{i^{(1+\delta)}} \ge \frac{e(i)(\frac{g(i+1)}{g(i)}-1)}{i^{(1+\delta)}} \xrightarrow[i\to\infty]{} \infty$$

by(4.4.41) and $g(i+1)-g(i) \xrightarrow[i\to\infty]{} \infty$, and

$$\frac{G(i)}{i^{(1+\delta)}} = \frac{e(i)(\frac{g(i)}{e(i)}-1)}{i^{(1+\delta)}} \xrightarrow[i\to\infty]{} \infty$$

by (4.4.41) and (4.4.42). Hence, conditions of Lemma 4.4.1 are satisfied. Without loss of generality, we may assume that (4.4.2)–(4.4.5) and the conclusion of Theorem 4.3.1 hold $\forall \omega \in S$. From now on assume that $\omega \in S$ is fixed.

It is clear that, for any constant $b > 0$

$$L(x_k) \leq L_{\min} + \frac{b}{2}, \quad \forall k \geq K_0, \tag{4.4.46}$$

if K_0 is large enough, since $x_k \longrightarrow x^0$ for $\omega \in S$.

Let

$$i(K_0) = \inf\{i : g(i) > K_0\}, \quad m_0 = \inf\{m : i_m > i(K_0)\}.$$

Rewrite (4.4.46) as

$$2(g(i(K_0)))^{2\gamma}(L(x_{g(i(K_0))}) - L_{\min}) \leq (g(i(K_0)))^{2\gamma} b. \tag{4.4.47}$$

Define

$$q \triangleq b(g(i(K_0)))^{2\gamma} \vee 1, \tag{4.4.48}$$

and

$$q_0 \triangleq (1+\gamma)q. \tag{4.4.49}$$

Noticing that there is no resetting between $i(K_0)$ and i_{m_0} and (4.4.47) corresponds to (4.4.8), by the same argument as that used in the proof of Lemma 4.4.3, we find that, for any $k : g(i(k_0)) < k \leq g(i_{m_0})$,

$$2k^{2\gamma}(L(x_k) - L_{\min}) \leq q_0. \tag{4.4.50}$$

Since $i_{m_0} - 1 \geq i(k_0)$, we have

$$\begin{aligned}\Lambda_{g(i_{m_0})} - L_{\min} &= \frac{1}{g(i_{m_0}) - g(i_{m_0} - 1)} \sum_{j=g(i_{m_0}-1)+1}^{g(i_{m_0})} (L(x_j) - L_{\min} + \xi_j^0) \\ &\leq \frac{q_0}{2(g(i_{m_0} - 1))^{2\gamma}} + \frac{1}{g(i_{m_0}) - g(i_{m_0} - 1)} \sum_{j=g(i_{m_0}-1)+1}^{g(i_{m_0})} \xi_j^0.\end{aligned}$$

By (4.4.3) (4.4.42) and (4.4.43) it follows that

$$\begin{aligned}\Lambda_{g(i_{m_0})} - L_{\min} &\leq \frac{q_0}{2(g(i_{m_0}-1))^{2\gamma}} + \frac{(i_{m_0}-1)^{\frac{1+\delta}{2}}}{\sqrt{g(i_{m_0}-1)}\sqrt{\frac{g(i_{m_0})}{g(i_{m_0}-1)}-1}} \\ &\leq \frac{q_0}{2(g(i_{m_0}-1))^{2\gamma}} + \frac{(i_{m_0}-1)^{\frac{1+\delta}{2}}}{\sqrt{e(i_{m_0}-1)}\cdot\sqrt{\frac{q_1}{i_{m_0}-1}}} \\ &\leq \frac{q_0}{2(g(i_{m_0}-1))^{2\gamma}} + \frac{(i_{m_0}-1)^{\frac{1+\delta}{2}}\cdot(i_{m_0}-1)^{\frac{1}{2}}}{(i_{m_0}-1)^{\frac{1}{2-4\mu-4\gamma}+\frac{\delta}{2}}\cdot\sqrt{q_1}},\end{aligned}$$

where for the last inequality (4.4.41) is used.

Thus, by (4.4.40)

$$\begin{aligned}\Lambda_{g(i_{m_0})} - L_{\min} &\leq \frac{q_0}{2(g(i_{m_0}-1))^{2\gamma}} + \frac{1}{\sqrt{q_1}}(i_{m_0}-1)^{-(\frac{4\mu+4\gamma-1}{2-4\mu-4\gamma})} \\ &\leq \frac{q_0}{2(g(i_{m_0}-1))^{2\gamma}} + \frac{1}{2}\lambda(i_{m_0}). \end{aligned} \tag{4.4.51}$$

By (4.4.33) it follows that

$$\begin{aligned}&\frac{1}{G(i_{m_0})}\sum_{j=1}^{G(i_{m_0})}(L(x_j^{(i_{m_0})}) - L_{\min}) \\ &\leq \Lambda_{g(i_{m_0})} - L_{\min} - 2\lambda(i_{m_0}) - \frac{1}{G(i_{m_0})}\sum_{j=1}^{G(i_{m_0})}\xi_{j+1}^{(i_{m_0},0)} \\ &\leq \Lambda_{g(i_{m_0})} - L_{\min} - 2\lambda(i_{m_0}) + \frac{1}{\sqrt{G(i_{m_0})}}\cdot i_{m_0}^{\frac{1+\delta}{2}}, \end{aligned} \tag{4.4.52}$$

provided $i(K_0)$ is large enough, where for the last inequality, (4.4.2) is used.

Since by (4.4.43)

$$\sqrt{G(i)} > \sqrt{g(i) - g(i-1)} > \sqrt{q_1 g(i-1)/(i-1)} > \sqrt{q_1 g(i-1)/i},$$

and since

$$e(i-1)i^{-\frac{1}{1-2\mu-2\gamma}-\delta} \longrightarrow \infty \quad \text{and} \quad g(i) > e(i),$$

we find

$$\frac{1}{\sqrt{G(i_{m_0})}} i_{m_0}^{\frac{1+\delta}{2}} \leq i_{m_0}^{\frac{1+\delta}{2}} \frac{\sqrt{i_{m_0}}}{\sqrt{q_1 g(i_{m_0}-1)}} \leq \frac{1}{\sqrt{q_1}} i_{m_0}^{1+\frac{\delta}{2}} \cdot \frac{1}{\sqrt{e(i_{m_0}-1)}}$$
$$\leq \frac{1}{\sqrt{q_1}} i_{m_0}^{1+\frac{\delta}{2}-\frac{1}{2(1-2\mu-2\gamma)}-\frac{\delta}{2}} = \frac{1}{\sqrt{q_1}} i_{m_0}^{1-\frac{1}{2-4\mu-4\gamma}} < \frac{1}{2}\lambda(i_{m_0}), \tag{4.4.53}$$

where the last inequality follows from (4.4.40).

Using (4.4.51) and (4.4.53), from (4.4.52) for sufficiently large K_0 we have

$$\frac{1}{G(i_{m_0})} \sum_{j=1}^{G(i_{m_0})} (L(x_j^{(i_{m_0})}) - L_{\min}) \leq \frac{q_0}{2(g(i_{m_0}-1))^{2\gamma}} - \lambda(i_{m_0}). \tag{4.4.54}$$

Using the second inequality of (4.4.43) and then observing that

$$(1 + \frac{q_2}{i_{m_0}-1})^{2\gamma} = 1 + \frac{2\gamma}{i_{m_0}} + o(1) \quad \text{with } o(1) \longrightarrow 0 \text{ as } K_0 \longrightarrow \infty$$

and

$$\frac{1}{(g(i_{m_0}))^{2\gamma} i_{m_0} \lambda(i_{m_0})} \leq \frac{1}{(e(i_{m_0}))^{2\gamma} i_{m_0} \lambda(i_{m_0})} \xrightarrow[i_{m_0} \longrightarrow \infty]{} 0$$

by (4.4.40) and (4.4.41) and $\mu + \gamma < \frac{1}{2}$, we find

$$\frac{1}{G(i_{m_0})} \sum_{j=1}^{G(i_{m_0})} (L(x_j^{(i_{m_0})}) - L_{\min}) \tag{4.4.55}$$
$$\leq \frac{q_0}{2(g(i_{m_0}))^{2\gamma}} (1 + \frac{q_2}{i_{m_0}-1})^{2\gamma} - \lambda(i_{m_0}) \leq \frac{q_0}{2(g(i_{m_0}))^{2\gamma}}.$$

We now show that there is $k \leq G(i_{m_0})$ such that

$$\max_{1 \leq j \leq G(i_{m_0})} 2(e(i_{m_0}) + k)^{2\gamma} (L(x_k^{(i_{m_0})}) - L_{\min}) < (1-\gamma) q. \tag{4.4.56}$$

Assume the converse:

$$\max_{1 \leq j \leq G(i_{m_0})} (e(i_{m_0}) + j)^{2\gamma} (L(x_j^{(i_{m_0})}) - L_{\min}) \geq \frac{(1-\gamma) q}{2}.$$

Then, we have

$$\begin{aligned}
&\frac{1}{G(i_{m_0})}\sum_{j=1}^{G(i_{m_0})}(L(x_j^{(i_{m_0})})-L_{\min}) \geq \frac{(1-\gamma)q}{2G(i_{m_0})}\sum_{j=1}^{G(i_{m_0})}\frac{1}{(e(i_{m_0})+j)^{2\gamma}}\\
&\geq \frac{(1-\gamma)q}{2(1-2\gamma)G(i_{m_0})}((g(i_{m_0})+1)^{1-2\gamma}-(e(i_{m_0})+1)^{1-2\gamma})\\
&\geq \frac{(1-\gamma)q(g(i_{m_0}))^{1-2\gamma}}{2(1-2\gamma)G(i_{m_0})}[1-(\frac{e(i_{m_0})+1}{g(i_{m_0})})^{1-2\gamma}]\\
&= \frac{(1+\gamma)q}{2(g(i_{m_0}))^{2\gamma}}\cdot\frac{(1-\gamma)}{1-\gamma-2\gamma^2}(\frac{g(i_{m_0})}{G(i_{m_0})})(1-(\frac{e(i_{m_0})+1}{g(i_{m_0})})^{1-2\gamma})\\
&> \frac{(1+\gamma)q}{2(g(i_{m_0}))^{2\gamma}}=\frac{q_0}{2(g(i_{m_0}))^{2\gamma}}
\end{aligned} \tag{4.4.57}$$

for large enough K_0, because

$$\begin{aligned}
&\frac{(1-\gamma)}{1-\gamma-2\gamma^2}(\frac{g(i_{m_0})}{G(i_{m_0})})(1-(\frac{e(i_{m_0})+1}{g(i_{m_0})})^{1-2\gamma})\\
&> \frac{(1-\gamma)}{1-\gamma-2\gamma^2}(1-(\frac{e(i_{m_0})+1}{g(i_{m_0})})^{1-2\gamma}) \xrightarrow[i_{m_0}\to\infty]{} \frac{(1-\gamma)}{1-\gamma-2\gamma^2}>1.
\end{aligned}$$

Inequality (4.4.57) contradicts (4.4.55). Consequently, (4.4.56) is true. In particular, for $k=G(i_{m_0})$, we have

$$2(g(i_{m_0}))^{2\gamma}(L(x_{g(i_{m_0})}^{(i_{m_0})})-L_{\min}) \leq (1+\gamma)(1-\gamma)q=(1-\gamma^2)q<q.$$

Completely by the same argument as that used for (4.4.47)–(4.4.50), by noticing that there is no resetting from $x_{g(i_{m_0})}$ to $x_{G(i_{m_0+1})}^{(i_{m_0+1})}$ we conclude that

$$2k^{2\gamma}(L(x_k)-L_{\min}) \leq (1+\gamma)q=q_0, \quad \forall k: g(i_{m_0})\leq k\leq G(i_{m_0+1}).$$

By the same treatment as that used for deriving (4.4.54) from (4.4.50), we obtain

$$\frac{1}{G(i_{m_0+1})}\sum_{j=1}^{G(i_{m_0+1})}(L(x_j^{(i_{m_0+1})})-L_{\min}) \leq \frac{q_0}{2(g(i_{m_0+1}-1))^{2\gamma}}-\lambda(i_{m_0+1}). \tag{4.4.58}$$

Comparing (4.4.58) with (4.4.54), we find that i_{m_0} has been changed to i_{m_0+1}, and this procedure can be continued if the number of resettings is infinite. Therefore, for any $m \geq m_0$, we have

$$\frac{1}{G(i_m)} \sum_{j=1}^{G(i_m)} (L(x_j^{(i_m)}) - L_{\min}) \leq \frac{q_0}{2(g(i_m - 1))^{2\gamma}} - \lambda(i_m). \tag{4.4.59}$$

From (4.4.40) we see

$$\lambda(i)i^{2\gamma} \geq \lambda(i)i^{(\frac{4\mu+4\gamma-1}{2-4\mu-4\gamma})\wedge 2\gamma} \longrightarrow \infty. \tag{4.4.60}$$

Since $0 < \gamma + \mu < \frac{1}{2}$, we have $\frac{2\gamma}{1-2\mu-2\gamma} > 2\gamma$ and hence by (4.4.60)

$$\lambda(i_m)i_m^{\frac{2\gamma}{1-2\mu-2\gamma}} \xrightarrow[m\longrightarrow\infty]{} \infty. \tag{4.4.61}$$

Consequently, by (4.4.41) the right-hand side of (4.4.59) can be estimated as follows:

$$\begin{aligned}\frac{q_0}{2(g(i_m - 1))^{2\gamma}} - \lambda(i_m) &< \frac{q_0}{2(e(i_m - 1))^{2\gamma}} - \lambda(i_m) \\ &\leq \frac{q_0}{(i_m - 1)^{\frac{2\gamma}{1-2\mu-2\gamma}+2\gamma\delta}} - \lambda(i_m) < 0\end{aligned}$$

by (4.4.61) if m is large enough.

However, the left-hand side of (4.4.59) is nonnegative. The obtained contradiction shows that m must be finite, and (4.4.44) is correct. □

By Theorem 4.4.1, our global optimization algorithm coincides with KW algorithm with randomized differences and expanding truncations for sufficiently large k. Therefore, theorems proved in Section 4.2 are applicable to the global optimization algorithm. By Theorems 4.2.1 and 4.2.2 we can derive convergence rate and asymptotic normality of the algorithm described by (GO1)–(GO5).

4.5. Application to Model Reduction

In this section we apply the global optimization algorithm to system modeling. A real system may be modeled by a high order system which, however, may be too complicated for control design. In control engineering the order reduction for a model is of great importance. In the linear system case, this means that a high order transfer function $F(z)$ is to be approximated by a lower order transfer function. For this one may use methods like the balanced truncation and the Hankel norm approximation. These methods are based on concept of the balanced realization. We are interested in recursively estimating the optimal coefficients of the reduced model by using the stochastic optimization algorithm presented in Section 4.3.

Let the high order transfer function $F(z)$ be

$$F(z) = \frac{\alpha_1 z^{n-1} + \alpha_2 z^{n-1} + \cdots + \alpha_{n-1} z + \alpha_n}{z^n + \beta_1 z^{n-1} + \cdots + \beta_{n-1} z + \beta_n} \tag{4.5.1}$$

and let it be approximated by a lower order transfer function $F_m(z) = \frac{C(z)}{D(z)}$. If $C(z)$ is of order $2s-1$ (or $2s$), then $D(z)$ is taken to be of order $2s$ (or $2s+1$). To be fixed, let us take $C(z)$ to be a polynomial of order $2s-1$ and $D(z)$ of order $2s$:

$$C(z) = c_1 z^{2s-1} + c_2 z^{2s-2} + \cdots + c_{2s-1} z + c_{2s}, \tag{4.5.2}$$

$$D(z) = (z^2 + d_{11} z + d_{21})(z^2 + d_{12} z + d_{22}) \cdots (z^2 + d_{1s} z + d_{2s}), \tag{4.5.3}$$

where coefficients c_i, $i = 1, \ldots, 2s$, should not be confused with step sizes used in Steps (GO1)-(GO5). Write $F_m(z)$ as $F_m(c,d,z)$, where c and d stand for coefficients of $C(z)$ and $D(z)$

$$c = [c_1, c_2, \ldots, c_{2s-1}, c_{2s}]^T \text{ and } d = [d_{11}, d_{21}, \ldots, d_{1s}, d_{2s}]^T.$$

It is natural to take

$$\begin{aligned} L(c,d) &\triangleq \|F(z) - F_m(c,d,z)\|_2^2 \\ &= \frac{1}{2\pi} \int_0^{2\pi} |F(e^{j\omega}) - F_m(c,d,e^{j\omega})|^2 d\omega \end{aligned} \tag{4.5.4}$$

as the performance index of approximation. The parameters c and d are to be selected to minimize $L(c,d)$ under the constraint that $F_m(c,d,z)$ is stable. For simplicity of notations we denote $x = \begin{bmatrix} c \\ d \end{bmatrix}$, and write $F_m(c,d,z)$ as $F_m(x,z)$.

Let us describe the x-set where $F_m(x,z)$ has the required property. Stability requires that

$$|D(z)| \neq 0, \quad \forall z : |z| \geq 1.$$

This implies that

$$|d_{1i}| < 2, \quad i = 1, \ldots, s, \tag{4.5.5}$$

because d_{1i} is the sum of two complex-conjugate roots of $D(z)$.

If $d_{1i} > 0$, then $\frac{-d_{1i} - \sqrt{d_{1i}^2 - 4d_{2i}}}{2} > -1$, which yields $d_{1i} - 1 < d_{2i}$. If $d_{1i} < 0$, then $-d_{1i} - 1 < d_{2i}$, and hence

$$|d_{1i}| - 1 < d_{2i} < 1 \qquad i = 1, \ldots, s. \tag{4.5.6}$$

Set

$$D = \{d_{1i}, d_{2i} : |d_{1i}| < 2, |d_{1i}| - 1 < d_{2i} < 1, i = 1, \dots, s\}. \tag{4.5.7}$$

Identify $L(x)$, x, $\mathbb{R}^l$, and l appeared in Section 4.3 to $L(c, d)$, $\begin{bmatrix} c \\ d \end{bmatrix}$, $\mathbb{R}^{4s}$, and $4s$ respectively for the present case.

We now apply the optimization algorithm (GO1)–(GO5) to minimizing $L(c, d)$ under constraint that the parameter x in $F_m(x, z)$ belongs to D. For this we first concretize Steps (GO1)–(GO5) described in Section 4.3.

Since $L(c, d)$ is convex in c for fixed d, we take the fixed initial value $c_0^{(i)} = (1, \dots, 1)$ for any search period i, and randomly select initial values only for d according to a distribution density $p(\cdot)$, which is defined as follows:

$$p(d) = \Pi_{l=1}^{s} p(d_{1l}, d_{2l}),$$

where $p(u, v) = q(v|u)q(u)$ with $q(u)$ and $q(v|u)$ being the uniform distributions over $[-2, 2]$ and $[|u| - 1, 1]$, respectively.

After $x_0^{(i)}$ having been selected in the ith search period, the algorithm (4.1.11) and (4.1.12) is calculated with $a_k \triangleq a_k^{(i)} = \frac{0.01}{e(i)+k+1}$ and $e(i) = (100+i)^{1.5}$. As to observations, in stead of (4.3.1) we will use information about gradient because in the present case the gradient $f(c, d) (\triangleq f(x))$ of $L(c, d)$ can explicitly be expressed:

$$\begin{aligned} f(x) \triangleq f(c, d) \triangleq \nabla L(c, d) &= \frac{1}{2\pi} \int_0^{2\pi} \nabla |F(e^{jw}) - F_m(c, d, e^{jw})|^2 dw \\ &= -\frac{1}{2\pi} \int_0^{2\pi} \operatorname{Re}[(F(e^{jw}) - F_m(c, d, e^{jw})) \nabla \overline{F_m(c, d, e^{jw})}] dw. \end{aligned} \tag{4.5.8}$$

In the ith search period the observation is denoted by $y_{k+1}^{(i)}$ and is given by

$$\begin{aligned} y_{k+1}^{(i)} = \frac{1}{100} \sum_{t=1}^{100} \operatorname{Re}[(F(e^{j(w_k + \frac{2\pi t}{100})}) - F_m(x_k^{(i)}, e^{j(w_k + \frac{2\pi t}{100})})) \\ \cdot \nabla \overline{F_m(x_k^{(i)}, e^{j(w_k + \frac{2\pi t}{100})})}], \end{aligned}$$

where w_k is independently selected from $[0, 2\pi]$ according to the uniform distribution, and $x_k^{(i)}$ stands for the estimate for $\begin{bmatrix} c \\ d \end{bmatrix}$ at time k in the ith search period. It is clear that $y_{k+1}^{(i)}$ is an approximation to the integral

(4.5.8) with $\begin{bmatrix} c \\ d \end{bmatrix} = x_k^{(i)}$. Therefore, we have observations in the form

$$y_{k+1}^{(i)} = f(x_k^{(i)}) + \varepsilon_{k+1}^{(i)}.$$

The expanding truncation method used in (4.1.11) and (4.1.12) requires projecting the estimated value to a fixed point, if the estimated value appears outside an expanding region. Let us denote it by Q_k. In (4.1.11) and (4.1.12) the spheres with expanding radiuses M_{σ_k} serve as the expanding regions Q_k, which are now modified as follows.

Let us write $x_k^{(i)} = \begin{bmatrix} c_k^{(i)} \\ d_k^{(i)} \end{bmatrix}$, where $d_k^{(i)} \in D$. Define

$$D_k^{(i)} \triangleq \{|d_{1i}| \le 2(1 - \frac{1}{\tau_k^{(i)}}),$$
$$(1 - \frac{1}{\tau_k^{(i)}})(|d_{1i}| - 1 + \frac{1}{\tau_k^{(i)}}) \le d_{2i} \le 1 - \frac{1}{\tau_k^{(i)}}\} \subset D, \quad (4.5.9)$$

$$Q_k^{(i)} \triangleq I\!R^{2s} \times D_k^{(i)}, \quad (4.5.10)$$

where

$$\tau_k^{(i)} = \sum_{j=0}^{k-1} I_{\{(x_j^{(i)} + a_j^{(i)} y_{j+1}^{(i)} \in (Q_j^{(i)})^c\}}, \quad \tau_0^{(i)} = 1. \quad (4.5.11)$$

The expanding truncations in (4.1.11) and (4.5.11) are also modified:

$$x_{k+1}^{(i)} = (x_k^{(i)} + a_k^{(i)} y_{k+1}^{(i)})) I_{[(x_k^{(i)} + a_k^{(i)} y_{k+1}^{(i)}) \in Q_k^{(i)}]}$$
$$+ (x_k^{(i)} + a_k^{(i)} y_{k+1}^{(i)})_p I_{[(x_k^{(i)} + a_k^{(i)} y_{k+1}^{(i)}) \in (Q_k^{(i)})^c]}$$

where $(x_k^{(i)} + a_k^{(i)} y_{k+1}^{(i)})_p$ means the projection of $x_k^{(i)} + a_k^{(i)} y_{k+1}^{(i)}$ to $Q_k^{(i)}$.

Take $g(i) = (100 + i)^2$. Then after $G(i) = (100 + i)^2 - (100 + i)^{1.5}$ steps, $x_{G(i)}^{(i)}$ will be obtained.

Concerning (GO2)–(GO4), the only change consists in observations. We replace $y_{k+1}^{(i,0)}$ in (GO2)–(GO4) by $Y_{k+1}^{(i)}$ which is defined by

$$Y_{k+1}^{(i)} = \frac{1}{100} \sum_{t=1}^{100} \left| F(e^{j(w_k + \frac{2\pi t}{100})}) - F_m(x_k^{(i)}, e^{j(w_k + \frac{2\pi t}{100})}) \right|^2,$$

where w_k are independently selected from $[0, 2\pi]$ according to the uniform distribution for each k. Clearly, $Y_{k+1}^{(i)}$ is an approximation to $L(x_k^{(i)})$ $(= L(c_k^{(i)}, d_k^{(i)})$. Finally, take $\lambda(i)$ equal to $\frac{c}{\ln(i)}$.

In control theory there are several well-known model reduction methods such as model reduction by balanced truncation, Hankel norm approximation among others. These methods depend on the balanced realization which is a state space realization method for a transfer matrix $F(s)$, keeping the Gramians for controllability and observability of the realized system balanced. In order to compare the proposed global optimization (GO) method, we take the commonly used model reduction methods by balanced truncation (BT) and Hankel norm approximation (HNA), which, are realized by using Matlab. For this, the discrete-time transfer functions $F(z)$ are transformed to the continuous time ones by using d2c provided in Matlab. Then the reduced systems are discretized to compute $L(c, d)$ for comparison.

As $F(z)$ we take a 10th order transfer function $F(z) = \frac{\alpha(z)}{\beta(z)}$ respectively for the following examples:

Example 4.5.1

$$\begin{aligned}\alpha(z) &= z^9 - 0.4z^8 + 0.08z^7 - 0.032z^6 + 0.0816z^5 - 0.0326z^4 \\ &\quad + 0.0288z^3 - 0.0115z^2 + 0.1296z - 0.0518, \\ \beta(z) &= z^{10} + 1.08z^8 + 0.8726z^6 + 0.6227z^4 + 0.4694z^2 + 0.1266;\end{aligned}$$

Example 4.5.2

$$\begin{aligned}\alpha(z) &= z^9 - 2.55z^8 + 4.62z^7 - 5.705z^6 + 6.1495z^5 - 5.9771z^4 \\ &\quad + 5.0659z^3 - 3.629z^2 + 1.7084z - 0.5523, \\ \beta(z) &= z^{10} - 3.55z^9 + 6.155z^8 - 5.688z^7 + 2.6317z^6 - 0.8835z^5 \\ &\quad + 2.5479z^4 - 4.714z^3 + 4.3881z^2 - 2.197z + 0.5194;\end{aligned}$$

Example 4.5.3

$$\begin{aligned}\alpha(z) &= z^9 + 1.1z^8 - 2.68z^7 - 2.858z^6 + 0.9821z^5 + 0.9453z^4 \\ &\quad - 0.1046z^3 - 0.828z^2 + 0.00858z + 0.002 \\ \beta(z) &= z^{10} - 3.6z^9 + 7.17z^8 - 10.836z^7 + 12.5713z^6 - 11.2381z^5 \\ &\quad + 7.9913z^4 - 4.3356z^3 + 1.5868z^2 - 0.3327z + 0.0296.\end{aligned}$$

Using the algorithm described in Section 4.3, for Examples 4.5.1-4.5.3 we obtain the approximate transfer functions of order 4, respectively, denoted by $F1_4(z)$, $F2_4(z)$ and $F3_4(z)$ with

$$F1_4(z) = \frac{0.9986z^3 + 0.0274z^2 - 0.7212z - 0.0865}{z^4 + 0.43z^3 + 0.4583z^2 + 0.1404z + 0.0757},$$

$$F2_4(z) = \frac{0.9435z^3 - 1.5672z^2 + 2.0739z - 1.4274}{z^4 - 2.7169z^3 + 3.3849z^2 - 2.1344z + 0.5807},$$

$$F3_4(z) = \frac{-2.9591z^3 + 8.2974z^2 + 3.5048z - 16.6678}{z^4 - 1.8622z^3 + 1.8829z^2 - 1.7667z + 0.7772}.$$

Using Matlab we also derive the 4th order approximations for Examples 4.5.1–4.5.3 by balanced truncation and Hankel norm approximation, which are as follows:

$$F1_b(z) = \frac{1.0168z^3 + 0.3238z^2 - 0.2054z - 0.1490}{z^4 + 0.6979z^3 + 0.9651z^2 + 0.3961z + 0.3682},$$

$$F1_H(z) = \frac{1.4257z^3 + 1.4368z^2 - 0.2350z - 0.3201}{z^4 + 1.1498z^3 + 0.76z^2 + 0.4191z + 0.1022},$$

$$F2_b(z) = \frac{0.728z^3 - 0.7624z^2 + 0.9906z - 0.361}{z^4 - 2.6733z^3 + 3.5289z^2 - 2.3527z + 0.7628},$$

$$F2_H(z) = \frac{2.5851z^3 - 4.4881z^2 + 4.8781z - 1.6623}{z^4 - 2.4062z^3 + 2.9866z^2 - 1.8512z + 0.5752},$$

$$F3_b(z) = \frac{-6.6681z^3 + 9.5183z^2 + 2.8167z - 10.7083}{z^4 - 1.925z^3 + 1.9375z^2 - 1.8604z + 0.8718},$$

$$F3_H(z) = \frac{110.9644z^3 - 131.9689z^2 + 139.8014z - 135.0758}{z^4 - 1.2098z^3 + 1.2268z^2 - 1.1031z + 0.2512},$$

where the subscripts b and H denote the results obtained by balanced truncation and Hankel norm approximation, respectively.

The approximation errors $L(c, d)$ are given in the following table:

	BT	HNA	GO
Example 1	0.1694	0.3136	0.0641
Example 2	6.8254	7.3206	2.9976
Example 3	1349.9	14820	761.8623

From this table we see that the algorithm presented in Section 4.3 gives less approximation errors in H_2-norm in comparison with other methods.

We now compare approximation errors in H_∞-norm and compare step responses between the approximate models and the true one by figures.

In the figures of step response

- the solid lines (——) denote the true high order systems;
- the dashed lines (- - -) denote the system reduced by Hankel norm approximation;

- the dotted lines $(\cdots\cdots)$ denote the system reduced by balanced truncation;
- The dotted-dashed lines $(- \cdot -)$ denote the systems reduced by the stochastic optimization method given in Section 3.

In the figures of the approximation error $|F(e^{jw}) - F_m(c, d, e^{jw})|^2$, $w \in [0, 2\pi]$

- the solid lines (——) denote the systems reduced by the stochastic optimization method;
- the dashed lines (- - -) denote the system reduced by Hankel norm approximation;
- the dotted lines $(\cdots\cdots)$ denote the system reduced by balanced truncation.

Example 4.5.1

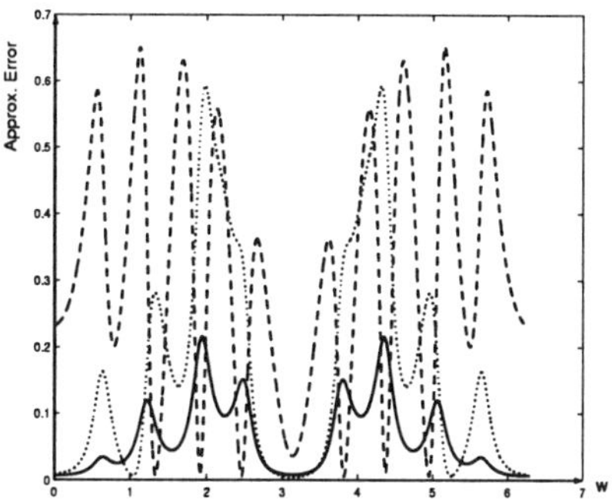

Approximation Errors

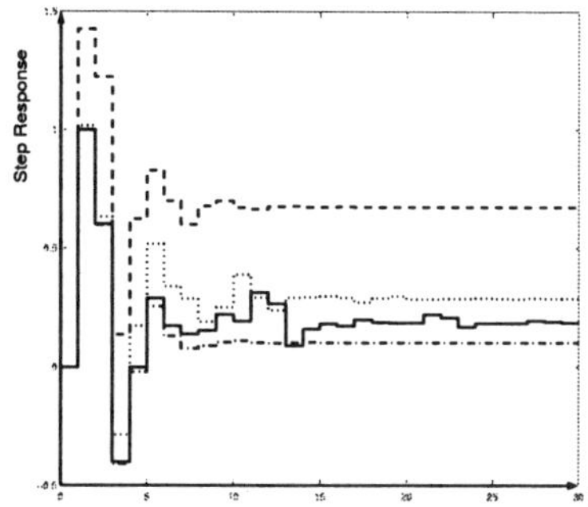

Step Responses

Example 4.5.2

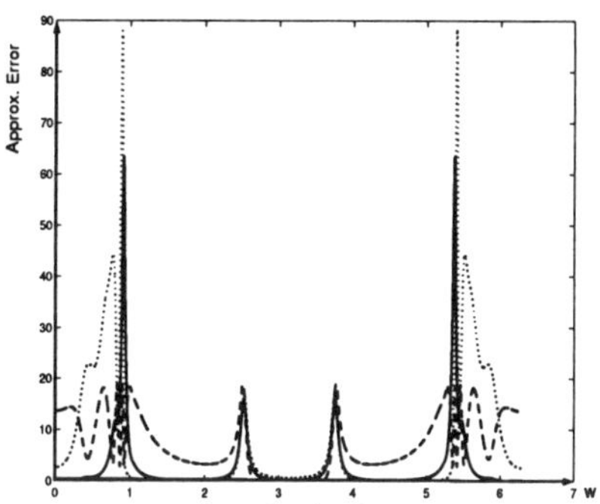

Approximation Errors

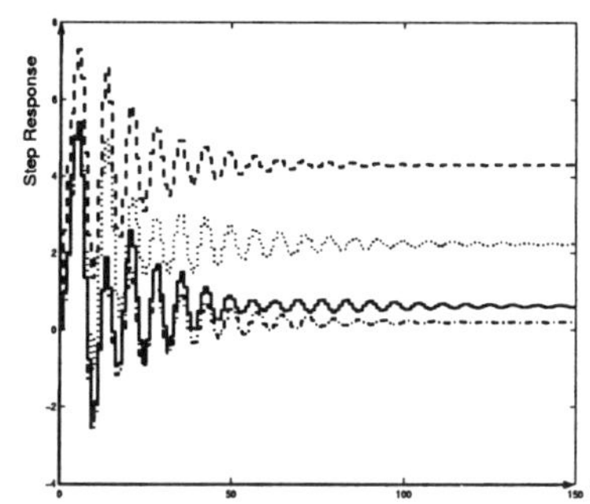

Step Responses

Example 4.5.3

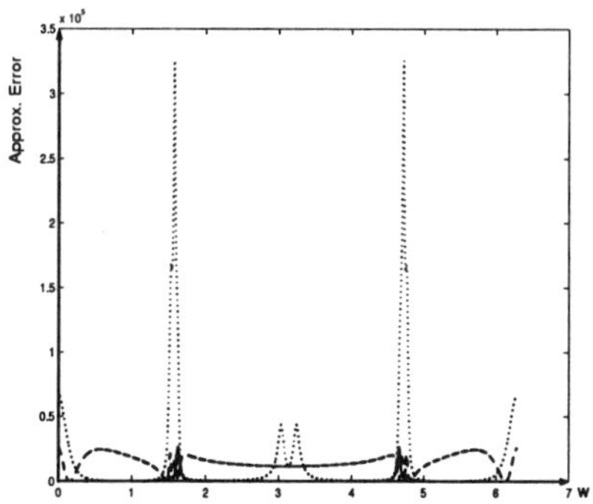

Approximation Errors

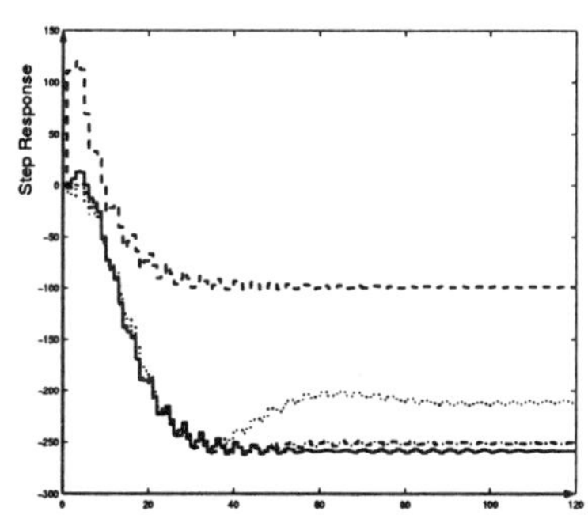

Step Responses

These figures show that the algorithm given in Section 4.3 gives less approximation error in H_∞-norm in comparison with other methods for Example 4.5.1 and the intermediate error in H_∞-norm for Examples 4.5.2 and 4.5.3. Concerning step responses, the algorithm given in Section 4.3 provides better approximation in comparison with other methods for all three examples.

4.6. Notes and References

The well-known paper [61] by Kiefer and Wolfowitz is the pioneer work using stochastic approximation method for optimization. The random version of KW algorithm was introduced in [63], and the random direction version of KW algorithm was dealt with in [85] by the ODE method. Theorems 2.4.1, 2.4.2 given in Section 4.1 are presented in [21], while Theorem 2.4.4 in [18]. The results on convergence rate and symptotic normality of KW algorithm presented in Section 4.2 can be found in [21].

Global optimization based on noisy observations by discrete-time simulated annealing is considered in [45, 52, 100]. Combination of the KW algorithm with a search method for global optimization is dealt with in [97]. A better combination given in [49] is presented in Section 4.3 and 4.4.

For model reduction we refer to [51, 102]. The global optimization method presented in Section 4.3 is applied to model reduction in Section 4.5, which is written based on [22].

Chapter 5

APPLICATION TO SIGNAL PROCESSING

The general convergence theorems developed in Chapter 2 can deal with noises containing not only random components but also structural errors. This property allows us to apply SA algorithms to parameter estimation problems arising from various fields. The general approach, roughly speaking, is as follows. First, the parameter estimation problem coming from practice is transformed to a root-seeking problem for a reasonable but unknown function $f(\cdot)$, which may not be directly observed. Then, the real observation y_{k+1} is artificially written in the standard form

$$y_{k+1} = f(x_k) + \epsilon_{k+1}$$

with $\epsilon_{k+1} = y_{k+1} - f(x_k)$. Normally, it is quite straightforward to arrive at this point. The main difficulty is to verify that the complicated noise $\epsilon_{k+1} = y_{k+1} - f(x_k)$ satisfies one of the noise conditions required in the convergence theorems. It is common that there is no standard method to complete the verification procedure, because ϵ_{k+1} for different problems are completely different from each other.

In Section 5.1, SA algorithms are applied to solve the blind channel identification problem, an active topic in communication. In Section 5.2, the principle component analysis used in pattern classification is dealt with by SA methods. Section 5.3 continues the problem discussed in Section 5.1, but in more general setting. Namely, unlike Section 5.1, the covariance matrix of the observation noise is no longer assumed to be known. In Section 5.4, adaptive filtering is considered: Very simple conditions for convergence of sign-algorithms are given. Section 5.5 discusses the asymptotic behavior of asynchronous SA algorithms, which take the possible communication delays between parallel processors into consideration.

5.1. Recursive Blind Identification

In system and control area, the unknown parameters are estimated on the basis of observed input and output data of the system. This is the subject of system identification. In contrast to this, for communication channels only the channel output is observed and the channel input is unavailable. The topic of blind channel identification is to estimate channel parameters by using the output data only. Blind channel identification has drawn much attention from researchers because of its potential applications in wireless communication. However, most existing estimation methods are "block" algorithms in nature, i.e., parameters are estimated after the entire block of data have been received.

By using the SA method, here a recursive approach is presented: Estimates are continuously improved while receiving new signals.

Consider a system consisting of p channels with L being the maximum order of the channels. Let s_k, $k = 0, 1, \dots, N$, be the one-dimensional input signal, and $[x_k^{(1)}, x_k^{(2)}, \dots, x_k^{(p)}]^T$ be the p-dimensional channel output at time k, $k = L, L+1, \dots, N$, where N is the number of samples and may not be fixed:

$$x_k = \sum_{i=0}^{L} h_i s_{k-i}, \quad k \geq L, \tag{5.1.1}$$

where

$$h_i = [h_i^{(1)}, \dots, h_i^{(p)}]^T, \quad i = 0, \dots, L, \tag{5.1.2}$$

are the unknown channel coefficients.

Let us denote by

$$h^{(i)} = [h_0^{(i)}, \dots, h_L^{(i)}]^T \tag{5.1.3}$$

the coefficients of the ith channel, and by

$$h^* = [(h_0)^T, \dots, (h_L)^T]^T \tag{5.1.4}$$

the coefficients of the whole system which compose a $p(L+1)$-dimensional vector.

The observations y_k may be corrupted by noise η_k:

$$y_k = x_k + \eta_k, \tag{5.1.5}$$

where η_k is a p-dimensional vector. The problem is to estimate h^* on the basis of observations.

Let us introduce polynomials in backward-shift operator z

$$h^{(j)}(z) \triangleq h_0^{(j)} + h_1^{(j)} z + \cdots + h_L^{(j)} z^L, \quad j = 1, \ldots, p, \tag{5.1.6}$$

where $zx_k = x_{k-1}$.

Write x_k, y_k and η_k in the component forms

$$x_k = [x_k^{(1)}, \ldots, x_k^{(p)}]^T, \quad y_k = [y_k^{(1)}, \ldots, y_k^{(p)}]^T, \quad \eta_k = [\eta_k^{(1)}, \ldots, \eta_k^{(p)}]^T,$$

respectively, and express the component $x_k^{(j)}$ via $h^{(j)}(z)$:

$$x_k^{(j)} = h^{(j)}(z) s_k, \qquad k = L, L+1, \ldots. \tag{5.1.7}$$

From this it is clear that

$$h^{(i)}(z) x_k^{(j)} = h^{(i)}(z) h^{(j)}(z) s_k = h^{(j)}(z) h^{(i)}(z) s_k = h^{(j)}(z) x_k^{(i)}, \tag{5.1.8}$$
$$\forall i, j = 1, \ldots, p, \quad k = 2L, 2L+1, \ldots$$

Define

$$\Phi_k = [X_k, \ldots, X_{k-L}], \tag{5.1.9}$$

where X_k is a $\frac{p(p-1)}{2} \times p$-matrix:

$$X_k \triangleq \begin{bmatrix} x_k^{(2)} & -x_k^{(1)} & 0 & \cdots & & 0 \\ x_k^{(3)} & 0 & -x_k^{(1)} & \ddots & & \vdots \\ \vdots & \vdots & \vdots & \ddots & & 0 \\ x_k^{(p)} & 0 & 0 & \cdots & & -x_k^{(1)} \\ 0 & x_k^{(3)} & -x_k^{(2)} & 0 & & 0 \\ & \vdots & & \ddots & & \vdots \\ & x_k^{(p)} & 0 & & & -x_k^{(2)} \\ \cdots & \cdots & & & & \\ & \ddots & & & & \ddots \\ \cdots & \cdots & & & & \\ 0 & \cdots & \cdots & 0 & x_k^{(p)} & -x_k^{(p-1)} \end{bmatrix}. \tag{5.1.10}$$

It is clear that Φ_k is a $\frac{p(p-1)}{2} \times p(L+1)$-matrix.

Similar to X_k and Φ_k let us define Y_k and Ψ_k, and N_k and Ξ_k, which have the same structure as X_k and Φ_k but with $x_k^{(i)}$ replaced by $y_k^{(i)}$ and $\eta_k^{(i)}$, respectively.

By (5.1.5) we have

$$\Psi_k = \Phi_k + \Xi_k. \tag{5.1.11}$$

From (5.1.8), (5.1.4), and (5.1.10) it is seen that

$$\Phi_k h^* = 0, \quad k = 2L, 2L+1, \dots \tag{5.1.12}$$

This means that the channel coefficient h^* satisfies the set of linear equations (5.1.12) with coefficients being the system outputs.

From the input sequence $\{s_0, s_1, \dots, s_N\}$, $N \geq 2L+1$, we form the $(N-2L+1) \times (2L+1)$-Hankel matrix $S_N^{(0)}(2L+1)$:

$$S_N^{(0)}(2L+1) \triangleq \begin{bmatrix} s_0 & s_1 & \cdots & s_{2L} \\ s_1 & s_2 & & s_{2L+1} \\ \vdots & \vdots & & \vdots \\ s_{N-2L} & s_{N-2L+1} & & s_N \end{bmatrix}. \tag{5.1.13}$$

It is clear that the maximal rank of $s_N^{(0)}(2L+1)$ is $2L+1$ as $N \longrightarrow \infty$. If $s_N^{(0)}(2L+1)$ is of full rank for some $N \geq 2L+1$, then $S_k^{(0)}(2L+1)$ will also be of full rank for any $k \geq N$.

Lemma 5.1.1 *Assume the following conditions hold:*

A5.1.1 $h^{(j)}(z)$ $j = 1, \dots, p$ *have no common root.*

A5.1.2 *The Hankel matrix* $s_N^{(0)}(2L+1)$ *composed of input signal is of full rank (rank*$=2L+1$*).*

Then h^* *is the unique up to a scalar multiple nonzero vector simultaneously satisfying*

$$\Phi_k h^* = 0, \quad k = 2L, 2L+1, \dots, N. \tag{5.1.14}$$

Proof. Assume there is another solution $\overline{h} \triangleq [(\overline{h}_0)^T, \dots, (\overline{h}_L)^T]^T$ to (5.1.14), which is different from h^*:

$$\Phi_k \overline{h} = 0, \quad k = 2L, 2L+1, \dots, N, \tag{5.1.15}$$

where $\overline{h}_i = [\overline{h}_i^{(1)}, \dots, \overline{h}_i^{(p)}]^T$ is p-dimensional, $\forall i = 0, 1, \dots, L$.

Denote

$$\overline{h}^{(i)} \triangleq [\overline{h}_0^{(i)}, \dots, \overline{h}_L^{(i)}]^T,$$
$$\overline{h}^{(i)}(z) \triangleq \overline{h}_0^{(i)} + \overline{h}_1^{(i)} z + \dots + \overline{h}_L^{(i)} z^L.$$

From (5.1.15) it follows that

$$\begin{aligned}&\overline{h}^{(i)}(z)x_k^{(j)} - \overline{h}^{(j)}(z)x_k^{(i)} = 0,\\&\forall i,j = 1,\dots,p; \quad k = 2L,\dots,N-(2L+1).\end{aligned}$$

By (5.1.7), we then have

$$\begin{aligned}&[\overline{h}^{(i)}(z)h^{k}(z) - \overline{h}^{(j)}(z)h^{(i)}(z)]s_k = 0,\\&\forall i,j = 1,\dots,p; \quad k = 2L,\dots,N-(2L+1),\end{aligned}$$

which implies

$$h^T(i,j)S_N^{(0)T}(2N+1) = 0, \quad \forall i,j = 1,\dots,p, \tag{5.1.16}$$

where by $h(i,j)$ we denote the $(2L+1)$-dimensional vector composed of coefficients of the polynomial $\overline{h}^{(i)}(z)h^{(j)}(z) - \overline{h}^{(j)}(z)h^{(i)}(z)$ written in the form of increasing orders of z.

Since $S_N^{(0)}(2L+1)$ is of full rank, $h(i,j) = 0$. In other words,

$$\overline{h}^{(i)}(z)h^{(j)}(z) - \overline{h}^{(j)}(z)h^{(i)}(z) \equiv 0, \quad \forall i,j = 1,\dots,p. \tag{5.1.17}$$

For a fixed j, (5.1.17) is valid for all $i = 1,\dots,p$, $i \neq j$. Therefore, all roots of $h^{(j)}(z)$ should be roots of $\overline{h}^{(j)}(z)h^{(i)}(z)$ for all $i \neq j$. By A5.1.1, all roots of $h^{(j)}(z)$ must be roots of $\overline{h}^{(j)}(z)$. Consequently, there is a constant α_j such that $\overline{h}^{(j)}(z) = \alpha_j h^{(j)}(z)$, $\forall j = 1,\dots,p$. Substituting this into (5.1.17) leads to

$$\alpha_i h^{(i)}(z)h^{(j)}(z) - \alpha_j h^{(j)}(z)h^{(i)}(z) = 0,$$

and hence $\alpha_i = \alpha_j \triangleq \alpha$, $\forall i,j = 1,\dots,p$. Thus, we conclude that

$$\overline{h} = \alpha h^*.$$

□

We first establish a convergence theorem for blind channel identification based on stochastic approximation methods for the case where a noise-free data sequence $\{x_L, x_{L+1},\dots,x_N\}$ is observed.

Then, we extend the results to the case where N is not fixed and observation is noise-corrupted.

Assume $\{x_L, x_{L+1},\dots,x_N\}$ is observed. In this case X_k, $L \le k \le N$ are available, and we have $\Phi_k, k = L, 2L+1,\dots,N$. We will repeatedly use the data by setting

$$\Phi_{k(N-2L+1)+i} = \Phi_i, \quad i = 2L,\dots,N, \quad k = 0,1,2,\dots. \tag{5.1.18}$$

Define estimate for h^* recursively by

$$h(k+1) = h(k) - a_k \Phi_{k+1}^T \Phi_{k+1} h(k), \quad k = 2L, 2L+1, \dots \tag{5.1.19}$$

with an initial value $h(2L-1) \neq 0$.

We need the following condition.

A5.1.3 $a_k > 0$, $a_{k+1} \leq a_k$, $\forall k$, $a_k \xrightarrow[k\to\infty]{} 0$ *and* $\sum_{k=1}^{\infty} a_k = \infty$.

Theorem 5.1.1 *Assume A5.1.1–A5.1.3 hold. Let $h(k)$ be given by (5.1.19) with any initial value $h(2L-1)$ with $h^T(2L-1)h^* \neq 0$. Then*

$$h(k) \xrightarrow[k\to\infty]{} \alpha h^*,$$

where $\alpha \triangleq \frac{h^T(2L-1)h^*}{\|h^*\|^2}$ *is a constant.*

Proof. Decompose $h(2L-1)$ and $h(k)$ respectively into orthogonal vectors:

$$h(2L-1) = \alpha h^* + h'(2L-1), \quad h(k) = \frac{h^T(k)h^*}{\|h^*\|^2} h^* + h'(k), \tag{5.1.20}$$

where $h'^T(k)h^* = 0$, $k = 2L-1, 2L, \dots$.

If αh^* serves as the initial value for (5.1.19), then by (5.1.14), $h(k) \equiv \alpha h^*$. Again, by (5.1.14) we have

$$h^{*T}(I - a_k \Phi_{k+1}^T \Phi_{k+1}) h'(k) = 0,$$

and we conclude that

$$h'(k+1) = h'(k) - a_k \Phi_{k+1}^T \Phi_{k+1} h'(k), \tag{5.1.21}$$

and

$$h(k) = \alpha h^* + h'(k). \tag{5.1.22}$$

Therefore, for proving the theorem it suffices to show that $h'(k) \longrightarrow 0$ as $k \longrightarrow \infty$.

Denote

$$\begin{aligned} F_{k+1} \triangleq &(I - a_{(k+1)(N-2L+1)-1} \Phi_{(k+1)(N-2L+1)}^T \Phi_{(k+1)(N-2L+1)}) \\ &\cdot (I - a_{(k+1)(N-2L+1)-2} \Phi_{(k+1)(N-2L+1)-1}^T \Phi_{(k+1)(N-2L+1)-1}) \\ &\cdots \\ &\cdot (I - a_{k(N-2L+1)} \Phi_{k(N-2L+1)+1}^T \Phi_{k(N-2L+1)+1}) \end{aligned}$$

and $g_k \triangleq h'(k(N-2L+1))$.

Then by (5.1.21) we have

$$g_{k+1} = F_{k+1} g_k. \tag{5.1.23}$$

Noticing that $a_k \longrightarrow 0$ and Φ_k is uniformly bounded with respect to k, for large k ($k \geq k_0$) we have

$$F_{k+1} \leq I - \frac{1}{2} \sum_{i=k(N-2L+1)}^{(k+1)(N-2L+1)-1} a_i \Phi_{i+1}^T \Phi_{i+1}, \tag{5.1.24}$$

and $F_k^2 \leq F_k$.

By (5.1.18)

$$\sum_{i=k(N-2L+1)}^{(k+1)(N-2L+1)-1} \Phi_{i+1}^T \Phi_{i+1} = \sum_{i=0}^{N-2L} \Phi_{i+1}^T \Phi_{i+1}, \tag{5.1.25}$$

and by Lemma 5.1.1, h^* is its unique up to a constant multiple eigenvector corresponding to the zero eigenvalue, and the rank of $\sum_{i=0}^{N-2L} \Phi_{i+1}^T \Phi_{i+1}$ is $p(L+1)-1$.

Denote by $\lambda_{\min}$ the minimal nonzero eigenvalue of $\sum_{i=0}^{N-2L} \Phi_{i+1}^T \Phi_{i+1}$.

Let h' be an arbitrary $p(L+1)$-dimensional vector orthogonal to h^*. Then h' can be expressed by

$$h' = \sum_{i=1}^{p(L+1)-1} \alpha_i u_i,$$

where u_i, $i = 1, \ldots, p(L+1)-1$, are the unit eigenvectors of $\sum_{i=0}^{N-2L} \Phi_{i+1}^T \cdot \Phi_{i+1}$ corresponding to its nonzero eigenvalues.

It is clear that

$$h'^T \sum_{i=0}^{N-2L} \Phi_{i+1}^T \Phi_{i+1} h' \geq \lambda_{\min} \|h'\|^2.$$

By this, from (5.1.23) and (5.1.24), it follows that for $k \geq k_0$

$$\|g_{k+1}\|^2 \leq g_k^T F_{k+1} g_k \leq \|g_k\|^2 - \frac{\lambda_{\min} a_{(k+1)(N-2L+1)-1} \|g_k\|^2}{2}$$

and

$$\|g_{k+1}\|^2 \le \prod_{i=k_0}^{k+1}(1-\frac{\lambda_{\min}}{2}a_{i(N-2L+1)})\|g_{k_0}\|^2.$$

Noticing that

$$\sum_{i=1}^{\infty} a_{i(N-2L+1)} \ge \frac{1}{N-2L+1}\sum_{i=N-2L+1}^{\infty} a_i = \infty,$$

we conclude

$$\prod_{i=k_0}^{k+1}(1-\frac{\lambda_{\min}}{2}a_{i(N-2L+1)})\|g_{k_0}\|^2 \xrightarrow[k\to\infty]{} 0,$$

and hence $g_k \xrightarrow[k\to\infty]{} 0$.

From (5.1.21) it is seen that $\|h'(k)\|$ is nonincreasing for $k \ge k_0$. Hence, the convergence $g_k \xrightarrow[k\to\infty]{} 0$ implies that $h'(k) \xrightarrow[k\to\infty]{} 0$.

The proof is completed. □

Remark 5.1.1 If the initial value $h(2L-1)$ is orthogonal to h^*, then $\alpha = 0$ and (5.1.20) is also true. But this is a non-interesting case giving no information about h^*.

Remark 5.1.2 Algorithm (5.1.19) is an SA algorithm with linear time-varying regression function $\Phi_{k+1}^T\Phi_{k+1}h$. The root set J for $\Phi_k^T\Phi_k$ is time-invariant: $J = \{0, \alpha h^*, \forall\alpha\}$. As mentioned above, $h(k)$ evolves in one of the subspaces $S_1 \triangleq \{0\}$ or $S_2 \triangleq \{\alpha h^*, \forall\alpha\}$ depending on the initial value: $h(2L-1) \in S_1$, or $h(2L-1) \in S_2$. In the proof of Theorem 5.5.1 we have actually verified that $h^T h$ may serve as the Lyapunov function $v(\cdot)$ satisfying A2.2.2^o for S_1. Then applying Remark 2.2.6 also leads to the desired conclusion.

We now assume the input signal $\{s_i\}$ is a sequence of infinitely many mutually independent random variables and that the observations do not contain noise, i.e., $\eta_k \equiv 0$ in (5.1.5).

Lemma 5.1.2 *Assume A5.1.1 holds and $\{s_i\}$ is a sequence of mutually independent random variables with $E|s_i|^2 \ne 0$. Then $\overline{h}^* \triangleq h^*/\|h^*\|$ is the unique unit eigenvector corresponding to the zero eigenvalue for the matrices*

$$B_{j,k} \triangleq \sum_{i=j+k(2L+1)}^{j+(k+1)(2L+1)-1} E\Phi_i^T\Phi_i, \quad \forall j \ge 0, \quad \forall k \ge 0, \tag{5.1.26}$$

and the rank of $B_{j,k}$ is $p(L+1)-1$.

Proof. Since $\{s_i\}$ is a sequence of mutually independent random variables and $E|s_i|^2 \neq 0$, it follows that

$$ES_{4L}^{(k)}(2L+1)S_{4L}^{(k)}(2L+1) > 0, \quad \forall k, \tag{5.1.27}$$

where

$$S_{4L}^{(k)}(2L+1) \triangleq \begin{bmatrix} s_k & s_{k+1} & \cdots & s_{k+2L} \\ s_{k+1} & s_{k+2} & \cdots & s_{k+2L+1} \\ & \cdots & & \\ s_{k+2L} & s_{k+2L+1} & \cdots & s_{k+4L} \end{bmatrix}. \tag{5.1.28}$$

Proceeding along the lines of the proof of Lemma 5.1.1., we arrive at the analogue of (5.1.16):

$$h^T(i,j)S_{4L}^{(k)}(2L+1) = 0, \qquad \forall i,j = 1,\dots,p, \quad \forall k \geq 2L,$$

which implies

$$h^T(i,j)ES_{4L}^{(k)}(2L+1)S_{4L}^{(k)}(2L+1)h(i,j) = 0, \tag{5.1.29}$$
$$\forall i,j = 1,\dots,p, \quad \forall k \geq 2L.$$

From (5.1.28) and (5.1.29) it follows that $h(i,j) = 0$. Then following the proof of Lemma 5.1.1, we conclude that $\overline{h}^*$ is the unique unit vector satisfying

$$E\Phi_i^T\Phi_i\overline{h}^* = 0,$$
$$\forall i : j + k(2L+1) \leq i \leq j + (k+1)(2L+1) - 1, \quad \forall j \geq 0.$$

This shows that $B_{j,k}$ is of rank $p(L+1)-1$, $\forall j \geq 0$, $\forall k \geq 0$, and $\overline{h}^*$ is its unique unit eigenvector corresponding to the zero eigenvalue. □

Let $\lambda_{\min}(k)$ denote the minimal nonzero eigenvalue of $B_{0,k}$. On $\{s_i\}$ we need the following condition.

A5.1.4 *$\{s_k\}$ is a sequence of mutually independent random variables with $E|s_k|^2 \neq 0$, $\sup_k E|s_k|^{2+\gamma} < \infty$ for some $\gamma > 0$ and such that*

$$\sum_{j=1}^{\infty} a_{(j+1)(2L+1)-1}\lambda_{\min}(j) = \infty. \tag{5.1.30}$$

Condition A5.1.3 is strengthened to the following A5.1.5.

A5.1.5 *A5.1.3 holds and $\sum_{i=1}^{\infty} a_i^{1+\frac{\gamma}{2}} < \infty$, where γ is given in A5.1.4.*

It is obvious that if $\{s_i\}$ is an iid sequence, then $\lambda_{\min}(j)$ is a positive constant, and (5.1.30) is automatically satisfied.

Theorem 5.1.2 *Assume A5.1.1, A5.1.4, and A5.1.5 hold, and $h(k)$ is given by (5.1.19) with initial value $h(2L-1)$. Then*

$$h(k) \longrightarrow \alpha h^* \quad a.s.,$$

where $\alpha = \frac{h^{*T}h(2L-1)}{\|h^*\|^2}$.

Proof. In the present situation we still have (5.1.21) and (5.1.22). So, it suffices to show $h'(k) \xrightarrow[k\to\infty]{} 0$.

With N replaced by $4L$ in the definitions of F_k and g_k we again arrive at (5.1.23).

Since $a_k \xrightarrow[k\to\infty]{} 0$, $\{E\Phi_{k+1}^T\Phi_{k+1}\}$ is bounded, and $\sum_{i=1}^{\infty} a_i(\Phi_{i+1}^T\Phi_{i+1} - E\Phi_{i+1}^T\Phi_{i+1})$ converges a.s. by A5.1.4 and A5.1.5, there is a large k_0 such that

$$\begin{aligned}F_{k+1} \leq & I - \sum_{i=k(2L+1)}^{(k+1)(2L+1)-1} a_i E\Phi_{i+1}^T\Phi_{i+1} - \sum_{i=k(2L+1)}^{(k+1)(2L+1)-1} a_i(\Phi_{i+1}^T\Phi_{i+1} \\ & - E\Phi_{i+1}^T\Phi_{i+1}) + o(a_{k(2L+1)}) \\ \leq & I - \frac{1}{2}\sum_{i=k(2L+1)}^{(k+1)(2L+1)-1} a_i E\Phi_{i+1}^T\Phi_{i+1}.\end{aligned}$$

Let h' be an arbitrary $p(L+1)$-dimensional vector such that $h'^T h^* = 0$. Then by Lemma 5.1.2,

$$h'^T B_{0,k} h' \geq \lambda_{\min}(k)\|h'\|^2,$$

and hence

$$\|g_{k+1}\|^2 \leq g_k^T F_{k+1} g_k \leq \|g_k\|^2 - \frac{a_{(k+1)(2L+1)-1}\lambda_{\min}(k)}{2}\|g_k\|^2, \quad k \geq k_0.$$

Therefore, $\|g_{k+1}\|^2 \leq \prod_{i=k_0}^{k+1}(1 - \frac{\lambda_{\min}(i)}{2}a_{(i+1)(2L+1)-1})\|g_{k_0}\|^2$, which tends to zero since $\sum_{i=1}^{\infty}\lambda_{\min}(i)a_{(i+1)(2L+1)-1} = \infty$. This implies $h'(k) \xrightarrow[k\to\infty]{} 0$. □

We now consider the noisy observation (5.1.5). By the definition (5.1.11), similar to (5.1.9) we have

$$\Psi_k = [Y_k, \dots, Y_{k-L}], \quad \Xi_k = [N_k, \dots, N_{k-L}], \tag{5.1.31}$$

where Y_k and N_k have the same structure as X_k given by (5.1.10) with $x_k^{(i)}$ replaced by $y_k^{(i)}$ and $\eta_k^{(i)}$, respectively.

The following truncated algorithm is used to estimate h^*:

$$\begin{aligned} h(k+1) = &\big(h(k) - a_k(\Psi_{k+1}^T\Psi_{k+1} - E\Xi_{k+1}^T\Xi_{k+1})h(k)\big) \\ &\cdot I_{[\|h(k)-a_k(\Psi_{k+1}^T\Psi_{k+1}-E\Xi_{k+1}^T\Xi_{k+1})h(k)\|<1]} \\ &+ h(2L-1)I_{[\|h(k)-a_k(\Psi_{k+1}^T\Psi_{k+1}-E\Xi_{k+1}^T\Xi_{k+1})h(k)\|\geq 1]} \end{aligned} \tag{5.1.32}$$

with initial value $h(2L-1)$: $h^T(2L-1)h^* \neq 0$ and $\|h(2L-1)\| = \frac{1}{4}$.

Introduce the following conditions.

A5.1.6 *$\{s_k\}$ and $\{\eta_k\}$ are mutually independent and each of them is a sequence of mutually independent random variables (vectors) such that $E\eta_k = 0$, $E(s_k - Es_k)^2 \neq 0$, and*

$$\sup_k\{|s_k| + \|\eta_k\|\} \leq \eta < \infty, \quad E\eta^{2+\gamma} < \infty$$

for some $\gamma > 0$.

A5.1.7 $a_k > 0$, $a_{k+1} \leq a_k$, $\forall k$, $\frac{a_k}{a_{k+1}} < c$, $\forall k$, $a_k \xrightarrow[k\to\infty]{} 0$, $\sum\limits_{k=0}^{\infty} a_k = \infty$, *and* $\sum\limits_{i=1}^{\infty} a_i^{1+\frac{\gamma}{2}} < \infty$, *where γ is given in A5.1.4.*

Set

$$R_k \triangleq \Phi_k^T\Phi_k - E\Phi_k^T\Phi_k + \Phi_k^T\Xi_k + \Xi_k^T\Phi_k + \Xi_k^T\Xi_k - E\Xi_k^T\Xi_k, \tag{5.1.33}$$

$$S_k \triangleq \Psi_k^T\Psi_k - E\Xi_k^T\Xi_k. \tag{5.1.34}$$

Then

$$S_k = E\Phi_k^T\Phi_k + R_k. \tag{5.1.35}$$

Denote by τ_k, $k = 1, 2, \dots$, the resetting times, i.e., $h(\tau_k) = h(2L-1)$. Then, we have

$$h(\tau_k + j) = h(2L-1) - \sum_{i=\tau_k}^{\tau_k+j-1} a_i(E\Phi_{i+1}^T\Phi_{i+1} + R_{i+1})h(i), \tag{5.1.36}$$

and

$$\|h(\tau_k + j)\| < 1, \quad j = 1, \ldots, \tau_{k+1} - \tau_k - 1, \quad \forall k.$$

Let $[V \vdots \overline{h}^*]$ be an orthogonal matrix, where

$$\overline{h}^* = \frac{h^*}{\|h^*\|}.$$

Denote

$$h'(k) \triangleq VV^T h(k), \quad \overline{h}^{*T} h(k) \triangleq \overline{h}(k). \tag{5.1.37}$$

Then

$$h(k) = h'(k) + \overline{h}(k)\overline{h}^*. \tag{5.1.38}$$

Noticing $\Phi_i h^* = 0$, $\forall i$ and $VV^T E\Phi_{i+1}^T \Phi_{i+1} = E\Phi_{i+1}^T \Phi_{i+1}$, we find that

$$\begin{aligned} h'(\tau_k + j) =& h'(2L-1) - \sum_{i=\tau_k}^{\tau_k+j-1} a_i E\Phi_{i+1}^T \Phi_{i+1} h'(i) \\ & - \sum_{i=\tau_k}^{\tau_k+j-1} a_i VV^T R_{i+1} h(i), \quad j = 1, \ldots, \tau_{k+1} - \tau_k - 1, \end{aligned} \tag{5.1.39}$$

$$\begin{aligned} \overline{h}(\tau_k + j) = \overline{h}(2L-1) - & \sum_{i=\tau_k}^{\tau_k+j-1} a_i \overline{h}^{*T} R_{i+1} h(i), \\ & j = 1, \ldots, \tau_{k+1} - \tau_k - 1. \end{aligned} \tag{5.1.40}$$

Lemma 5.1.3 *Assume A5.1.6 and A5.1.7 hold. Then for $h(j)$ given by (5.1.32),*

$$\sum_{j=1}^{\infty} a_j R_{j+1} h(j) < \infty \ a.s. \tag{5.1.41}$$

Proof. Setting

$$D_k = \Phi_k \Xi_k + \Xi_k^T \Phi_k + \Xi_k^T \Xi_k - E\Xi_k^T \Xi_k,$$

we have

$$R_k = \Phi_k^T \Phi_k - E\Phi_k^T \Phi_k + D_k.$$

By A5.1.6, $D_{k+1}h(k)$ is a martingale difference sequence with $\sup_k E\|D_k\|^{1+\frac{\gamma}{2}} < \infty$. Noticing $\sum_{i=1}^{\infty} a_i^{1+\frac{\gamma}{2}} < \infty$ and $\|h(i)\| < 1$, we find that

$$\sum_{i=1}^{\infty} a_i D_{i+1} h(i) < \infty \quad \text{a.s.} \tag{5.1.42}$$

by the convergence theorem for martingale difference sequences.

Since $\Phi_k^T \Phi_k - E\Phi_k^T \Phi_k$ is independent of $\Phi_{k+2L+1}^T \Phi_{k+2L+1} - E\Phi_{k+2L+1}^T \cdot \Phi_{k+2L+1}$ and $\sup_k E\|\Phi_k^T \Phi_k - E\Phi_k^T \Phi_k\|^{1+\frac{\gamma}{2}} < \infty$, we also have

$$\sum_{i=1}^{\infty} a_i (\Phi_{i+1}^T \Phi_{i+1} - E\Phi_{i+1}^T \Phi_{i+1}) h(i) < \infty \quad \text{a.s.,}$$

which together with (5.1.42) implies (5.1.41). □

Lemma 5.1.4 *Under the condition A5.1.6, if $\tau_k \leq j$, $m(j,T) \leq \tau_{k+1} - \tau_k - 1$, then there is a constant $c_0 > 0$, possibly depending on sample path, such that*

$$\|h(i+1) - h(j)\| \leq c_0 t, \quad \forall i : j \leq i \leq m(j,t), \quad \forall t \in [0,T], \tag{5.1.43}$$

where

$$m(j,T) = \max\{k : \sum_{i=j}^{k} a_i \leq T\}.$$

Proof. By A5.1.6 there is a constant $c_0 > 0$, possibly depending on sample path, such that

$$\|E\Phi_{i+1}^T \Phi_{i+1} + R_{i+1}\| < c_0, \quad \forall i. \tag{5.1.44}$$

Then the lemma follows from (5.1.36) by noticing $\|h(i)\| \leq 1$. □

Lemma 5.1.5 *Assume A5.1.1 and A5.1.6 hold. Then for any j and any k, the matrix*

$$\sum_{i=j+k(2L+1)}^{j+(k+1)(2L+1)-1} E\Phi_i^T \Phi_i \tag{5.1.45}$$

has rank $p(L+1) - 1$, and $\overline{h}^*$ serves as its unique unit eigenvector corresponding to the zero eigenvalue.

Proof. Since $\{s_i\}$ is a sequence of mutually independent nondegenerate random variables, $ES_{4L}^{(j)}(2L+1)S_{4L}^{(j)T}(2L+1) \neq 0,\ \forall j \geq 0$ where

$$S_{4L}^{(j)}(2L+1) \triangleq \begin{bmatrix} s_j & s_{j+1} & \cdots & s_{j+2L} \\ s_{j+1} & s_{j+2} & \cdots & s_{j+2L+1} \\ \cdots & & & \\ s_{j+2L} & s_{j+2L+1} & \cdots & s_{j+4L} \end{bmatrix}. \tag{5.1.46}$$

Notice that $S_{4L}^{(j)}(2L+1)$ coincides with $S_N^{(0)}(2L+1)$ given by (5.1.13) if setting $N = 4L$ and $j = 0$ in (5.1.13).

Proceeding as the proof of Lemma 5.1.1, we again arrive at (5.1.16). Then, we have $h^T(i,j)ES_{4L}^{(j)}(2L+1)S_{4L}^{(j)T}(2L+1)h(i,j) = 0$. Since $ES_{4L}^{(j)}(2L+1)S_{4L}^{(j)T}(2L+1) > 0$, we find that $h(i,j) = 0$. Then by the same argument as that used in the proof of Lemma 5.1.1, we conclude that for any j, $\overline{h}^*$ is the unique unit nonzero vector simultaneously satisfying

$$E\Phi_i^T\Phi_i\overline{h}^* = 0, \quad \forall i : j + k(2L+1) \leq i \leq j + (k+1)(2L+1) - 1. \tag{5.1.47}$$

Since $E\Phi_i^T\Phi_i$ is a $p(L+1) \times p(L+1)$ matrix, the above assertion proves that the rank of $\sum\limits_{i=j+k(2L+1)}^{j+(k+1)(2L+1)-1} E\Phi_i^T\Phi_i$ is $p(L+1)-1$, and also proves that $\overline{h}^*$ is its unique unit eigenvector corresponding to the zero eigenvalue. □

Denote by $\lambda_{\min}(j,k) > 0$ the minimal nonzero eigenvalue of $\sum\limits_{i=j+k(2L+1)}^{j+(k+1)(2L+1)-1} E\Phi_i^T\Phi_i$. We need the following condition.

A5.1.8 *There is a* $\lambda > 0$ *such that*

$$\lambda_{\min}(j,k) \geq \lambda, \quad \forall j \geq 0, \forall k \geq 0.$$

It is clear that if $\{s_i\}$ is an iid sequence, then $\lambda_{\min}(j,k)$ is independent of j and k and A5.1.8 is automatically satisfied.

Lemma 5.1.6 *Assume A5.1.1 and A5.1.6–A5.1.8 hold. Then for any* $h' \triangleq VV^Th \neq 0$

$$-h'^T \sum_{i=j}^{m(j,t)} a_i E\Phi_{i+1}^T\Phi_{i+1}h' \leq -\epsilon t\|h'\|^2, \quad \forall j \geq N, \forall t \in [0,T], \tag{5.1.48}$$

if N is large enough, where $\epsilon = \frac{\lambda}{2(2L+1)c^{2L+1}} > 0$ with c and λ given in A5.1.7 and A5.1.8, respectively.

Proof. Let $[u_1^{(j,k)}, \ldots, u_{p(L+1)-1}^{(j,k)}, \overline{h}^*]$ be the orthogonal matrix composed of eigenvectors of $\sum\limits_{i=j+k(2L+1)}^{j+(k+1)(2L+1)-1} E\Phi_{i+1}^T \Phi_{i+1}$. By Lemma 5.1.4, $\overline{h}^*$ is the only eigenvector corresponding to the zero eigenvalue.

Since $h'^T \overline{h}^* = 0$, h' can be expressed as

$$h' = \sum_{i=1}^{p(L+1)-1} \alpha_i u_i^{(j,k)}.$$

Then

$$-h'^T \sum_{i=j+k(2L+1)}^{j+(k+1)(2L+1)-1} E\Phi_{i+1}^T \Phi_{i+1} h' \le -\sum_{i=1}^{p(L+1)-1} \lambda |\alpha_i|^2 = -\lambda \|h'\|^2, \forall j, \forall k. \tag{5.1.49}$$

By A5.1.4 $\|E\Phi_{i+1}^T \Phi_{i+1}\|$ is bounded with respect to i, and hence by (5.1.48) and the nonincreasing property of $\{a_i\}$ we have

$$\begin{aligned}
-h'^T &\sum_{i=j}^{m(j,t)} a_i E\Phi_{i+1}^T \Phi_{i+1} h' \\
&\le -h'^T \sum_{k=0}^{[\frac{m(j,t)-j+1}{2L+1}]} \sum_{i=j+k(2L+1)}^{j+(k+1)(2L+1)-1} a_i E\Phi_{i+1}^T \Phi_{i+1} h' \\
&\le -\lambda \|h'\|^2 \sum_{k=0}^{[\frac{m(j,t)-j+1}{2L+1}]} a_{j+(k+1)(2L+1)-1} \\
&= -\lambda \|h'\|^2 \sum_{k=0}^{[\frac{m(j,t)-j+1}{2L+1}]} \frac{1}{2L+1} \sum_{i=0}^{2L} a_{j+(k+1)(2L+1)-1}
\end{aligned} \tag{5.1.50}$$

where $[x]$ denotes the integer part of x.

Since $\frac{a_{k+1}}{a_k} \ge \frac{1}{c}$ and $c \ge 1$, we have

$$\sum_{i=0}^{2L} a_{j+(k+1)(2L+1)-1} = \sum_{i=0}^{2L} \frac{a_{j+(k+1)(2L+1)-1}}{a_{j+k(2L+1)} + i} \cdot a_{j+k(2L+1)+i}$$

$$\geq \frac{1}{c^{2L+1}} \sum_{i=0}^{2L} a_{j+k(2L+1)+i},$$

which incorporating with (5.1.44) leads to

$$-h'^T \sum_{i=j}^{m(j,t)} a_i E\Phi_{i+1}^T \Phi_{i+1} h'$$

$$\leq -\frac{\lambda}{(2L+1)c^{2L+1}} \|h'\|^2 \sum_{k=0}^{[\frac{m(j,t)-j+1}{2L+1}]} \sum_{i=0}^{2L} a_{j+k(2L+1)+i}$$

$$= -\frac{\lambda}{(2L+1)c^{2L+1}} \|h'\|^2 \sum_{i=0}^{[\frac{m(j,t)-j+1}{2L+1}](2L+1)+2L} a_{j+i}$$

$$= -\frac{\lambda}{(2L+1)c^{2L+1}} \|h'\|^2 \sum_{i=j}^{m(j,t)} a_i + o(1)$$

$$\leq -\epsilon t \|h'\|^2 \quad \text{for large enough } j,$$

where $\epsilon = \frac{\lambda}{2(2L+1)c^{2L+1}}$ and $o(1) \longrightarrow 0$ as $j \longrightarrow \infty$. □

Theorem 5.1.3 *Assume A5.1.1 and A5.1.6–A5.1.8 hold. Then for $h(k)$ given by (5.1.32) with initial value $h(2L-1)$: $h^T(2L-1)h^* \neq 0$ and $\|h(2L-1)\| = \frac{1}{4}$,*

$$h(k) \underset{k \longrightarrow \infty}{\longrightarrow} \alpha h^*, \tag{5.1.51}$$

where α is a random variable expressed by (5.1.60).

Proof. We first prove that the number of truncations is finite, i.e., $\lim_{k\longrightarrow\infty} \tau_k \overset{\Delta}{=} \tau < \infty$ a.s.

Assume the converse: $\tau_k \underset{k\longrightarrow\infty}{\longrightarrow} \infty$.

By Lemma 5.1.3, for any given $\delta > 0$

$$\|\sum_{i=\tau_k}^{\tau_k+j} a_i R_{i+1} h(i)\| < \delta$$

and

$$\|h^T(2L-1)\overline{h}^* - \overline{h}^{*T} \sum_{i=\tau_k}^{\tau_k+j} a_i R_{i+1} h(i)\| < \frac{1}{4} + \delta, \quad \forall j = 1, 2, \ldots, \tag{5.1.52}$$

if k is large enough, say, $k \geq K$.

By the definition of τ_{k+1}, we have

$$\|h(2L-1) - \sum_{i=\tau_k}^{\tau_{k+1}-1} a_i S_{i+1} h(i)\| \geq 1,$$

which incorporating with (5.1.52) implies

$$\|h'(2L-1) - \sum_{i=\tau_k}^{\tau_{k+1}-1} a_i E\Phi_{i+1}^T \Phi_{i+1} h'(i)\| \geq \frac{3}{4} - \delta,$$

and

$$\begin{aligned}\|h'(2L-1) - \sum_{i=\tau_k}^{\tau_{k+1}-1} a_i E\Phi_{i+1}^T \Phi_{i+1} h'(i) - \sum_{i=\tau_k}^{\tau_{k+1}-1} a_i VV^T R_{i+1} h(i)\| \\ > \frac{3}{4} - 2\delta. \qquad (5.1.53)\end{aligned}$$

Define

$$\begin{aligned} j(k) \triangleq \{\min j : j < \tau_{k+1} - \tau_k, \|h'(2L-1) \\ - \sum_{i=\tau_k}^{\tau_k+j-1} a_i E\Phi_{i+1}^T \Phi_{i+1} h'(i) - \sum_{i=\tau_k}^{\tau_k+j-1} a_i VV^T R_{i+1} h(i)\| \geq \frac{3}{4} - 2\delta\}. \\ (5.1.54)\end{aligned}$$

Since $a_i E\Phi_{i+1}^T \Phi_{i+1} h'(i) - a_i VV^T R_{i+1} h(i) \xrightarrow[i\to\infty]{} 0$, $j(k)$ is well-defined by (5.1.54). Notice that from τ_k to $\tau_k + j(k)$ there is no truncation. Consequently,

$$h(\tau_k + j(k)) = h(2L-1) - \sum_{i=\tau_k}^{\tau_k+j(k)-1} a_i S_{i+1},$$

and

$$\begin{aligned} h'(\tau_k + j(k)) = h'(2L-1) - \sum_{i=\tau_k}^{\tau_k+j(k)-1} a_i E\Phi_{i+1}^T \Phi_i h'(i) \\ - \sum_{i=\tau_k}^{\tau_k+j(k)-1} a_i VV^T R_{i+1} h(i), \quad \forall k \geq K. \qquad (5.1.55)\end{aligned}$$

To be fixed, let us take $\delta < \frac{1}{8}$.

From (5.1.52) and (5.1.54) it follows that sequences $\{\|h'(\tau_k + j)\|, j = 0, 1, \ldots, j(k)\}$ starting from $\|h'(2L-1)\| < \frac{1}{4}$ cross the interval $[\frac{1}{4}, \frac{1}{2}]$ for each $k \geq K$. This means that $\{\|h'(\tau_k + i)\|^2, i = 0, 1, \ldots, j(k)\}$ crosses interval $[\frac{1}{16}, \frac{1}{4}]$ for each $k \geq K$.

Here, we call that the sequence $\{\|h'(\tau_k + i)\|^2, i = l_k, l_k + 1, \ldots, m_k\}$ crosses an interval $[a, b]$ with $0 < a < b < \infty$, if $\|h'(\tau_k + l_k)\|^2 \leq a, \|h'(\tau_k + m_k)\|^2 \geq b$, $a < \|h'(\tau_k + i)\|^2 < b$, $\forall i : l_k < i < m_k$, and there is no truncation in the algorithm (5.1.32) for $\forall i : l_k \leq i \leq m_k$.

Without loss of generality, we may assume $h'(\tau_k + l_k)$ converges: $h'(\tau_k + l_k) \xrightarrow[k\to\infty]{} h'$.

It is clear that $\|h'\| = \frac{1}{4}$ and $h'^T h^* = 0$.

By Lemma 5.1.4, there is no truncation for $h(i)$, $i = \tau_k + l_k$, $\tau_k + l_k + 1, \ldots, m(\tau_k + l_k, T) + 1$, if T is small enough.

Then, similar to (2.2.24), for large k, by Lemmas 5.1.3 and 5.1.4 we have

$$\begin{aligned}
&\|h'(m(\tau_k + l_k, T) + 1)\|^2 - \|h'(\tau_k + l_k)\|^2 \\
&= -\sum_{i=\tau_k+l_k}^{m(\tau_k+l_k,T)} h'^T a_i (E\Phi_{i+1}^T \Phi_{i+1} h'(i) + VV^T R_{i+1} h(i)) + o(T) \\
&= -\sum_{i=\tau_k+l_k}^{m(\tau_k+l_k,T)} h'^T a_i E\Phi_{i+1}^T \Phi_{i+1} (h' + h'(i) - h') \\
&\quad - \sum_{i=\tau_k+l_k}^{m(\tau_k+l_k,T)} h'^T a_i VV^T R_{i+1} h(i) + o(T) \\
&= -\sum_{i=\tau_k+l_k}^{m(\tau_k+l_k,T)} h'^T a_i E\Phi_{i+1}^T \Phi_{i+1} h' + o(1) + o(T),
\end{aligned} \tag{5.1.56}$$

where $o(1) \xrightarrow[k\to\infty]{} 0$ and $o(T) \xrightarrow[T\to 0]{} 0$.

By Lemma 5.1.6, for large k and small T we have

$$\|h'(m(\tau_k + l_k, T) + 1)\|^2 - \|h'(\tau_k + l_k)\|^2 \leq -\frac{\epsilon T}{2}\|h'\|^2. \tag{5.1.57}$$

By Lemma 5.1.4 $h'(m(\tau_k + l_k, T) + 1) \xrightarrow[T\to 0]{} h'(\tau_k + l_k)$. Noticing that $\|h'(\tau_k + l_k)\|^2 \leq \frac{1}{16}$ and $\|h'(\tau_k + m_k)\|^2 \geq \frac{1}{4}$, by definition of crossing we see that for small enough T, $m(\tau_k + l_k, T) + 1 < \tau_k + m_k$.

This implies that

$$\|h'(m(\tau_k + l_k, T) + 1)\|^2 > \frac{1}{16}. \tag{5.1.58}$$

Letting $k \longrightarrow \infty$ in (5.1.57), we find that

$$\limsup_{k \longrightarrow \infty} \|h'(m(\tau_k + l_k, T) + 1)\|^2 \leq \frac{1}{16} - \frac{\epsilon T}{32},$$

which contradicts (5.1.58). The contradiction shows that

$$\lim_{k \longrightarrow \infty} \tau_k \triangleq \tau < \infty \quad \text{a.s.}$$

Thus, starting from τ, the algorithm (5.1.32) suffers from no truncation.

If $\|h'(k)\|^2$, $k \geq \tau$, did not converge as $k \longrightarrow \infty$, then $\liminf_{k \longrightarrow \infty} \|h'(k)\|^2 < \limsup_{k \longrightarrow \infty} \|h'(k)\|^2$ and $\|h'(k)\|^2$ would cross a nonempty interval $[a, b]$ infinitely often. But this leads to a contradiction as shown above. Therefore, $\|h'(k)\|$ converges as $k \longrightarrow \infty$.

If $\lim_{k \longrightarrow \infty} \|h'(k)\|$ were not zero, then there would exist a convergent subsequence $h'(k_j) \longrightarrow h' \neq 0$. Replacing $\tau_k + l_k$ in (5.1.56) by k_j, from (5.1.57) it follows that

$$\|h'(m(k_j, T) + 1)\|^2 - \|h'(k_j)\|^2 \leq -\frac{\epsilon T}{2}\|h'\|^2. \tag{5.1.59}$$

Since $\|h'(k)\|^2$ converges, the left-hand side of (5.1.59) tends to zero, which makes (5.1.59) a contradictory inequality. Thus, we have proved $h'(k) \xrightarrow[k \longrightarrow \infty]{} 0$ a.s.

Since $\tau_k \longrightarrow \tau < \infty$, from (5.1.40) it follows that

$$\lim_{k \longrightarrow \infty} \bar{h}(k) = h^T(2L - 1)\bar{h}^* - \sum_{i=\tau}^{\infty} a_i \bar{h}^{*T} R_{i+1} h(i) \triangleq \alpha. \tag{5.1.60}$$

By (5.1.38) and the fact that $h'(k) \xrightarrow[k \longrightarrow \infty]{} 0$, we finally conclude that

$$\lim_{k \longrightarrow \infty} h(k) = \alpha \bar{h}^* \quad \text{a.s.}$$

□

The difficulty of applying the algorithm (5.1.32) consists in that the second moment $E\Xi_k^T \Xi_i$ of the noise may not be available. Identification of channel coefficients without using $E\Xi_i^T \Xi_i$ will be discussed in Section 5.3, by using the principal component analysis to be described in the next section.

5.2. Principal Component Analysis

The principal component analysis (PCA) is one of the basic methods used in feature extraction, signal processing and other areas. Roughly speaking, PCA gives recursive algorithms for finding eigenvectors of a symmetric matrix A based on the noisy observations on A.

Let $\{A_i\}$ be a sequence of observed $l \times l$ symmetric matrices, and $EA_i \xrightarrow[i\to\infty]{} A$. The problem is to find eigenvectors of A, in particular, the one corresponding to the maximal eigenvalue.

Define

$$\tilde{u}_{k+1}^{(1)} = u_k^{(1)} + a_k A_{k+1} u_k^{(1)}, \tag{5.2.1}$$

$$u_{k+1}^{(1)} = \tilde{u}_{k+1}^{(1)} / \|\tilde{u}_{k+1}^{(1)}\|, \quad \text{if } \|\tilde{u}_{k+1}^{(1)}\| \neq 0, \tag{5.2.2}$$

with initial value $u_0^{(1)}$ being a nonzero unit vector. $u_k^{(1)}$ serves as an estimate for unit eigenvector of A.

If $\|\tilde{u}_{k+1}^{(1)}\| = 0$, then $u_k^{(1)}$ is reset to a different vector with norm equal to 1.

Assume $u_k^{(i)}$, $i = 1, \dots, j$, have been defined as estimates for j unit eigenvectors of A. Denote $V_k^{(j)} \triangleq [u_k^{(1)}, P_k^{(1)} u_k^{(2)}, \dots, P_k^{(j-1)} u_k^{(j)}]$ which is an $l \times j$-matrix, where

$$P_k^{(i)} \triangleq I - V_k^{(i)} V_k^{(i)+}, \quad i = 1, \dots, j-1, \tag{5.2.3}$$

where $V_k^{(j)+}$ denotes the pseudo-inverse of $V_k^{(j)}$. Since for large k, $V_k^{(j)}$ is a full-rank $l \times j$ matrix, $l > j$,

$$V_k^{(j)+} = (V_k^{(j)T} V_k^{(j)})^{-1} V_k^{(j)T}. \tag{5.2.4}$$

Define

$$\tilde{u}_{k+1}^{(j+1)} = P_k^{(j)} u_k^{(j+1)} + a_k P_k^{(j)} A_{k+1} P_k^{(j)} u_k^{(j+1)}, \tag{5.2.5}$$

$$u_{k+1}^{(j+1)} = \tilde{u}_{k+1}^{(j+1)} / \|\tilde{u}_{k+1}^{(j+1)}\|, \tag{5.2.6}$$

if $\|\tilde{u}_{k+1}^{(j+1)}\| \geq \epsilon$ with $0 < \epsilon < \frac{1}{4}$, $j = 1, \dots, l-1$.

If $\|\tilde{u}_{k+1}^{(j+1)}\| < \epsilon$, we redefine an $u_k^{(j+1)}$ with $\|u_k^{(j+1)}\| = 1$ such that $\|P_k^{(j)} u_k^{(j+1)}\| = 1$.

Define the estimate $\lambda_k^{(j)}$ for the eigenvalue $\lambda^{(j)}$ corresponding to the eigenvector whose estimate at time k is $u_k^{(j)}$ by the following recursion.

Take an increasingly diverging to infinity sequence $M_k > M_{k-1} > 0$ and define $\lambda_k^{(j)}$ by the SA algorithm with expanding truncations:

$$\begin{aligned}\lambda_{k+1}^{(j)} =& [\lambda_k^{(j)} - a_k(\lambda_k^{(j)} - u_k^{(j)T} A_{k+1} u_k^{(j)})] \\ & \cdot I_{[|\lambda_k^{(j)} - a_k(\lambda_k^{(j)} - u_k^{(j)T} A_{k+1} u_k^{(j)}| > M_{\sigma_k}]}, \qquad j = 1, \dots, l,\end{aligned} \tag{5.2.7}$$

where

$$\sigma_k = \sum_{i=0}^{k-1} I_{[|\lambda_i^{(j)} - a_i(\lambda_i^{(j)} - u_i^{(j)T} A_{i+1} u_i^{(j)}| \geq M_{\sigma_i}]}, \quad \sigma_0 = 0$$

We will use the following conditions:

A5.2.1 $a_k > 0$, $a_k \xrightarrow[k\to\infty]{} 0$, *and* $\sum_{k=1}^{\infty} a_k = \infty$;

A5.2.2 $A_k, k = 1, 2, \dots$ *are symmetric, and* $EA_k \xrightarrow[k\to\infty]{} A$;

A5.2.3 $\sup_k \|A_k\| = \zeta < \infty$, *and*

$$\lim_{T\to 0} \limsup_{k\to\infty} \frac{1}{T} \| \sum_{i=k}^{m(k,T_k)} a_i(A_{i+1} - EA_{i+1}) \| = 0, \quad \forall T_k \in [0, T], \tag{5.2.8}$$

where $m(k, T)$ is given by (1.3.2).

Examples for which (5.2.8) is satisfied are given in Chapters 1 and 2. We now give one more example.

Example 5.2.1 Assume $\{A_i\}$ is stationary and ergodic, $EA_1 = A < \infty$. If $a_i = \frac{a}{i}$, then $\{A_i\}$ satisfies (5.2.8). Set $\xi_k = \sum_{i=1}^{k}(A_i - A)$. By ergodicity, we have $\frac{1}{k}\xi_k \longrightarrow 0$ a.s. By a partial summation it follows that $\sum_{i=k}^{m(k,T)} a_i(A_{i+1} - A) = \sum_{i=k}^{m(k,T)} a_i(\xi_{i+1} - \xi_i) = a_{m(k,T)}\xi_{m(k,T)+1} - a_k\xi_k + \sum_{i=k+1}^{m(k,T)} \xi_i(a_{i-1} - a_i) = \frac{a}{m(k,T)}\xi_{m(k,T)+1} - \frac{a}{k}\xi_k + \sum_{i=k+1}^{m(k,T)} \xi_i \frac{a}{i(i-1)} \xrightarrow[k\to\infty]{} 0$, which implies (5.2.8).

Let ϕ_i be the unit eigenvector of A corresponding to eigenvalue $\lambda^{(i)}$, $i = 1, \dots, l$, where $\lambda^{(i)}$ may not be different.

Theorem 5.2.1 *Assume A5.2.1 and A5.2.2 hold. Then $u_k^{(j)}$ given by (5.2.1)-(5.2.6) converges at those samples (ω) for which A5.2.3 holds, $\forall j : j = 1, \dots, l$, and the limits of $\{u_k^j\}$ coincide with $\{\phi_i, i = 1, \dots, l\}$. Let ϕ_j denote the limit of $u_k^{(j)}$ as $k \longrightarrow \infty$. Then $\lambda_k^{(j)} \longrightarrow \lambda^{(j)}$.*

Proof. Consider those ω for which A5.2.3 holds. We first prove convergence of $u_k^{(1)}$. Note that $\|\tilde{u}_{k+1}^{(1)}\| = 0$ may happen only for a finite number of steps because $a_k \longrightarrow 0$ as $k \longrightarrow \infty$ and $\|u_k^{(1)}\| \equiv 1$. By boundedness of $\{A_k\}$ we expand $u_k^{(1)}$ into the power series of a_k:

$$\begin{aligned}
u_{k+1}^{(1)} =&(u_k^{(1)} + a_k A_{k+1} u_k^{(1)})(1 + 2a_k u_k^{(1)T} A_{k+1} u_k^{(1)} + a_k^2 u_k^{(1)T} A_{k+1}^2 u_k^{(1)})^{-\frac{1}{2}} \\
=&(u_k^{(1)} + a_k A_{k+1} u_k^{(1)})\{1 - a_k u_k^{(1)T} A_{k+1} u_k^{(1)} - \frac{a_k^2}{2} u_k^{(1)T} A_{k+1}^2 u_k^{(1)} \\
&+ \frac{3}{8}[4a_k^2 (u_k^{(1)T} A_{k+1} u_k^{(1)})^2 + 4a_k^3 u_k^{(1)T} A_{k+1} u_k^{(1)} u_k^{(1)T} A_{k+1}^2 u_k^{(1)}] \\
&- \frac{5}{16} \cdot 8a_k^3 (u_k^{(1)T} A_{k+1} u_k^{(1)})^3 + O(a_k^4)\} \\
=&u_k^{(1)} + a_k A_{k+1} u_k^{(1)} - a_k (u_k^{(1)T} A_{k+1} u_k^{(1)}) u_k^{(1)} + a_k \nu_{k+1}^{(1)},
\end{aligned} \tag{5.2.9}$$

where

$$\begin{aligned}
\nu_{k+1}^{(1)} \triangleq &\Bigg[- \frac{1}{2} (u_k^{(1)T} A_{k+1}^2 u_k^{(1)}) u_k^{(1)} \\
&+ \frac{3}{2} (u_k^{(1)T} A_{k+1} u_k^{(1)})^2 u_k^{(1)} - (u_k^{(1)T} A_{k+1} u_k^{(1)}) A_{k+1} u_k^{(1)} \Bigg] a_k \\
&+ \Bigg[\frac{3}{2} (u_k^{(1)T} A_{k+1} u_k^{(1)}) (u_k^{(1)T} A_{k+1}^2 u_k^{(1)}) u_k^{(1)} - \frac{5}{2} (u_k^{(1)T} A_{k+1} u_k^{(1)})^3 u_k^{(1)} \\
&- \frac{1}{2} (u_k^{(1)T} A_{k+1}^2 u_k^{(1)}) A_{k+1} u_k^{(1)} + \frac{3}{2} (u_k^{(1)T} A_{k+1} u_k^{(1)})^2 A_{k+1} u_k^{(1)} \Bigg] a_k^2 \\
&+ O(a_k^3)
\end{aligned} \tag{5.2.10}$$

Further, we rewrite (5.2.9) as

$$u_{k+1}^{(1)} = u_k^{(1)} + a_k (A u_k^{(1)} - (u_k^{(1)T} A u_k^{(1)}) u_k^{(1)}) + a_k (\mu_{k+1}^{(1)} + \nu_{k+1}^{(1)}), \tag{5.2.11}$$

where

$$\mu_{k+1}^{(1)} \triangleq (A_{k+1} - A) u_k^{(1)} + (u_k^{(1)T} (A - A_{k+1}) u_k^{(1)}) u_k^{(1)}. \tag{5.2.12}$$

Denote

$$\epsilon_{k+1}^{(1)} = \mu_{k+1}^{(1)} + \nu_{k+1}^{(1)}.$$

From (5.2.10) and the boundedness of $\{A_k\}$ and $\{u_k^{(1)}\}$ it is seen that $\nu_k^{(1)} \longrightarrow 0$ as $k \longrightarrow \infty$. Therefore, in order to show that $\{\epsilon_i\}$ satisfies A2.2.3 it suffices to show

$$\lim_{T\longrightarrow 0} \limsup_{k\longrightarrow\infty} \frac{1}{T} \| \sum_{i=n_k}^{m(n_k,T_k)} a_i \mu_{i+1}^{(1)} \| = 0, \quad \forall T_k \in [0,T], \tag{5.2.13}$$

for any convergent subsequence $u_{n_k}^{(1)} \longrightarrow \bar{u}$.

By boundedness of $\{A_k\}$ and $\{u_k^{(1)}\}$, it is clear that

$$\|u_i^{(1)} - u_{n_k}^{(1)}\| \le ct, \quad \forall i : n_k \le i \le m(n_k, t) + 1, \tag{5.2.14}$$

where c is a constant for a fixed sample. For any $t \in [0, T]$ there is a k_T such that

$$\|u_{n_k}^{(1)} - \bar{u}\| \le t, \quad \forall k \ge k_T.$$

Consequently, we have

$$\|u_i^{(1)} - \bar{u}\| \le \|u_i^{(1)} - u_{n_k}^{(1)}\| + \|u_{n_k}^{(1)} - \bar{u}\| \le (c+1)t, \tag{5.2.15}$$

$\forall i : n_k \le i \le m(n_k, t) + 1$, $\forall k \ge k_T$.

Expressing the first part of μ_{k+1} as

$$\begin{aligned}(A_{k+1} - A)u_k^{(1)} =&(A_{k+1} - EA_{k+1} + EA_{k+1} - A)(u_k^{(1)} - \bar{u} + \bar{u})\\ =&(A_{k+1} - EA_{k+1})\bar{u} + (A_{k+1} - EA_{k+1})(u_k^{(1)} - \bar{u})\\ &+ (EA_{k+1} - A)u_k^{(1)},\end{aligned} \tag{5.2.16}$$

we find that

$$\lim_{T\longrightarrow 0} \limsup_{k\longrightarrow\infty} \frac{1}{T} \| \sum_{i=n_k}^{m(n_k,T_k)} a_i (A_{i+1} - A) u_i^{(1)} \| = 0, \quad \forall T_k \in [0,T]. \tag{5.2.17}$$

This is because (5.2.8) is applied for the first term on the right-hand side of (5.2.16), while for the other two terms we have used (5.2.15), $EA_{k+1} \xrightarrow[k\longrightarrow\infty]{} A$, and the boundedness of $\{A_k\}$, $\{EA_k\}$, and $\{u_k^{(1)}\}$.

Similar treatment can also be applied to the second part of μ_k. Thus, we have verified (5.2.13), and A2.2.3 too.

Denote by S the unit sphere in $\mathbb{R}^l$. Then $u_k^{(1)}$ defined by (5.2.2) evolves on S.

Define

$$f(u) = Au - (u^T Au)u, \quad u \in S. \tag{5.2.18}$$

The root set of $f(\cdot)$ on S is

$$J \triangleq \{\phi_i, i = 1, \dots, l\}. \tag{5.2.19}$$

Defining $v(u) \triangleq -\frac{1}{2}u^T Au$, we find for $u \in S$

$$\begin{aligned} v_u^T(u)f(u) &= -u^T A[Au - (u^T Au)u] \\ &= -u^T A^2 u + (u^T Au)^2 \\ &\begin{cases} < \|Au\|^2\|u\|^2 - u^T A^2 u = 0, \text{ if } u \notin J, \\ = 0, \text{ if } u \in J. \end{cases} \end{aligned} \tag{5.2.20}$$

Thus, Condition A2.2.2(S) introduced in Remark 2.2.6 is satisfied. Since $\{u_k^{(1)}\}$ is bounded, no truncation is needed. Then, by Remark 2.2.6 we conclude that $u_k^{(1)}$ converges to one of $\{\phi_i\}$, say ϕ_1.

Denote

$$V^{(i)} \triangleq [\phi_1, \dots, \phi_i], \quad P^{(i)} \triangleq I - V^{(i)}V^{(i)+}, \quad P^{(0)} = I.$$

Inductively, we now assume

$$P_k^{(i-1)}u_k^{(i)} \longrightarrow \phi_i, \quad P^{(i-1)}\phi_i = \phi_i, \quad i = 1, \dots, j-1. \tag{5.2.21}$$

We then have

$$V_k^{(i)} \xrightarrow[k\to\infty]{} V^{(i)}, \quad P_k^{(i)} \xrightarrow[k\to\infty]{} P^{(i)}, \quad i = 1, \dots, j-1. \tag{5.2.22}$$

Since $a_k \xrightarrow[k\to\infty]{} 0$ and $\|A_k\| \le \zeta < \infty$, from (5.2.21) and (5.2.5) it follows that $\tilde{u}_k^{(i)} \longrightarrow \phi_i$, and by (5.2.6)

$$u_k^{(i)} \longrightarrow \phi_i, \quad i = 1, \dots, j-1.$$

We now proceed to show that $u_k^{(j)}$ converges to one of unit eigenvectors contained in $J \backslash \{\phi_1, \dots, \phi_{j-1}\}$.

From (5.2.5) we see that the last term in the recursion

$$\tilde{u}_{k+1}^{(j)} = P_k^{(j-1)}u_k^{(j)} + a_k P_k^{(j-1)} A_{k+1} u_k^{(j)}$$

tends to zero as $k \longrightarrow \infty$. So, by (5.2.22) we need to reset $u_k^{(j)}$ with $\|u_k^{(j)}\| = 1$ and $\|P_k^{(j-1)} u_k^{(j)}\| = 1$ at most for a finite number of times.

Replacing $u_k^{(1)}$ by $P_k^{(j-1)} u_k^{(j)}$ in (5.2.9)–(5.2.11), we again arrive at (5.2.11) for j. Precisely,

$$
\begin{aligned}
u_{k+1}^{(j)} = P_k^{(j-1)} u_k^{(j)} &+ a_k (P_k^{(j-1)} A P_k^{(j-1)} u_k^{(j)} \\
&- (u_k^{(j)} P_k^{(j-1)} A P_k^{(j-1)} u_k^{(j)}) P_k^{(j-1)} u_k^{(j)}) + a_k (\mu_{k+1}^{(j)} + \nu_{k+1}^{(j)}), \quad (5.2.23)
\end{aligned}
$$

where

$$
\begin{aligned}
\mu_{k+1}^{(j)} &= (A_{k+1} - A) P_k^{(j-1)} u_k^{(j)} \\
&+ (u_k^{(j)T} P_k^{(j-1)} (A - A_{k+1}) P_k^{(j-1)} u_k^{(j)}) P_k^{(j-1)} u_k^{(j)}, \qquad (5.2.24)
\end{aligned}
$$

and

$$\nu_{k+1}^{(j)} = O(a_k). \qquad (5.2.25)$$

By noticing

$$P_k^{(j-1)} P_k^{(j-1)} = P_k^{(j-1)}$$

and using (5.2.22), (5.2.23) can be rewritten as

$$
\begin{aligned}
P_{k+1}^{(j-1)} u_{k+1}^{(j)} =& P_k^{(j-1)} u_k^{(j)} - V_{k+1}^{(j-1)} V_{k+1}^{(j-1)+} P_k^{(j-1)} u_k^{(j)} + a_k P^{(j-1)} \\
&\cdot \Big[P^{(j-1)} A P^{(j-1)} P_k^{(j-1)} u_k^{(j)} \\
&- (u_k^{(j)T} P_k^{(j-1)} P^{(j-1)} A P^{(j-1)} P_k^{(j-1)} u_k^{(j)}) P_k^{(j-1)} u_k^{(j)} \Big] \\
&+ a_k (\mu_{k+1}^{(j)} + \nu_{k+1}^{(j)} + o(1)), \qquad (5.2.26)
\end{aligned}
$$

where $o(1) \longrightarrow 0$ as $k \longrightarrow \infty$.

Since $u_k^{(1)}$ tends to an eigenvector of A, from (5.2.11) it follows that

$$
\begin{aligned}
V_{k+1}^{(1)} V_{k+1}^{(1)+} - V_k^{(1)} V_k^{(1)+} =& (V_{k+1}^{(1)} - V_k^{(1)}) V_{k+1}^{(1)+} + V_k^{(1)} (V_{k+1}^{(1)+} - V_k^{(1)+}) \\
=& (u_{k+1}^{(1)} - u_k^{(1)}) u_{k+1}^{(1)T} + u_k^{(1)} (u_{k+1}^{(1)} - u_k^{(1)})^T \\
=& o(a_k) + a_k \gamma_{k+1}^{(1)},
\end{aligned}
$$

where

$$\gamma_{k+1}^{(1)} = \epsilon_{k+1}^{(1)} u_{k+1}^{(1)T} + u_k^{(1)} \epsilon_{k+1}^{(1)T}.$$

Since $u_k^{(1)}$ converges, from (5.2.13) and $\nu_k^{(1)} \xrightarrow[k \longrightarrow \infty]{} 0$ it follows that

$$\lim_{T \longrightarrow 0} \limsup_{n \longrightarrow \infty} \frac{1}{T} \| \sum_{i=n}^{m(n,t)} a_i \gamma_{i+1}^{(1)} \| = 0, \quad \forall t \in [0, T]. \qquad (5.2.27)$$

Inductively, assume that

$$V_{k+1}^{(j-1)}V_{k+1}^{(j-1)+} - V_k^{(j-1)}V_k^{(j-1)+} = o(a_k) + a_k\gamma_{k+1}^{(j-1)} \tag{5.2.28}$$

with $\{\gamma_k^{j-1}\}$ satisfying (5.2.27), i.e.,

$$\lim_{T\longrightarrow 0}\limsup_{n\longrightarrow\infty}\frac{1}{T}\|\sum_{i=n}^{m(n,t)} a_i\gamma_{i+1}^{(j-1)}\| = 0, \quad \forall t\in[0,T]. \tag{5.2.29}$$

Noticing that $VV^+VV^+ = VV^+$ for any matrix V, we have

$$\begin{aligned}
&V_{k+1}^{(j-1)}V_{k+1}^{(j-1)+}P_k^{(j-1)}\\
=&V_{k+1}^{(j-1)}V_{k+1}^{(j-1)+}(I - V_{k+1}^{(j-1)}V_{k+1}^{(j-1)+} + V_{k+1}^{(j-1)}V_{k+1}^{(j-1)+}\\
&\quad - V_k^{(j-1)}V_k^{(j-1)+})\\
=&V_{k+1}^{(j-1)}V_{k+1}^{(j-1)+}(V_{k+1}^{(j-1)}V_{k+1}^{(j-1)+} - V_k^{(j-1)}V_k^{(j-1)+})\\
=&o(a_k) + a_kV_{k+1}^{(j-1)}V_{k+1}^{(j-1)+}\gamma_{k+1}^{(j-1)}
\end{aligned}$$

by (5.2.28).

Since $V_{k+1}^{(j-1)}V_{k+1}^{(j-1)+} \longrightarrow V^{(j-1)}V^{(j-1)+}$ by (5.2.24), denoting by $\delta_{k+1}^{(j)}$ the term $(-\frac{1}{a_k}V_{k+1}^{(j-1)}V_{k+1}^{(j-1)+}P_k^{(j-1)}u_k^{(j)})$ we have

$$\lim_{T\longrightarrow 0}\limsup_{k\longrightarrow\infty}\frac{1}{T}\|\sum_{i=n_k}^{m(n_k,T_k)} a_i\delta_{i+1}^{(j)}\| = 0, \quad \forall T_k\in[0,T], \tag{5.2.30}$$

for any convergent subsequence $u_{n_k}^{(j)}$.

Denoting

$$w_k^{(j)} = P_k^{(j-1)}u_k^{(j)},$$

from (5.2.26) we see

$$\begin{aligned}
w_{k+1}^{(j)} =&w_k^{(j)} + a_k\Big[P^{(j-1)}AP^{(j-1)}w_k^{(j)} - (w_k^{(j)T}P^{(j-1)}AP^{(j-1)}w_k^{(j)})w_k^{(j)}\Big]\\
&+ a_k(\delta_{k+1}^{(j)} + \mu_{k+1}^{(j)} + \nu_{k+1}^{(j)} + o(1)).
\end{aligned} \tag{5.2.31}$$

By (5.2.8) and (5.2.30), similar to (5.2.18)–(5.2.20), by Remark 2.2.6 $w_k^{(j)}$ converges to an unit eigenvector of $P^{(j-1)}AP^{(j-1)}$. From (5.2.5) it is seen that $\tilde{u}_k^{(j)}$ converges since $a_k \longrightarrow 0$ and $\|A_k\| \le \zeta < \infty$. Then from (5.2.6) it follows that $u_k^{(j)}$ itself converges as $k \longrightarrow \infty$: $u_k^{(j)} \xrightarrow[k\longrightarrow\infty]{} u^{(j)}$. Thus, we have

$$w_k^{(j)} \xrightarrow[k\longrightarrow\infty]{} P^{(j-1)}u^{(j)}. \tag{5.2.32}$$

From (5.2.5) it follows that

$$V^{(j-1)}V^{(j-1)+}u_{k+1}^{(j)} \xrightarrow[k \to \infty]{} 0,$$

which implies that $V^{(j-1)}V^{(j-1)+}u^{(j)} = 0$, and consequently,

$$P^{(j-1)}u^{(j)} = u^{(j)}. \tag{5.2.33}$$

Since the limit of $w_k^{(j)}$, $P^{(j-1)}u^{(j)}$, is an unit eigenvector of $P^{(j-1)}AP^{(j-1)}$, we have

$$P^{(j-1)}Au^{(j)} - (u^{(j)T}Au^{(j)})u^{(j)} = 0. \tag{5.2.34}$$

By (5.2.33) it is clear that $u^{(j)}$ can be expressed as a linear combination of eigenvectors $\phi_j, \dots, \phi_l$. Consequently, $P^{(j-1)}Au^{(j)} = Au^{(j)}$, which incorporating with (5.2.34) implies that

$$Au^{(j)} - (u^{(j)T}Au^{(j)})u^{(j)} = 0.$$

This means that $u^{(j)}$ is an eigenvector of A, and $u^{(j)}$ is different from $\phi_1, \dots, \phi_{j-1}$ by (5.2.33).

Thus, we have shown (5.2.21) for $i = j$. To complete the induction it remains to show (5.2.28) for j.

As have just shown, $[P^{(i-1)}AP^{(i-1)}w_k^{(i)} - (w_k^{(i)T}P^{(i-1)}AP^{(i-1)}w_k^{(i)})w_k^{(i)}]$ tends to zero as $k \longrightarrow \infty$, from (5.2.31) we have

$$w_{k+1}^{(i)} - w_k^{(i)} = o(a_k) + a_k\alpha_{k+1}^{(i)}, \quad i = 1, \dots, j, \tag{5.2.35}$$

where $\{\alpha_{k+1}^{(i)}\}$ satisfies (5.2.29) with $\gamma_{k+1}^{(j-1)}$ replaced by $\alpha_{k+1}^{(i)}$, by taking notice of that (5.2.30) is fulfilled for whole sequence $\{u_k^{(i)}\}$ because which has been shown to be convergent.

Elementary manipulation leads to

$$\begin{aligned}
&V_{k+1}^{(j)}V_{k+1}^{(j)+} - V_k^{(j)}V_k^{(j)+}\\
&= (V_{k+1}^{(j)} - V_k^{(j)})V_{k+1}^{(j)+} + V_k^{(j)}(V_{k+1}^{(j)+} - V_k^{(j)+})\\
&= (V_{k+1}^{(j)} - V_k^{(j)})V_{k+1}^{(j)+} + V_k^{(j)}[(V_{k+1}^{(j)T}V_{k+1}^{(j)})^{-1}V_{k+1}^{(j)T}\\
&\quad - (V_k^{(j)T}V_k^{(j)})^{-1}V_k^{(j)T}]\\
&= (V_{k+1}^{(j)} - V_k^{(j)})V_{k+1}^{(j)+} + V_k^{(j)}(V_{k+1}^{(j)T}V_{k+1}^{(j)})^{-1}(V_k^{(j)T}V_k^{(j)}\\
&\quad - V_{k+1}^{(j)T}V_{k+1}^{(j)})V_k^{(j)+}\\
&= (V_{k+1}^{(j)} - V_k^{(j)})V_{k+1}^{(j)+} + V_k^{(j)}(V_{k+1}^{(j)T}V_{k+1}^{(j)})^{-1}\\
&\quad \cdot \big((V_k^{(j)T} - V_{k+1}^{(j)T})V_k^{(j)} + V_{k+1}^{(j)T}(V_k^{(j)} - V_{k+1}^{(j)})\big)V_k^{(j)+}.
\end{aligned}$$

This expression incorporating with (5.2.35) proves (5.2.28) for j.

Thus, we have proved that $u_k^{(i)}, i = 1, \ldots, l$, given by (5.2.1)–(5.2.6) converge to different unit eigenvectors of A, respectively.

To complete the proof of the theorem it remains to show $\lambda_k^{(j)} \longrightarrow \lambda^{(j)}$.

Rewrite the untruncated version of (5.2.7) as follows

$$\begin{aligned}\lambda_{k+1}^{(j)} =& \lambda_k^{(j)} + a_k[\lambda^{(j)} - \lambda_k^{(j)} + u_k^{(j)T}((A_{k+1} - EA_{k+1}) + (EA_{k+1} \\ & - A))u_k^{(j)} + u_k^{(j)T} A u_k^{(j)} - \lambda^{(j)}]. \end{aligned} \tag{5.2.36}$$

We have just proved that $u_k^{(j)} A u_k^{(j)} - \lambda^{(j)} \xrightarrow[k \longrightarrow \infty]{} 0$. Then by (5.2.8) and noticing the fact that $u_k^{(j)}$ converges and $EA_k \xrightarrow[k \longrightarrow \infty]{} A$, we see that

$$\epsilon_{k+1} \triangleq u_k^{(j)T} A_{k+1} u_k^{(j)} - \lambda^{(j)}$$

satisfies A2.2.3.

The regression function in (5.2.36) is linear:

$$f(x) = \lambda^{(j)} - x.$$

Applying Theorem 2.2.1 leads to $\lambda_k^{(j)} \longrightarrow \lambda^{(j)}$, $j = 1, \ldots, l$. □

Remark 5.2.1 If in (5.2.1) and (5.2.3) $+a_k$ is replaced by $-a_k$, Theorem 5.2.1 remains valid. In this case $f(u)$ given by (5.2.18) should change to $-Au + (u^T A u)u$, and correspondingly $v(u)$ changes to $\frac{1}{2} u^T A u$. As a result, the limit of $u_k^{(j)}$, $j = 1, \ldots, l$ changes to the opposite sign, from ϕ_j to $-\phi_j$.

5.3. Recursive Blind Identification by PCA

As mentioned in Section 5.1, the algorithm (5.1.32) for identifying channel coefficients uses the second moment $E\Xi_{k+1}^T \Xi_{k+1}$ of the observation noise. This causes difficulty in possible applications, because $E\Xi_{k+1}^T \Xi_{k+1}$ may not be available.

We continue to consider the problem stated in Section 5.1 with notations introduced there. In particular, (5.1.1)–(5.1.12), and (5.1.31) will be used without explanation.

In stead of (5.1.32) we now consider the following normalized SA algorithm:

$$\tilde{h}(k+1) = h(k) - a_k \Psi_{k+1}^T \Psi_{k+1} h(k), \tag{5.3.1}$$

$$h(k+1) = \tilde{h}(k+1) / \|\tilde{h}(k+1)\|. \tag{5.3.2}$$

Comparing (5.3.1) and (5.3.2) with (5.2.1) and (5.2.2), we find that the channel parameter identification algorithm coincides with the PCA algorithm with $A_{k+1} = \Psi_{k+1}^T \Psi_{k+1}$. By Remark 5.2.1, Theorem 5.2.1 can be applied to (5.3.1) and (5.3.2) if conditions A5.2.1, A5.2.2, and A5.2.3 hold.

The following conditions will be used.

A5.3.1 *The input* $\{s_k\}$ *is a* ϕ*-mixing sequence, i.e., there exist a constant* $M \geq 0$ *and a function* $\phi(m) \xrightarrow[m \to \infty]{} 0$ *such that for any* $n \geq 1$

$$\sup_{V \in \mathcal{F}_1^n, U \in \mathcal{F}_{n+m}^\infty} |P(U|V) - P(U)| \leq \phi(m), \quad \forall m \geq M,$$

where $\mathcal{F}_i^j = \sigma\{s_k, i \leq k \leq j\}$.

A5.3.2 *There exists a distribution function* $F_0(\cdot)$ *over* $\mathbb{R}^{2L+1}$ *such that*

$$|\sup_{S \in \mathcal{B}^{2L+1}} P\{(s_{k-2L}, \ldots, s_k) \in S\} - \int_S dF_0(\boldsymbol{\omega})| \xrightarrow[k \to \infty]{} 0,$$

where $\mathcal{B}^{2l+1}$ *denotes the Borel* σ*-algebra in* $\mathbb{R}^{2L+1}$ *and* $\boldsymbol{\omega} = (\omega_1, \ldots, \omega_{2L+1})^T$.

A5.3.3 *The* $(2L+1) \times (2L+1)$*-matrix* $Q = \{q_{ij}\}$ *is nondegenerate, where* $q_{ij} = \int_{\mathbb{R}^{2L+1}} \omega_i \omega_j dF_0(\boldsymbol{\omega})$.

A5.3.4 *The signal* $\{s_k\}$ *is independent of* $\{\eta_k\}$ *and* $|s_k(\omega)| \leq \zeta(\omega) < \infty$ *a.s., where* $\zeta(\omega)$ *is a random variable with* $E\zeta^{2+r} < \infty$.

A5.3.5 *All components of* $\{\eta_k^{(i)}, i = 1, \ldots, p, k = 1, 2, \ldots\}$ *of* $\{\eta_k\}$ *are mutually independent with* $E\eta_k = 0$, $E(\eta_k^{(i)})^3 = 0$, $E(\eta_k^{(i)})^2 = c > 0$, *and* $E((\eta_k^{(i)})^2 - c)^2 > 0$, $\forall i, k$, *and* $\{\eta_k^{(i)}\}$ *is bounded* $\|\eta_k(\omega)\| < \eta < \infty$ *where* η *is a constant.*

A5.3.6 $h^{(j)}(z)$, $j = 1, \ldots,$ *have no common root.*

A5.3.7 $a_k > 0$, $\sum_{k=1}^{\infty} a_k = \infty$, $\sum_{k=1}^{\infty} a_k^2 < \infty$, $a_{k+1}/a_k = 1 + O(a_k)$, *and*

$$\limsup_k a_k k^{\frac{(1+\mu)}{2}} \triangleq \chi < \infty, \quad \mu \in (0, \frac{1}{2}).$$

For Theorem 5.1.1, $\{s_k\}$ is assumed to be a sequence of mutually independent random variables (Condition A5.1.6), while in A5.3.1 the independence is weakened to a ϕ-mixing property, but the distribution of

$(s_{k-L}, \ldots, s_k)$ is additionally required to be convergent. Although there is no requirement on distribution of $\{s_k\}$ in Theorem 5.1.1, we notice that (5.1.30) is satisfied if s_k $k = 1, 2, \ldots$, are identically distributed.

In the sequel, $I_{n\times n}$ denotes the n-dimensional identity matrix.

Define $H = (H_0, \cdots, H_L)$ with

$$H_j = \begin{pmatrix} H_j^{(2)} & -H_j^{(1)} & 0 & 0 & 0 \\ H_j^{(3)} & 0 & -H_j^{(1)} & 0 & 0 \\ \vdots & \vdots & \vdots & \ddots & \vdots \\ H_j^{(p)} & 0 & 0 & 0 & -H_j^{(1)} \\ 0 & H_j^{(3)} & -H_j^{(2)} & 0 & 0 \\ \vdots & \vdots & \vdots & \ddots & \vdots \\ 0 & H_j^{(p)} & 0 & 0 & -H_j^{(2)} \\ \cdots & \cdots & \cdots & \cdots & \cdots \\ \vdots & \vdots & \vdots & \vdots & \vdots \\ \cdots & \cdots & \cdots & \cdots & \cdots \\ 0 & 0 & 0 & H_j^{(p)} & -H_j^{(p-1)} \end{pmatrix}$$

and $H_j^{(i)} = (\underbrace{0, \ldots, 0}_{j}, h_0^{(i)}, \ldots, h_L^{(i)}, \underbrace{0, \ldots, 0}_{L-j})^T$.

In what follows $\otimes$ denotes the Kronecker product.

Theorem 5.3.1 *Assume A5.3.1–A5.3.7 hold. Then*

$$E\{\Phi_k^T \Phi_k\} \xrightarrow[k\to\infty]{} H^T (I_{\frac{p(p-1)}{2}\times\frac{p(p-1)}{2}} \otimes Q) H \triangleq C, \tag{5.3.3}$$

where C is a $p(L+1)\times p(L+1)$-matrix and Q is given in A5.3.3, and for $h(k)$ given by (5.3.1) and (5.3.2),

$$d(h(k), J) \xrightarrow[k\longrightarrow\infty]{} 0 \quad a.s., \tag{5.3.4}$$

where J denotes the set of unit eigenvectors of C.

Proof. By the definition of Φ_k, we have

$$\Phi_k = (I_{\frac{p(p-1)}{2}\times\frac{p(p-1)}{2}}) \otimes (s_k, \ldots, s_{k-2L}))H. \tag{5.3.5}$$

Since

$$(I_{\frac{p(p-1)}{2}\times\frac{p(p-1)}{2}}) \otimes (s_k, \ldots, s_{k-2L}))^T (I_{\frac{p(p-1)}{2}\times\frac{p(p-1)}{2}}) \otimes (s_k, \ldots, s_{k-2L}))$$
$$= I_{\frac{p(p-1)}{2}\times\frac{p(p-1)}{2}}) \otimes ((s_k, \ldots, s_{k-2L})^T (s_k, \cdots, s_{k-2L})),$$

and $E\{(s_k,\dots,s_{k-2L})^T(s_k,\dots,s_{k-2L})\} \xrightarrow[k\to\infty]{} Q$ by A5.3.2, (5.3.3) immediately follows.

From the definition (5.1.31) for Ξ_i, by A5.3.5 it is clear that $E\Xi_k^T\Xi_k$ is a $p(L+1)\times p(L+1)$-identity matrix multiplied by $c(p-1)$ with $c = E(\eta_k^{(i)})^2$. Then by A5.3.4 and A5.3.5

$$E\Psi_k^T\Psi_k = E\Phi_k^T\Phi_k + E\Xi_k^T\Xi_k \xrightarrow[k\longrightarrow\infty]{} C + c(p-1)I_{p(L+1)\times p(L+1)} \triangleq B. \tag{5.3.6}$$

Identifying A_{k+1} in Theorem 5.2.1 to $\Psi_{k+1}^T\Psi_{k+1}$, we find that Theorem 5.2.1 can be applied to the present algorithm, if we can show (5.2.8), which, in the present case, is expressed as

$$\lim_{T\longrightarrow 0}\limsup_{k\longrightarrow\infty}\frac{1}{T}\|\sum_{i=k}^{m(k,T_k)} a_i(\Psi_{i+1}^T\Psi_{i+1} - B)\| = 0 \quad \text{a.s.} \quad \forall T_k \in [0,T], \tag{5.3.7}$$

where $m(k,T)$ is given by (1.3.2), and B is given by (5.3.6).

Notice, by the notation introduced by (5.1.33),

$$\Psi_{k+1}^T\Psi_{k+1} - E\Phi_{k+1}^T\Phi_{k+1} - E\Xi_{k+1}^T\Xi_{k+1} = R_{k+1}.$$

Since

$$\begin{aligned}
&\Psi_{k+1}^T\Psi_{k+1} - E\Phi_{k+1}^T\Phi_{k+1} - E\Xi_{k+1}^T\Xi_{k+1}\\
&\quad = \Phi_{k+1}^T\Phi_{k+1} - E\Phi_{k+1}^T\Phi_{k+1} + \Phi_{k+1}^T\Xi_{k+1} + \Xi_{k+1}^T\Phi_{k+1} + \Xi_{k+1}^T\Xi_{k+1}\\
&\qquad - E\Xi_{k+1}^T\Xi_{k+1},
\end{aligned}$$

$\sup_k E\|\Phi_{k+1}^T\Xi_{k+1} + \Xi_{k+1}^T\Phi_{k+1} + \Xi_{k+1}^T\Xi_{k+1} - E\Xi_{k+1}^T\Xi_{k+1}\|^{1+\frac{\gamma}{2}} < \infty$, and $\sum_{k=1}^{\infty} a_k^{1+\frac{\gamma}{1}} < \infty$, by the convergence theorem for martingale difference sequences, for (5.3.7) it suffices to show

$$\lim_{T\longrightarrow 0}\limsup_{k\longrightarrow\infty}\frac{1}{T}\|\sum_{i=k}^{m(k,T_k)} a_i(\Phi_{i+1}^T\Phi_{i+1} - E\Phi_{i+1}^T\Phi_{i+1})\| = 0, \quad \text{a.s.} \quad \forall T_k \in [0,T]. \tag{5.3.8}$$

Identifying ξ_{i+1} and $f(x_i,\xi_{i+1})$ in Lemma 2.5.2 to $[s_{i-2L},\dots,s_i]^T$, and $\Phi_{i+1}^T\Phi_{i+1}$, respectively, we find that conditions required there are satisfied. Then (5.3.8) follows from Lemma 2.5.2, and hence (5.3.7) is fulfilled.

By Theorem 5.2.1 $h(k)$ given by (5.3.1) and (5.3.2) converges to an unit eigenvector of B, which clearly is an eigenvector of C. □

Lemma 5.3.1 *h^* is the unique up to a scalar multiple nonzero vector simultaneously satisfying*

$$\Phi_k h^* = 0, \quad \forall k = 2L+1, 2L+2, \ldots. \tag{5.3.9}$$

Proof. Since it is known that h^* satisfies (5.3.9), it suffices to prove the uniqueness.

As in the proof of Lemma 5.1.1, assume $\overline{h} \triangleq [(\overline{h}_0)^T, \ldots, (\overline{h}_L)^T]^T$ is also a solution to (5.3.9). Then, along the lines of the proof of Lemma 5.1.1, we obtain the analogue of (5.1.16), which implies (5.1.29):

$$h^T(i,j) E S_{4L}^{(k)}(2L+1) S_{4L}^{(k)}(2L+1) h(i,j) = 0, \quad \forall i,j = 1,\ldots,p, \quad \forall k \geq 2L,$$

where $S_{4L}^{(h)}(2L+1)$ is given by (5.1.28) while $h(i,j)$ by (5.1.16).

By A5.3.3 $E S_{4L}^{(k)}(2L+1) S_{4L}^{(k)}(2L+1) \xrightarrow[k\to\infty]{} Q$, which is nondegenerate. Then we have $h(i,j) = 0$. The rest of proof for uniqueness coincides with that given in Lemma 5.1.1. □

By Lemma 5.3.1 zero is an eigenvalue of C with multiplicity one and the corresponding eigenvector is h^*, $h^*/\|h^*\| \in J$. Theorem 5.3.1 guarantees that the estimate $h(k)$ approaches to J, but it is not clear if $h(k)$ tends to the direction of h^*.

Let $0 = \lambda_1 < \lambda_2 < \cdots < \lambda_m$, $m \leq p(L+1)$ be all different eigenvalues of C. J is composed of disconnected sets $J_s = \{h \in \mathbb{R}^p, \|h\| = 1$ and $Ch = \lambda_s h\}$, $s = 1, \ldots, m$, where $J_1 = \{h^*/\|h^*\|, -h^*/\|h^*\|\}$. Note that the limit points of $h(k)$ are in a connected set, so $h(k)$ converges to a J_s for some s. Let $\Gamma_s = \{\omega, d(h(k), J_s) \xrightarrow[k\to\infty]{} 0\}$. We want to prove that $d(h(k), J_1) \xrightarrow[k\to\infty]{} 0$ a.s. or $P\{\Gamma_1\} = 1$. This is the conclusion of Theorem 5.3.2, which is essentially based on the following lemma, proved in [9].

Lemma 5.3.2 *Let $\{\mathcal{F}_k\}$ be a family of nondecreasing $\sigma-$algebras and $\{\varepsilon_k, \mathcal{F}_k\}$ be a martingale difference sequence with*

$$E\{\|\varepsilon_{k+1}\|^2 | \mathcal{F}_k\} < \infty, \; E\{\varepsilon_{k+1} | \mathcal{F}_k\} = 0.$$

Let $\{\Theta_k, \mathcal{F}_k\}$ be an adapted random sequence and $\{c_k\}$ be a real sequence such that $c_k > 0$, $\sum_k c_k = +\infty$ and $\sum c_k^2 < \infty$. Suppose that on $\Gamma \subset \Omega$, the following conditions 1, 2 and 3 hold.

1) $$\limsup_{k\to\infty} E\{\|\varepsilon_{k+1}\|^2 | \mathcal{F}_k\} < \infty, \quad \liminf_{k\to\infty} E\{\|\varepsilon_{k+1}\| | \mathcal{F}_k\} > 0; \tag{5.3.10}$$

2) Θ_k can be decomposed into two adapted sequences $\{r_k, \mathcal{F}_k\}$ and $\{R_k, \mathcal{F}_k\}$: $\Theta_k = r_k + R_k$ such that

$$\sum_k \|r_k\|^2 < \infty \text{ and } E\{I_\Gamma \sum_{k=n}^{\infty} \|c_k R_k\|\} = o\left(\sum_{k=n}^{\infty} c_k^2\right)^{1/2} \quad \text{as } n \to \infty. \tag{5.3.11}$$

3) $\sum_{k=n}^{\infty} c_k(\Theta_k + \varepsilon_k)$ coincides with an $\mathcal{F}_n$-measurable random variable for some n.

Then $P(\Gamma) = 0$.

Theorem 5.3.2 *Assume A5.3.1–A5.3.7 hold. Then $h(k)$ defined by (5.3.1) and (5.3.2) converges to h^* up-to a constant multiple:*

$$h(k) \xrightarrow[k \longrightarrow \infty]{} \alpha h^* \quad a.s.$$

where α equals either $\|h^*\|^{-1}$ or $-\|h^*\|^{-1}$.

Proof. Assume the contrary: $P\{\Gamma_s\} > 0$ for some $s > 1$, $\lambda_s > 0$. Since C is a symmetric matrix, $h^{*T}h(k) \to 0$ for $\omega \in \Gamma_s$, where and hereafter a possible set with zero probability in Γ_s is ignored. The proof is completed by four steps.

Step 1. We first explicitly express $\theta_n \triangleq h^{*T}h(n)$.

Expanding $h(k)$ defined by (5.3.2) to the power series of a_k, we derive

$$\begin{aligned} h(k+1) = h(k+1) - a_k(&Bh(k) \\ &- (h^T(k)\Psi_{k+1}^T\Psi_{k+1}h(k))h(k) + \mu_{k+1} + \beta_{k+1}), \end{aligned} \tag{5.3.12}$$

where

$$\mu_{k+1} = (\Psi_{k+1}^T\Psi_{k+1} - B)h(k), \quad B = C + (p-1)cI_{p(L+1)\times p(L+1)}, \tag{5.3.13}$$

$$\beta_{k+1} = O(a_k). \tag{5.3.14}$$

Noting $h^{*T}C = 0$ and $h^{*T}\Phi_{k+1} = 0$, we derive

$$\begin{aligned} \theta_{k+1} = \theta_k + a_k((&h^T(k)\Psi_{k+1}^T\Psi_{k+1}h(k) - (p-1)c)\theta_k \\ &- h^{*T}\mu_{k+1} - h^{*T}\beta_{k+1}), \end{aligned} \tag{5.3.15}$$

and

$$\begin{aligned} h^{*T}\mu_{k+1} &= h^{*T}(\Psi_{k+1}^T\Psi_{k+1} - B)h(k) \\ &= h^{*T}(\Xi_{k+1}^T\Xi_{k+1} + \Xi_{k+1}^T\Phi_{k+1})h(k) - (p-1)c\theta_k \end{aligned}$$

$$= (\sum_{i=0}^{L} h_i^T N_{k+1-i}^T)(\sum_{i=0}^{L} N_{k+1-i} h_i(k))$$
$$+ (\sum_{i=0}^{L} h_i^T N_{k+1-i}^T)(\sum_{i=0}^{L} X_{k+1-i} h_i(k)) - (p-1)c\theta_k, \tag{5.3.16}$$

where h_i, $i = 0, \ldots, L$, is defined by (5.1.4), N_k is given by (5.1.10) with $x_k^{(i)}$ replaced by $\eta_k^{(i)}$, the observation noise, and $h_i(k)$ denotes the estimate for h_i at time k.

By (5.3.4) and (5.3.5), there exists $\alpha(\omega) < \infty$ a.s. such that $\left\|\Psi_{k+1}^T \Psi_{k+1} - (p-1)cI_{p(L+1)\times p(L+1)}\right\| < \alpha(\omega)$ a.s.

For any integers m and n define $\Gamma_m = \{\omega, \alpha(\omega) < m\} \cap \Gamma_s$ and

$$B_n = \prod_{k=n_0}^{n} \{1 + a_k(h^T(k)(\Psi_{k+1}^T \Psi_{k+1} - (p-1)cI_{p(L+1)\times p(L+1)})h(k))\}. \tag{5.3.17}$$

Note that for $\omega \in \Gamma_m$,

$$h^T(k)Ch(k) \xrightarrow[k\to\infty]{} \lambda_s > 0,$$

and by the convergence of $h(k)$ from (5.3.12) it follows that $\|h(j) - h(k)\| < c_0 T$, $\forall j : k \le j \le m(k,T)$, where c_0 is a constant for all ω in Γ_m. By (5.3.7) we then have

$$\left|\sum_{k=j}^{m(j,T)} a_k\big(h^T(k)\big(\Psi_{k+1}^T\Psi_{k+1} - B\big)h(k)\big)\right|$$
$$\le \left\|\sum_{k=j}^{m(j,T)} a_k(\Psi_{k+1}^T\Psi_{k+1} - B)\right\| + 2c_0 Tm \sum_{k=j}^{m(j,T)} a_k = o(T)$$

as $T \longrightarrow 0$, where and hereafter T should not be confused with the superscript T for transpose.

Choose large enough n_0 and sufficiently small T such that $o(T)/T < \lambda_s/4$, $\forall j \ge n_0$. Let $k_0 = n_0$, $k_1 = m(n_0, T) + 1$, $k_2 = m(k_1, T) + 1$, $\ldots$, $k_{j+1} = m(k_j, T) + 1, \ldots$, and $m(k_l, T) \le n \le m(k_{l+1}, T)$. It then follows that for $\omega \in \Gamma_m$

$$\ln B_n$$
$$= \ln \prod_{k=n_0}^{n} \{1 + a_k(h(k)^T(\Psi_{k+1}^T\Psi_{k+1} - (p-1)cI_{p(L+1)\times p(L+1)})h(k))\}$$

$$= \sum_{k=n_0}^{n} a_k (h^T(k)(\Psi_{k+1}^T \Psi_{k+1} - (p-1)cI_{p(L+1)\times p(L+1)})h(k))$$
$$+ O(\sum_{k=n_0}^{n} a_k^2)$$
$$= \sum_{k=n_0}^{n} h^T(k)Ch(k)a_k + \sum_{k=n_0}^{n} a_k(h^T(k)(\Psi_{k+1}^T\Psi_{k+1} - B)h(k))$$
$$+ O(\sum_{k=n_0}^{n} a_k^2)$$
$$\geq \sum_{j=0}^{l} \sum_{k=k_j}^{m(k_j,T)} \frac{\lambda_s}{2} a_k > \frac{\lambda_s}{3} \sum_{k=n_0}^{n} a_k \tag{5.3.18}$$

for n_0 sufficiently large.

Consequently, for $\omega \in \Gamma_m$ with fixed m

$$B_n \geq e^{\frac{\lambda_s}{3} \sum_{k=n_0}^{n} a_k} \tag{5.3.19}$$

and hence

$$B_n / (\sum_{k=n_0}^{n} a_k)^2 \xrightarrow[n\to\infty]{} \infty. \tag{5.3.20}$$

Define

$$\Lambda_l = \{\omega, B_n > \big(\sum_{k=1}^{n} a_k\big)^2, \forall n \geq l\}.$$

From (5.3.15) it follows that

$$\theta_n = B_{n-1}(\theta_{n_0} - \sum_{j=n_0}^{n-1} B_j^{-1} a_j (h^{*T}\mu_{j+1} + h^{*T}\beta_{j+1})). \tag{5.3.21}$$

Tending $n \to \infty$ in (5.3.21) and replacing n_0 by n in the resulting equality, by (5.3.19) we have

$$\theta_n = \sum_{j=n}^{\infty} B_j^{-1} a_j (h^{*T}\mu_{j+1} + h^{*T}\beta_{j+1}),\ \forall n,\ \omega \in \Gamma_m \cap \Lambda_l. \tag{5.3.22}$$

Thus, we have expressed θ_n in two ways: (5.3.21) shows that θ_n is $\mathcal{F}_n$-measurable, while (5.3.22) is in the form required in 5.3.2, where
$\mathcal{F}_k = \sigma\{\xi_l, l = 0, \ldots, k, s_l, l = 0, \ldots, k + 2L + 1\}$.

Step 2. In order to show that the summand in (5.3.22) can be expressed as that required in Lemma 5.3.2 we first show that the series

$$S_n \triangleq \sum_{k=n}^{\infty} a_k (h^{*T} \mu_{k+1} + h^{*T} \beta_{k+1}) \tag{5.3.23}$$

is convergent on Γ_s. By (5.3.14) and (5.3.7) it suffices to show $\sum_{k=n}^{\infty} a_k \mu_{k+1}$ is convergent on Γ_s.

Define

$$\varepsilon_{k+1}^{(1)} = \sum_{i=0}^{L} (h_i^T N_{k+1}^T)(N_{k+1} h_i(k)) - (p-1)c\theta_k, \tag{5.3.24}$$

$$\begin{aligned} \varepsilon_{k+1}^{(2)} = \sum_{i=0}^{L-1} \Big[& (h_i^T N_{k+1}^T)(\sum_{l=i+1}^{L} N_{k+i+1-l} h_l(k)) \\ & + (\sum_{l=i+1}^{L-1} h_l^T N_{k+i+1-l}^T)(N_{k+1} h_i(k)) \Big], \end{aligned} \tag{5.3.25}$$

$$\varepsilon_{k+1}^{(3)} = \sum_{i=0}^{L} [(h_i^T N_{k+1}^T)(\sum_{l=0}^{L} X_{k+i+1-l} h_l(k)), \tag{5.3.26}$$

and

$$\delta_{k+1} = \sum_{i=1}^{3} \varepsilon_{k+1}^{(i)}. \tag{5.3.27}$$

Clearly, δ_k is measurable with respect to $\mathcal{F}_k$ and $E\{\delta_{k+1} | \mathcal{F}_k\} = 0$. Then by the convergence theorem for martingale difference sequences,

$$\sum_{k=n}^{\infty} a_k \delta_{k+1} < \infty. \tag{5.3.28}$$

By (5.3.16) it follows that

$$\begin{aligned} & \sum_{k=n}^{\infty} a_k [h^{*T} \mu_{k+1} + (p-1)c\theta_k] \\ & = \sum_{k=n}^{\infty} a_k \Big[\sum_{i=0}^{L} (h_i^T N_{k+1-i}^T)(\sum_{s=0}^{L} N_{k+1-s} h_s(k)) \\ & \qquad + (\sum_{i=0}^{L} h_i^T N_{k+1-i}^T)(\sum_{s=0}^{L} X_{k+1-s} h_s(k)) \Big] \end{aligned}$$

$$= \sum_{i=0}^{L}\sum_{k=n}^{\infty}[a_k h_i^T N_{k+1-i}^T(\sum_{s=0}^{L} N_{k+1-s}h_s(k))$$

$$+ a_k h_i^T N_{k+1-i}^T(\sum_{s=0}^{L} X_{k+1-s}h_s(k))]$$

$$= \sum_{i=0}^{L}\sum_{l=n-i}^{\infty}[a_{l+i}h_i^T N_{l+1}^T(\sum_{s=0}^{L} N_{l+i+1-s}h_s(l+i))$$

$$+ a_{l+i}h_i^T N_{l+1}^T(\sum_{s=0}^{L} X_{l+i+1-s}h_s(l+i))]. \tag{5.3.29}$$

The first term on the right-hand side of the last equality of (5.3.29) can be expressed in the following form:

$$\sum_{i=0}^{L}\sum_{l=n-i}^{\infty} a_{l+i}(h_i^T N_{l+1}^T)(N_{l+1}h_i(l+i))$$

$$+\sum_{i=0}^{L-1}\sum_{l=n-i}^{\infty} a_{l+i}(h_i^T N_{l+1}^T)(\sum_{s=i+1}^{L} N_{l+i+1-s}h_s(l+i))$$

$$+\sum_{i=1}^{L}\sum_{l=n-i}^{\infty} a_{l+i}(h_i^T N_{l+1}^T)(\sum_{s=0}^{i-1} N_{l+i+1-s}h_s(l+i)), \tag{5.3.30}$$

where the last term equals

$$\sum_{s=0}^{L-1}\sum_{i=s+1}^{L-1}\sum_{l=n-i}^{\infty} a_{l+i}(h_i^T N_{l+1}^T)(N_{l+i+1-s}h_s(l+i))$$

$$= \sum_{s=0}^{L-1}\sum_{i=s+1}^{L-1}\sum_{m=n-s}^{\infty} a_{m+s}(h_i^T N_{m-i+s+1}^T)(N_{m+1}h_s(m+s)). \tag{5.3.31}$$

Combining (5.3.30) and (5.3.31) we derive that the first term on the right-hand side of the last equality of (5.3.29) is

$$\sum_{i=0}^{L}\sum_{l=n-i}^{\infty} a_{l+i}(h_i^T N_{l+1}^T)(N_{l+1}h_i(l+i))$$

$$+\sum_{i=0}^{L-1}\sum_{l=n-i}^{\infty}\sum_{s=i+1}^{L-1} a_{l+i}[(h_i^T N_{l+1}^T)(N_{l+i+1-s}h_s(l+i))$$

$$+ (h_s^T N_{l+i+1-s}^T)(N_{l+1}h_i(l+i))]. \tag{5.3.32}$$

By A5.3.4, A5.3.5, and A5.3.7 it is clear that $\|h(l+i)-h(l)\| = O(a_l)$, $\forall i,\ 0 \le i \le L$. Hence replacing $h(l+i)$ by $h(l)$ in (5.3.29) results in producing an additional term of magnitude $O(a_l)$. Thus, by (5.3.24)–(5.3.26) we can rewrite (5.3.29) as

$$\sum_{k=n}^{\infty} a_k h^{*T}\mu_{k+1} = \sum_{k=n}^{\infty} a_k\Big(\sum_{i=1}^{3}\varepsilon_{k+1}^{(i)} + \nu_{k+1}\Big) = \sum_{k=n}^{\infty} a_k(\delta_{k+1}+\nu_{k+1}), \tag{5.3.33}$$

where $\nu_{k+1} = O(a_{k+1})$ and is $\mathcal{F}_{k+1}$-measurable. By (5.3.28) and A5.3.7 the series (5.3.33) is convergent, and hence S_n given by (5.3.23) is a convergent series.

Step 3. We now define sequences corresponding to r_k, R_k, and ϵ_k in Lemma 5.3.2.

Let $B_{n-1} = I$. We have

$$\begin{aligned}
\theta_n &= \sum_{k=n}^{\infty} B_k^{-1}(S_k - S_{k+1}) = \sum_{k=n}^{\infty}(B_k^{-1} - B_{k-1}^{-1})S_k + S_{n_0}\\
&= \sum_{k=n}^{\infty}[(B_k^{-1} - B_{k-1}^{-1})S_k + a_k(h^{*T}\mu_{k+1} + h^{*T}\beta_{k+1})]\\
&= \sum_{j=0}^{\infty}\Big[\sum_{l=j(L+1)+n}^{(j+1)(L+1)+n-1} R_l^1 + a_{j(L+1)+n}\sum_{l=j(L+1)+n}^{(j+1)(L+1)+n-1}(\delta_{l+1}+\tilde{\nu}_{l+1}\\
&\quad + h^{*T}\beta_{l+1})\Big],
\end{aligned}$$

where $R_j^1 = (B_j^{-1} - B_{j-1}^{-1})S_j$,

$$\begin{aligned}
\tilde{\nu}_{l+1} &= \Big(\frac{a_l}{a_{j(L+1)+n}} - 1\Big)(\delta_{l+1} + \nu_{l+1} + h^{0T}\beta_{l+1}) + \nu_{l+1}\\
&= O(a_{j(L+1)+n}),\quad \forall l : j(L+1)+n \le l < (j+1)(L+1)+n.
\end{aligned}$$

Denote

$$\begin{aligned}
R_j &= \frac{1}{c_j}\sum_{l=j(L+1)+n}^{(j+1)(L+1)+n-1} R_l^1,\\
r_j &= \sum_{l=j(L+1)+n}^{(j+1)(L+1)+n-1}(\tilde{\nu}_{l+1} + h^{*T}\beta_{l+1}),\ \varepsilon_j = \sum_{l=j(L+1)+n}^{(j+1)(L+1)+n-1}\delta_{l+1},\\
c_j &= a_{j(L+1)+n},\ \text{and } \mathcal{F}_j' \triangleq \mathcal{F}_{(j+1)(L+1)+n}.
\end{aligned}$$

Then $\{R_j, \mathcal{F}'_j\}$ and $\{r_j, \mathcal{F}'_j\}$ are adapted sequences, $\{\varepsilon_j, \mathcal{F}'_j\}$ is a martingale difference sequence, and θ_n is written in the form of Lemma 5.3.2: $\theta_n = \sum_{j=n}^{\infty} c_j(R_j + r_j + \varepsilon_j)$.

It remains to verify (5.3.10) and (5.3.11).

From (5.3.23) and (5.3.33) it follows that there is a constant $\eta > 0$ such that $E\{|S_n|^2\} \leq \eta \sum_{k=n}^{\infty} a_k^2$. Then for $n > l$ noticing

$$\left|B_j^{-1} - B_{j-1}^{-1}\right| \leq B_j^{-1} a_j \left|(h^T(j)\Psi_{j+1}^T \Psi_{j+1} h(j)) - (p-1)c\right|,$$

and

$$\begin{aligned}\sum_{j=n}^{\infty}(E\{I_{\Gamma_m \cap \Lambda_l} \left|B_j^{-1} - B_{j-1}^{-1}\right|^2\})^{1/2} &\leq \sum_{j=n}^{\infty}(E\{1_{\Gamma_m \cap \Lambda_l} B_j^{-2} a_j^2 (m)^2\})^{1/2} \\ &\leq m \sum_{j=n}^{\infty} \frac{a_j}{(\sum_{k=1}^{j} a_k)^2} \leq \int_{\sum_{k=1}^{n-1} a_k}^{\infty} \frac{1}{x^2} dx < \infty,\end{aligned}$$

we have

$$\begin{aligned}E\{I_{\Gamma_m \cap \Lambda_l} \sum_{j=n}^{\infty} |c_j R_j|\} &\leq E\{I_{\Gamma_m \cap \Lambda_l} \sum_{k=n}^{\infty} |R_k^1|\} \\ &= E\{I_{\Gamma_m \cap \Lambda_l} \sum_{k=n}^{\infty} |B_k^{-1} - B_{k-1}^{-1}| \, |S_k|\} \\ &\leq \sum_{k=n}^{\infty} (E\{I_{\Gamma_m \cap \Lambda_l} |B_k^{-1} - B_{k-1}^{-1}|^2\} E\{I_{\Gamma_m \cap \Lambda_l} |S_k|^2\})^{1/2} \\ &\leq o\left(\sum_{k=n}^{\infty} a_k^2\right)^{1/2} = o\left(\sum_{j=n}^{\infty} c_j^2\right)^{1/2} \quad \text{as } n \longrightarrow \infty.\end{aligned}$$

By A5.3.4 and A5.3.5 it follows that

$$\limsup_{k\to\infty} E\{\delta_{k+1}^{2+\gamma} | \mathcal{F}_k\} < \infty \text{ for some } \gamma > 0. \tag{5.3.34}$$

As in Step 4 it will be shown that

$$\liminf_{k\to\infty} E\left\{ | \sum_{l=k}^{k+L} \delta_{l+1} |^2 | \mathcal{F}_k \right\} > 0. \tag{5.3.35}$$

From this it follows that

$$\liminf_{k\to\infty} E\{|\varepsilon_{k+1}|^2 \,|\mathcal{F}'_k\}$$
$$= \liminf_{k\to\infty} E\{|\sum_{l=(k+1)(L+1)+n}^{(k+2)(L+1)+n} \delta_{l+1}|^2|\mathcal{F}_{(k+1)(L+1)+n}\} > 0. \qquad (5.3.36)$$

Then from the following inequality

$$E\{|\varepsilon_{k+1}|^2 \,|\mathcal{F}'_k\} < E\{|\varepsilon_{k+1}|^{2+\gamma} \,|\mathcal{F}'_k\}^{\frac{1}{1+\gamma}} E\{|\varepsilon_{k+1}| \,|\mathcal{F}'_k\}^{\frac{\gamma}{1+\gamma}}$$

by (5.3.34) and (5.3.36) it follows that

$$\liminf_{k\to\infty} E\{|\varepsilon_{k+1}| \,|\mathcal{F}'_k\} > 0.$$

Therefore all conditions required in Lemma 5.3.2 are met, and we conclude $P\{\Gamma_m \cap \Lambda_l\} = 0$. Since $\Gamma_s = \bigcup_{m,l} \Gamma_m \cap \Lambda_l$, it follows that $P\{\Gamma_s\} = 0$, and $h(k)$ must converge to αh^*, a.s.

Step 4. To complete the proof we have to show (5.3.35).

Proof. If (5.3.35) were not true, then there would exist a subsequence $\{k_n\}$ such that

$$E\Big\{|\sum_{l=k_n}^{k_n+L} \delta_{l+1}|^2 \Big|\mathcal{F}_{k_n}\Big\} \xrightarrow[n\to\infty]{} 0. \qquad (5.3.37)$$

For notational simplicity, let us denote the subsequence k_n still by k.

Since by A5.3.5 $E\{\varepsilon^{(s)}_{j+1}\varepsilon^{(t)}_{i+1}|\mathcal{F}_k\} = 0$ for j, $i \geq k$ if $s \neq t$, and for any j, $i \geq k$ but $j \neq i$ if $s = t$, we then have

$$E\{|\sum_{l=k}^{k+L} \delta_{l+1}|^2|\mathcal{F}_k\} = \sum_{l=k}^{k+L}[E\{(\varepsilon^{(1)}_{l+1})^2 \,|\mathcal{F}_k\} + E\{(\varepsilon^{(2)}_{l+1})^2 \,|\mathcal{F}_k\}$$
$$+ E\{(\varepsilon^{(3)}_{l+1})^2 \,|\mathcal{F}_k\}],$$

which incorporating with (5.3.37) implies that

$$E\{(\varepsilon^{(1)}_{k+1})^2 \,|\mathcal{F}_k\} \xrightarrow[k\to\infty]{} 0, \qquad (5.3.38)$$

and

$$E\{(\varepsilon^{(2)}_{k+L+1})^2 \,|\mathcal{F}_k\} \xrightarrow[k\to\infty]{} 0. \qquad (5.3.39)$$

Noticing that $\theta_k \xrightarrow[k\to\infty]{} 0$ and $|h_i(j-L)-h_i(j)| = O(a_j)$, from (5.3.38) and (5.3.24) it follows that

$$E\{(\sum_{i=0}^{L}(h_i^T N_{k+1}^T)(N_{k+1}h_i(k)))^2|\mathcal{F}_k\} \xrightarrow[k\to\infty]{} 0.$$

On the other hand, we have

$$
\begin{aligned}
&\sum_{i=0}^{L}(h_i^T N_{k+1}^T)(N_{k+1}h_i(k)) \\
&= \sum_{i=0}^{L}\sum_{n=1}^{p}\sum_{m=n+1}^{p}(h_i^{(n)}\xi_{k+1}^{(m)} - h_i^{(m)}\xi_{k+1}^{(n)})(h_i^{(n)}(k)\xi_{k+1}^{(m)} - h_i^{(m)}(k)\xi_{k+1}^{(n)}) \\
&= \sum_{i=0}^{L}\sum_{n=1}^{p}\sum_{\substack{m=1\\ m\neq n}}^{p}(h_i^{(n)}h_i^{(n)}(k)\xi_{k+1}^{(m)}\xi_{k+1}^{(m)} - h_i^{(m)}h_i^{(n)}(k)\xi_{k+1}^{(n)}\xi_{k+1}^{(m)}),
\end{aligned}
$$

and hence,

$$
\begin{aligned}
E\{[&\sum_{i=0}^{L}\sum_{m=1}^{p}\sum_{\substack{n=1\\ n\neq m}}^{p}(h_i^{(n)}h_i^{(n)}(k)\xi_{k+1}^{(m)}\xi_{k+1}^{(m)} \\
&- h_i^{(m)}h_i^{(n)}(k)\xi_{k+1}^{(n)}\xi_{k+1}^{(m)})]^2|\mathcal{F}_k\} \xrightarrow[k\to\infty]{} 0,
\end{aligned}
\tag{5.3.40}
$$

where $h_i^{(n)}(k)$ denotes the estimate provided by $h(k)$ for $h_i^{(n)}$ at time k.

Since for any $s \neq t$,

$$E\{(h_i^{(n)}h_i^{(n)}(j)\xi_{j+1}^{(m)}\xi_{j+1}^{(m)})(h_i^{(t)}h_i^{(s)}(j)\xi_{j+1}^{(s)}\xi_{j+1}^{(t)})\,|\mathcal{F}_j\} = 0,$$

we have

$$
\begin{aligned}
E\{[&\sum_{m=1}^{p}\sum_{\substack{n=1\\ n\neq m}}^{p}((\sum_{i=0}^{L}h_i^{(n)}h_i^{(n)}(k))\xi_{k+1}^{(m)}\xi_{k+1}^{(m)} \\
&- (\sum_{i=0}^{L}h_i^{(m)}h_i^{(n)}(k))\xi_{k+1}^{(n)}\xi_{k+1}^{(m)})]^2\,|\mathcal{F}_k\} \\
={}& E\{[\sum_{m=1}^{p}\sum_{\substack{n=1\\ n\neq m}}^{p}(\sum_{i=0}^{L}h_i^{(n)}h_i^{(n)}(k))\xi_{k+1}^{(m)}\xi_{k+1}^{(m)}]^2\,|\mathcal{F}_k\} \\
&+ E\{[\sum_{m=1}^{p}\sum_{\substack{n=1\\ n\neq m}}^{p}(\sum_{i=0}^{L}h_i^{(m)}h_i^{(n)}(k))\xi_{k+1}^{(n)}\xi_{k+1}^{(m)})]^2\,|\mathcal{F}_k\}.
\end{aligned}
$$

Hence (5.3.40) implies that

$$E\{[\sum_{m=1}^{p}\sum_{\substack{n=1\\ n\neq m}}^{p}(\sum_{i=0}^{L}h_i^{(n)}h_i^{(n)}(k))(\xi_{k+1}^{(m)})^2]^2\,|\mathcal{F}_k\} \xrightarrow[k\to\infty]{} 0, \tag{5.3.41}$$

and

$$E\{[\sum_{m=1}^{p}\sum_{\substack{n=1\\n\neq m}}^{p}(\sum_{i=0}^{L}h_i^{(m)}h_i^{(n)}(k))\xi_{k+1}^{(n)}\xi_{k+1}^{(m)})]^2\,|\mathcal{F}_k\}\xrightarrow[k\to\infty]{}0. \tag{5.3.42}$$

By A5.3.4 the left-hand side of (5.3.41) equals

$$\begin{aligned}
&E\{[\sum_{m=1}^{p}(\sum_{n=1}^{p}\sum_{i=0}^{L}h_i^{(n)}h_i^{(n)}(k)-\sum_{i=0}^{L}h_i^{(m)}h_i^{(m)}(k))(\xi_{k+1}^{(m)})^2]^2\,|\mathcal{F}_k\}\\
&=E\{[\sum_{m=1}^{p}(\theta_k-\sum_{i=0}^{L}h_i^{(m)}h_i^{(m)}(k))(\xi_{k+1}^{(m)})^2]^2\,|\mathcal{F}_k\}\\
&=\sum_{m=1}^{p}(\theta_k-\sum_{i=0}^{L}h_i^{(m)}h_i^{(m)}(k))^2E\{((\xi_{k+1}^{(m)})^2-c)^2\}\\
&\quad+c^2(\sum_{m=1}^{p}(\theta_k-\sum_{i=0}^{L}h_i^{(m)}h_i^{(m)}(k)))^2\\
&=\sum_{m=1}^{p}(\theta_k-\sum_{i=0}^{L}h_i^{(m)}h_i^{(m)}(k))^2E\{((\xi_{k+1}^{(m)})^2-c)^2\}+(p-1)^2c^2\theta_k^2.
\end{aligned}$$

Since $\theta_k\to 0$, it follows that for any m,

$$\sum_{i=0}^{L}h_i^{(m)}h_i^{(m)}(k)\xrightarrow[k\to\infty]{}0. \tag{5.3.43}$$

The left side of (5.3.42) equals

$$\begin{aligned}
&E\{\sum_{m=1}^{p}\sum_{\substack{n=1\\n\neq m}}^{p}(\sum_{i=0}^{L}(h_i^{(m)}h_i^{(n)}(k)+h_i^{(n)}h_i^{(m)}(k)))^2(\xi_{k+1}^{(n)}\xi_{k+1}^{(m)})^2\,|\mathcal{F}_k\}\\
&=c^2\sum_{m=1}^{p}\sum_{n=1}^{m-1}(\sum_{i=0}^{L}h_i^{(m)}h_i^{(n)}(k)+h_i^{(n)}h_i^{(m)}(k))^2.
\end{aligned}$$

Thus (5.3.42) implies that for any $m\neq n$,

$$\sum_{i=0}^{L}(h_i^{(m)}h_i^{(n)}(k)+h_i^{(n)}h_i^{(m)}(k))\xrightarrow[k\to\infty]{}0. \tag{5.3.44}$$

Noticing $|h_{k+i} - h(k)| = O(a_k)$ $\forall i = 1, \ldots, L$, from (5.3.25) we have

$$\begin{aligned}
\varepsilon^{(2)}_{k+L+1} &= \sum_{i=0}^{L-1} \sum_{l=0}^{L-i-1} [(h_i^T N_{k+L+1}^T)(N_{k+L-l} h_{l+i+1}(k)) \\
&\quad + (h_{l+i+1}^T N_{k+L-l}^T)(N_{k+1} h_i(k))] \\
&= \sum_{l=0}^{L-1} \sum_{i=0}^{L-l-1} [(h_i^T N_{k+L+1}^T)(N_{k+L-l} h_{l+i+1}(k)) \\
&\quad + (h_{l+i+1}^T N_{k+L-l}^T)(N_{k+L+1} h_i(k))] + O(a_k).
\end{aligned}$$

Then by A5.3.5, (5.3.39) implies that for any l

$$\begin{aligned}
E\Bigg\{ \Big[\sum_{i=0}^{L-l-1} & [(h_i^T N_{k+L+1}^T)(N_{k+L-l} h_{l+i+1}(k)) \\
& + (h_{l+i+1}^T N_{k+L-l}^T)(N_{k+L+1} h_i(k))] \Big]^2 | \mathcal{F}_k \Bigg\} \xrightarrow[k\to\infty]{} 0. \qquad (5.3.45)
\end{aligned}$$

Notice that

$$\begin{aligned}
&(h_i^T N_{k+L+1}^T)(N_{k+L-l} h_{l+i+1}(k)) \\
&= \sum_{m=1}^{p} \sum_{n=1}^{m-1} (h_i^{(n)} \xi_{k+L+1}^{(m)} - h_i^{(m)} \xi_{k+L+1}^{(n)})(h_{l+i+1}^{(n)}(k) \xi_{k+L-l}^{(m)} \\
&\qquad - h_{l+i+1}^{(m)}(k) \xi_{k+L-l}^{(n)}) \\
&= \sum_{m=1}^{p} \sum_{\substack{n=1 \\ n\neq m}}^{p} (h_i^{(n)} h_{l+i+1}^{(n)}(k) \xi_{k+L+1}^{(m)} \xi_{k+L-l}^{(m)} - h_i^{(n)} h_{l+i+1}^{(m)}(k) \xi_{k+L+1}^{(m)} \xi_{k+L-l}^{(n)}),
\end{aligned} \qquad (5.3.46)$$

and

$$\begin{aligned}
&(h_{l+i+1}^T N_{k+L-l}^T)(N_{k+L+1} h_i(k)) \\
&= \sum_{m=1}^{p} \sum_{\substack{n=1 \\ n\neq m}}^{p} (h_{l+i+1}^{(n)} h_i^{(n)}(k) \xi_{k+1}^{(m)} \xi_{k+L-l}^{(m)} - h_{l+i+1}^{(m)} h_i^{(n)}(k) \xi_{k+L+1}^{(m)} \xi_{k+L-l}^{(n)}).
\end{aligned} \qquad (5.3.47)$$

Then by A5.3.5, from (5.3.45)–(5.3.47) it follows that

$$E\Big\{\Big[\sum_{i=0}^{L-l-1}[(h_i^T N_{k+L+1}^T)(N_{k+L-l}h_l(k))$$

$$+(h_{l+i+1}^T N_{k+L-l}^T)(N_{k+L+1}h_i(k))]\Big]^2\Big\}$$

$$=c^2\sum_{m=1}^{p}[(\sum_{i=0}^{L-l-1}\sum_{\substack{n=1\\n\neq m}}^{p}(h_i^{(n)}h_{l+i+1}^{(n)}(k)+h_{l+i+1}^{(n)}h_i^{(n)}(k)))^2$$

$$+\sum_{\substack{n=1\\n\neq m}}^{p}(\sum_{i=0}^{L-l-1}(h_i^{(m)}h_{l+i+1}^{(n)}(k)+h_{l+i+1}^{(n)}h_i^{(m)}(k)))^2]\xrightarrow[k\to\infty]{}0,$$

and hence for any $l=0,\dots,L-1$

$$\sum_{i=0}^{L-l-1}(h_i^{(m)}h_{l+i+1}^{(n)}(k)+h_{l+i+1}^{(n)}h_i^{(m)}(k))\xrightarrow[k\to\infty]{}0 \tag{5.3.48}$$

and

$$\sum_{i=0}^{L-l-1}\sum_{\substack{n=1\\n\neq m}}^{p}(h_i^{(n)}h_{l+i+1}^{(n)}(k)+h_{l+i+1}^{(n)}h_i^{(n)}(k))\xrightarrow[k\to\infty]{}0. \tag{5.3.49}$$

Notice that (5.3.49) means that

$$p\sum_{n=1}^{p}\sum_{i=0}^{L-l-1}(h_i^{(n)}h_{l+i+1}^{(n)}(k)+h_{l+i+1}^{(n)}h_i^{(n)}(k))$$

$$-\sum_{m=1}^{p}\sum_{i=0}^{L-l-1}(h_i^{(m)}h_{l+i+1}^{(m)}(k)+h_{l+i+1}^{(m)}h_i^{(m)}(k))\xrightarrow[k\to\infty]{}0.$$

However, the above expression equals

$$(p-1)\sum_{n=1}^{p}\sum_{i=0}^{L-l-1}(h_i^{(n)}h_{l+i+1}^{(n)}(k)+h_{l+i+1}^{(n)}h_i^{(n)}(k)).$$

Therefore,

$$\sum_{i=0}^{L-l-1}(h_i^{(n)}h_{l+i+1}^{(n)}(k)+h_{l+i+1}^{(n)}h_i^{(n)}(k))\xrightarrow[k\to\infty]{}0. \tag{5.3.50}$$

In the sequel, it will be shown that (5.3.43), (5.3.44), (5.3.48), and (5.3.50)) imply that $h(k) \xrightarrow[k\to\infty]{} 0$, which contradicts with $\|h(k)\| = 1$. This means that the converse assumption (5.3.37) is not true.

For any $m \neq n$, since $h^{(n)}(z)$, $h^{(m)}(z)$ are coprime, where $h^{(n)}(z)$ is given in (5.1.6), there exist polynomials $d_1(z)$, $d_2(z)$ such that

$$d_1(z)h^{(n)}(z) + d_2(z)h^{(m)}(z) = 1. \tag{5.3.51}$$

Let r_1 and r_2 be the degrees of $d_1(z)$ and $d_2(z)$, respectively. Set $q = 4(r_1 + r_2) + 5L + 1$. Introduce the q-dimensional vector g_k^s and $q \times q$ square matrices W and A as follows:

$$g_k^s = (\underbrace{0,\dots,0}_{2(r_1+r_2+L)}, h_0^{(s)}(k),\dots,h_L^{(s)}(k), \underbrace{0,\dots,0}_{2(r_1+r_2+L)})^T,$$

$$W = \begin{pmatrix} 0 & \cdots & 0 & 1 \\ \vdots & & / & 0 \\ 0 & / & & \vdots \\ 1 & 0 & \cdots & 0 \end{pmatrix}, \quad A = \begin{pmatrix} 0 & \cdots & \cdots & 0 \\ 1 & \ddots & \ddots & \vdots \\ 0 & \ddots & \ddots & \vdots \\ 0 & 0 & 1 & 0 \end{pmatrix}.$$

Note that $Wg = (g_q,\dots,g_1)^T$, where $g = (g_1,\dots,g_q)^T$ and $Ag = (0, g_1,\dots,g_{q-1})^T$, $A^Tg = (g_2,\dots,g_q,0)^T$. Then (5.3.43), (5.3.44), (5.3.48), and (5.3.50) can be written in the following compact form:

$$h^{(s)}(A)Wg_k^t + h^{(t)}(A^T)A^Lg_k^t \underset{k\to\infty}{\to} 0, \ \forall s,t = 1,\dots,p. \tag{5.3.52}$$

To see this, note that for any fixed s and t, on the left hand sides of (5.3.48) and (5.3.50) there are $2L$ different sums when l varies from 0 to $L-1$ and s, t replace roles each other. These together with (5.3.43) and (5.3.44) give us $2L+1$ sums, and each of them tends to zero. Explicitly expressing (5.3.52), we find that there are $2L+1$ nonzero rows and each row corresponds to one of the relationships in (5.3.43), (5.3.44), (5.3.48), and (5.3.50).

Since we have put enough zeros in the definition of g_k^s, after multiplying the left hand side of (5.3.52) by A^i, $\forall i \leq r_1 + r_2$, $A^i(h^{(s)}(A)Wg_k^t + h^{(t)}(A^T)A^Lg_k^s)$ has only shifted nonzero elements in $h^{(s)}(A)Wg_k^t + h^{(t)}(A^T)A^Lg_k^s$.

From (5.3.52) it follows that for any l, $l = 1,\dots,p$, and m, n in (5.3.51)

$$\big(d_1(A)(h^{(n)}(A)Wg_k^l + h^{(l)}(A^T)A^Lg_k^n) + d_2(A)(h^{(m)}(A)Wg_k^l$$

$$
\begin{aligned}
&+ h^{(l)}(A^T)A^L g_k^m) = d_1(A)h^{(n)}(A) + d_2(A)h^{(m)}(A))Wg_k^l \\
&+ h^{(l)}(A^T)A^L(d_1(A)g_k^n + d_2(A)g_k^m) \\
&= Wg_k^l + h^{(l)}(A^T)A^L(d_1(A)g_k^n + d_2(A)g_k^m) \xrightarrow[k\to\infty]{} 0. \qquad (5.3.53)
\end{aligned}
$$

From (5.3.53) it follows that

$$
\begin{aligned}
&d_1(A^T)[Wg_k^n + h^{(n)}(A^T)A^L(d_1(A)g_k^n + d_2(A)g_k^m)] \\
&+ d_2(A^T)[Wg_k^m + h^{(m)}(A^T)A^L(d_1(A)g_k^n + d_2(A)g_k^m)] \xrightarrow[k\to\infty]{} 0.
\end{aligned} \tag{5.3.54}
$$

Note that for any polynomial $d(z)$ of degree r, $d(A^T)Wy = Wd(A)y$, if the last r elements of y are zeros. From (5.3.54) it follows that

$$
\begin{aligned}
&d_1(A^T)Wg_k^n + d_2(A^T)Wg_k^m + (d_1(A^T)h^{(n)}(A^T) \\
&\qquad + d_2(A^T)h^{(m)}(A))z^L(d_1(A)g_k^n + d_2(A)g_k^m) \\
&= W(d_1(A)g_k^n + d_2(A)g_k^m) + A^L(d_1(A)g_k^n + d_2(A)g_k^m) \xrightarrow[k\to\infty]{} 0.
\end{aligned} \tag{5.3.55}
$$

Denoting

$$
g_k = (g_{k,1}, \dots, g_{k,q})^T \triangleq d_1(A)g_k^n + d_2(A)g_k^m,
$$

from (5.3.55) we find that

$$
Wg_k + A^L g_k \xrightarrow[k\to\infty]{} 0. \tag{5.3.56}
$$

By the definition of g_k^n, the first $2(r_1+r_2+L)$ elements of g_k are zeros, i.e., $g_{k,i} = 0$, $i = 1, \dots, 2(r_1+r_2+L)$. This means that the last $2(r_1+r_2+L)$ elements of Wg_k are zeros, i.e.,

$$
Wg_k = (\underbrace{g_{k,q}, g_{k,q-1}, \dots, g_{k,2(r_1+r_2+L)+1}}_{2(r_1+r_2+L)+L}, \underbrace{0, \dots, 0}_{2(r_1+r_2+L)})^T. \tag{5.3.57}
$$

On the other hand,

$$
A^L g_k = (\underbrace{0, \dots, 0,}_{2(r_1+r_2+L)+L} \underbrace{g_{k,2(r_1+r_2+L)+1}, \dots, g_{k,q-L}}_{2(r_1+r_2+L)}). \tag{5.3.58}
$$

By (5.3.56), from (5.3.57) and (5.3.58) it is seen that $g_k \xrightarrow[k\to\infty]{} 0$, i.e.,

$$
d_1(A)g_k^n + d_2(A)g_k^m \xrightarrow[k\to\infty]{} 0.
$$

From (5.3.53) it then follows that

$$g_k^l \xrightarrow[k\to\infty]{} 0, \quad \forall l = 1, \dots, p,$$

i.e., $h^{(l)}(k) \xrightarrow[k\to\infty]{} 0$, $\forall l = 1, \dots p$. But this is impossible, because $h(k)$ are unit vectors. Consequently, (5.3.37) is impossible and this completes the proof of Theorem 5.3.2. □

5.4. Constrained Adaptive Filtering

We now apply SA methods to adaptive filtering, which is an important topic in signal processing. We consider the constrained problem, while the unconstrained problem is only a special case of the constrained one as to be explained.

Let $\{Y_k\}$ and $\{S_k\}$ be two observed sequences, where Y_k and S_k are $r \times l$-and $m \times l$-matrices, respectively. Assume $\begin{bmatrix} Y_k \\ S_k \end{bmatrix}$ is stationary and ergodic with

$$E \begin{bmatrix} Y_k \\ S_k \end{bmatrix} [\, Y_k^T S_k^T \,] \triangleq \begin{bmatrix} R_y & R_{ys} \\ R_{sy} & R_{ss} \end{bmatrix}, \tag{5.4.1}$$

which, however, is unknown.

It is required to design the optimal weighting $m \times r$-matrix X, which minimizes

$$E(XY_1 - S_1)(XY_1 - S_1)^T, \tag{5.4.2}$$

under constraint

$$XC = \Phi, \tag{5.4.3}$$

where C and Φ are $r \times l$-and $m \times l$-given matrices, respectively. In the case where $C = 0$, the problem is reduced to the unconstrained one.

It is clear that (5.4.3) is solvable with respect to X if and only if $\Phi C^+ C = \Phi$, and in this case the solution to (5.4.3) is

$$X = \Phi C^+ + Z(I - CC^+), \tag{5.4.4}$$

where Z is any $m \times r$-matrix.

For notational simplicity, denote

$$P = I - CC^+. \tag{5.4.5}$$

Let $L(C)$ denote the vector space spanned by the columns of matrix C, and let the columns of matrix V_1 be an orthogonally normalized basis of $L(C)$. Then there is a full-rank decomposition $C = V_1 K$, $KK^T > 0$. Noticing $C^+ = K^T(KK^T)^{-1}V_1^T$, we have $CC^+ = V_1 V_1^T$. Let $[V_1, V_2]$ be an orthogonal matrix. Then

$$P = I - CC^+ = I - V_1 V_1^T = V_2 V_2^T, \tag{5.4.6}$$

and hence

$$P = P^+, \quad P^2 = P. \tag{5.4.7}$$

From this it follows that

$$\begin{aligned} &E[(PR_yPR_y^+ - I)PY][(PR_yPR_y^+ - I)PY]^T \\ &= (PR_yPR_y^+ - I)PR_yP(R_y^+PR_yP - I) = 0, \end{aligned}$$

and hence $PR_yPR_y^+PY = PY$ a.s. This implies that

$$PR_yPR_y^+PR_y = PR_y, \quad PR_yPR_y^+PR_{ys} = PR_{ys}. \tag{5.4.8}$$

Let us express the optimal X minimizing (5.4.2) via R_y, R_{ys}. By (5.4.8) substituting (5.4.4) into (5.4.2) leads to

$$\begin{aligned} &E(XY_1 - S_1)(XY_1 - S_1)^T \\ &= E((\Phi C^+ + ZP)Y_1 - S_1)(\Phi C^+ + ZP)Y_1 - S_1)^T \\ &= E(ZPY_1 + (\Phi C^+ Y_1 - S_1))(ZPY_1 + (\Phi C^+ Y_1 - S_1))^T \\ &= ZPR_yPZ^T + ZP(R_yC^{+T}\Phi^T - R_{ys}) + (\Phi C^+ R_y - R_{sy})PZ^T \\ &\quad + \Phi C^+ R_y C^{+T}\Phi^T - R_{sy}C^{+T}\Phi^T - \Phi C^+ R_{ys} + R_s \\ &= [Z - (R_{sy} - \Phi C^+ R_y)PR_y^+P]PR_yP[Z - (R_{sy} - \Phi C^+ R_y)PR_y^+P]^T \\ &\quad - (R_{sy} - \Phi C^+ R_y)PR_y^+P(R_{ys} - R_yC^{+T}\Phi^T) \\ &\quad + \Phi C^+ R_y C^{+T}\Phi^+ - R_{sy}C^{+T}\Phi^T - \Phi C^+ R_{ys} + R_s. \end{aligned} \tag{5.4.9}$$

On the right-hand side of (5.4.9) only the first term, which is quadratic, depends on Z. Therefore, the optimal Z_{opt} should be the solution of

$$[Z - (R_{sy} - \Phi C^+ R_y)PR_y^+P]PR_yP = 0, \tag{5.4.10}$$

i.e.,

$$Z_{opt} = (R_{sy} - \Phi C^+ R_y)PR_y^+P + Z_a, \tag{5.4.11}$$

where Z_a is any $m \times r$-matrix satisfying

$$Z_a PR_yP = 0.$$

Combining (5.4.4) with (5.4.11), we find that

$$X_{opt} = \Phi C^+ + (R_{sy} - \Phi C^+ R_y) P R_y^+ P + Z_a P. \tag{5.4.12}$$

Using the ergodic property of $[Y_k^T S_k^T]$, we may replace R_{sy} and R_y by their sample averages to obtain the estimate for X_{opt}. And, the estimate can be updated by using new observations. However, to update the estimate, it involves taking the pseudo-inverse of the updated estimate for R_y, which may be of high dimension. This will slow down the computation speed. Instead, we now use an SA algorithm to approach X_{opt}.

By (5.4.8), we can rewrite (5.4.10) as

$$ZPR_yP - (R_{sy}P - \Phi C^+ R_y P) = 0,$$

or

$$(\Phi C^+ + ZP) R_y P - R_{sy} P = 0. \tag{5.4.13}$$

We now face to the standard root-seeking problem for a linear function

$$f(Z) \triangleq (\Phi C^+ + ZP) R_y P - R_{sy} P. \tag{5.4.14}$$

As before, let $M_k > 0$, $M_{k+1} > M_k$, $\forall k$, and $M_k \xrightarrow[k \to \infty]{} \infty$. The following algorithm is used to estimate Z_{opt} given by (5.4.12), which in the notations used in previous chapters is the root set J for the linear function $f(Z)$ given by (5.4.14):

$$J = \{(R_{sy} - \Phi C^+ R_y) P R_y^+ P + Z_a, \quad \forall Z_a : Z_a P R_y P = 0\}, \tag{5.4.15}$$

$$\begin{aligned} Z_{k+1} =& (Z_k - a_k((\Phi C^+ + Z_k P) Y_{k+1} Y_{k+1}^T P - S_{k+1} Y_{k+1}^T P)) \\ & \cdot I_{[\|Z_k - a_k((\Phi C^+ + Z_k P) Y_{k+1} Y_{k+1}^T P - S_{k+1} Y_{k+1}^T P)\| \le M_{\sigma_k}]}, \end{aligned} \tag{5.4.16}$$

$$\sigma_k = \sum_{i=1}^{k-1} I_{[\|Z_i - a_i((\Phi C^+ + Z_i P) Y_{i+1} Y_{i+1}^T P - S_{i+1} Y_{i+1}^T P)\| > M_{\sigma_i}]}$$

with initial value Z_0 such that $Z_0 P = Z_0$, and $a_k = \frac{1}{k}$.

Theorem 5.4.1 *Assume that* $\begin{bmatrix} Y_k \\ S_k \end{bmatrix}$ *is stationary and ergodic with second moment given by (5.4.1) and that* $\sup_k (\|Y_k\| + \|S_k\|) \le \eta < \infty$ *a.s. Then, after a finite number of steps, say* $k \ge k_0$, *(5.4.16) has no more truncations, i.e.,*

$$Z_{k+1} = Z_k - a_k[(\Phi C^+ + Z_k P) Y_{k+1} Y_{k+1}^T P - S_{k+1} Y_{k+1}^T P], \quad k \ge k_0, \tag{5.4.17}$$

and

$$d(Z_k, J) \xrightarrow[k\to\infty]{} 0 \quad a.s. \tag{5.4.18}$$

i.e.,

$$d(\Phi C^+ + Z_k, X_{opt}) \xrightarrow[k\to\infty]{} 0 \quad a.s., \tag{5.4.19}$$

where X_{opt} given by (5.4.12) solves the stated constrained optimization problem.

Proof. We first note that (5.4.16) is a matrix recursion. However, if in lieu of Z_k we consider $Z_k^T a$ with a being an arbitrary constant vector, then we have a conventional vector recursion, and by (5.4.9) $a^T E(XY_1 - S_1)(XY_1 - S_1)^T a$ may serve as a Lyapunov function for the corresponding regression function obtained from (5.4.14):

$$a^T f(Z) = (a^T \Phi C^+ + a^T ZP)R_y P - a^T R_{sy} P.$$

Therefore, in order to apply Theorem 2.2.1, we need only to verify the noise condition.

Denoting

$$\epsilon_{k+1} = (\Phi C^+ + Z_k P)(Y_{k+1}Y_{k+1}^T - R_y)P - (S_{k+1}Y_{k+1}^T - R_{sy})P, \tag{5.4.20}$$

then from (5.4.16) we have

$$\begin{aligned} Z_{k+1} =& (Z_k - a_k((\Phi C^+ + Z_k P)R_y P - R_{sy}P \\ &+ \epsilon_{k+1}))I_{[\|Z_k - a_k((\Phi C^+ + Z_k P)R_y P - R_{sy}P + \epsilon_{k+1})\| \le M_{\sigma_k}]}. \end{aligned} \tag{5.4.21}$$

We now show that for a fixed sample if $Z_{n_k} \xrightarrow[k\to\infty]{} \overline{Z}$, then

$$\begin{aligned} Z_{m+1} = Z_m - a_m((\Phi C^+ + Z_m P)R_y P - R_{sy}P + \epsilon_{m+1}),& \\ \forall m : n_k \le m \le m(n_k, T) + 1,& \end{aligned} \tag{5.4.22}$$

and there is a constant $c_0 > 0$ such that

$$\|Z_m - Z_{n_k}\| \le c_0 T, \quad \forall m : n_k \le m \le m(n_k, T) + 1, \tag{5.4.23}$$

if k is sufficiently large and T is small enough, where $m(k, T) = \max\{m : \sum_{i=k}^{m} \frac{1}{i} \le T\}$.

We need the following fact, which is an extension of Example 5.2.1. Assume the process $\{A_i\}$ is stationary and ergodic with $E\|A_1\| < \infty$, and $\{\Psi_i\}$ is a convergent sequence of random matrices $\Psi_i \xrightarrow[i \longrightarrow \infty]{} \Psi < \infty$ a.s. Then

$$\lim_{k \longrightarrow \infty} \left(\sum_{i=k}^{m(k,T_k)} \frac{\Psi_i}{i} A_{i+1} - \Psi E A_1 T_k \right) = 0 \quad \text{a.s.} \tag{5.4.24}$$

Let $Q_k = \sum_{i=1}^{k} \Psi_i A_{i+1}$. Then by ergodicity of both $\{A_i\}$ and $\{\|A_i\|\}$ we have

$$\frac{Q_k}{k} = \frac{\Psi \sum_{i=1}^{k} A_{i+1}}{k} + \frac{\sum_{i=1}^{k} (\Psi_i - \Psi) A_{i+1}}{k} \xrightarrow[k \longrightarrow \infty]{} \Psi E A_1 \quad \text{a.s.,} \tag{5.4.25}$$

because the second term on the right-hand side of the equality can be estimated as follows

$$\begin{aligned} \|\frac{1}{k} \sum_{i=1}^{k} (\Psi_i - \Psi) A_{i+1}\| \leq & \frac{1}{k} \| \sum_{i=1}^{k_1} (\Psi_i - \Psi) A_{i+1}\| + \max_{k_1 \leq i \leq k} \|\Psi_i - \Psi\| \\ & \cdot \frac{1}{k} \sum_{k=k_1}^{k} \|A_{i+1}\|, \end{aligned}$$

which tends to zero as $k \longrightarrow \infty$ and then $k_1 \longrightarrow \infty$. By a partial summation and by using (5.4.25) we have

$$\begin{aligned} \sum_{i=k}^{m(k,T_k)} \frac{\Psi_i A_{i+1}}{i} = & \sum_{i=k}^{m(k,T_k)} \frac{1}{i} (Q_i - Q_{i-1}) \\ = & \frac{Q_{m(k,T_k)}}{m(k,T_k)} - \frac{Q_{k-1}}{k} + \sum_{i=k+1}^{m(k,T_k)} Q_{i-1} (\frac{1}{i-1} - \frac{1}{i}), \end{aligned}$$

which implies (5.4.24) by (5.4.25).

Let us consider the following algorithm starting from n_k without truncation:

$$Z'^T_{k+1} = (I - \frac{1}{k} P Y_{k+1} Y^T_{k+1} P) Z'^T_k - \frac{1}{k} P (Y_{k+1} Y^T_{k+1} C^{+T} \Phi^T - Y_{k+1} S^T_{k+1}), \tag{5.4.26}$$

Set

$$Z'_{n_k} = Z_{n_k}, \quad \Phi_{i,i} = I, \text{ and}$$

$$\Phi_{k+1,i} = (I - \frac{1}{k}PY_{k+1}Y_{k+1}^T P)(I - \frac{1}{k-1}PY_kY_k^TP)$$
$$\cdots (I - \frac{1}{i}PY_{i+1}Y_{i+1}^T P).$$

Then from (5.4.26) it follows that

$$Z'^T_{m(n_k,t)+1} = \Phi_{m(n_k,t)+1,k} Z^T_{n_k} - \sum_{j=n_k}^{m(n_k,t)} \frac{1}{j}\Phi_{m(n_k,t),j} P(Y_{j+1}Y_{j+1}^T C^{+T}\Phi^T - Y_{j+1}S_{j+1}^T) \tag{5.4.27}$$

Denote

$$G_{j+1} = P(Y_{j+1}Y_{j+1}^T C^{+T}\Phi^T - Y_{j+1}S_{j+1}^T) \tag{5.4.28}$$

and

$$F_k = \sum_{j=1}^{k} G_{j+1}.$$

Since $\{G_j\}$ is stationary and ergodic, $\frac{1}{k}F_k \xrightarrow[k\to\infty]{} EG_1$ a.s., and $\frac{1}{k}\sum_{i=1}^{k}\|G_i\| \xrightarrow[k\to\infty]{} E\|G_1\|$. Then by a partial summation, we have

$$\begin{aligned}
\sum_{j=n_k}^{m(n_k,t)} \frac{1}{j}\Phi_{m(n_k,T),j} G_{j+1} &= \sum_{j=n_k}^{m(n_k,t)} \frac{1}{j}\Phi_{m(n_k,t),j}(F_j - F_{j-1}) \\
&= \frac{1}{m(n_k,t)}F_{m(n_k,t)} - \frac{1}{n_k}\Phi_{m(n_k,t),n_k}F_{n_k-1} \\
&\quad + \sum_{j=n_k+1}^{m(n_k,t)} (\frac{1}{j-1}\Phi_{m(n_k,t),j-1} - \frac{1}{j}\Phi_{m(n_k,t),j} F_{j-1} \\
&= \frac{1}{m(n_k,t)}F_{m(n_k,t)} - \frac{1}{n_k}\Phi_{m(n_k,T),n_k}F_{n_k-1} \\
&\quad + \sum_{j=n_k+1}^{m(n_k,t)} \Phi_{m(n_k,t),j}(\frac{1}{j-1}(I - \frac{1}{j-1}PY_jY_j^TP) - \frac{1}{j})F_{j-1} \\
&= \frac{1}{m(n_k,t)}F_{m(n_k,t)} - \frac{1}{n_k}\Phi_{m(n_k,t),n_k}F_{n_k-1} \\
&\quad + \sum_{j=n_k+1}^{m(n_k,t)} \Phi_{m(n_k,T),j}(\frac{1}{j(j-1)} - \frac{1}{(j-1)^2}PY_jY_j^TP)F_{j-1}.
\end{aligned} \tag{5.4.29}$$

Notice that $\frac{1}{k}Y_{k+1}Y_{k+1}^T \xrightarrow[k\to\infty]{} 0$ a.s. by ergodicity. Then for large k, $\|\Phi_{m(n_k,t),j}\| \leq 1$, $\forall j \geq k$, and from (5.4.29) it follows that

$$\begin{aligned}
&\limsup_{k\to\infty} \| \sum_{j=n_k}^{m(n_k,t)} \frac{1}{j}\Phi_{m(n_k,T),j}G_{j+1}\| \\
&\leq \|EG_1\| + E\|G_1\| + \limsup_{k\to\infty} \sum_{j=n_k+1}^{m(n_k,t)} (\frac{1}{j(j-1)} \\
&\quad - \frac{1}{(j-1)^2}\operatorname{tr} Y_jY_j^T)\|F_{j-1}\| \\
&\leq \|EG_1\| + E\|G_1\| + t\|EG_1\| + t\operatorname{tr} R_y\|EG_1\|, \qquad (5.4.30)
\end{aligned}$$

where (5.4.24) is used incorporating with the fact that $\frac{\|F_j\|}{j} \xrightarrow[j\to\infty]{} \|EG_1\|$ and $\operatorname{tr} Y_jY_j^T$ is stationary with $E\operatorname{tr} Y_jY_j^T = \operatorname{tr} R_y$.

From (5.4.27)–(5.4.30) by convergence of Z_{n_k} it follows that

$$\|Z'_{m(n_k,t)+1}\| \leq c_1 + c_2T, \quad \forall t \in [0,T] \qquad (5.4.31)$$

for large k and small T, where c_1 and c_2 are constants independent of k and t.

Consequently, in the case $\sigma_k \xrightarrow[k\to\infty]{} \infty$, i.e., $M_{\sigma_k} \xrightarrow[k\to\infty]{} \infty$ in (5.4.16), $Z'_{m(n_k,t)+1}$ will never reach the truncation bound M_{σ_k} for Z_m, $\forall m : n_k \leq m \leq m(n_k,T)+1$ if k is large enough and T is small enough. Then Z'_m coincides with Z_m, $\forall m : n_k \leq m \leq m(n_k,T)+1$. This verifies (5.4.22), while (5.4.23) follows from (5.4.16) because for a fixed ω, $\{\|Y_k\|\}$ and $\{\|S_k\|\}$ are bounded, and $\{\|Z_m\|, m = n_k, n_k+1, \ldots, m(n_k,T)+1\}$ are also bounded by (5.4.31) and the convergence $Z_{n_k} \xrightarrow[k\to\infty]{} \bar{Z}$. In the case $\sigma_k \xrightarrow[k\to\infty]{} \sigma < \infty$, i.e., $\sigma_k = \sigma$, $\forall k \geq k_1$ for some k_1, $\{Z_k\}$ is bounded, and hence (5.4.22) and (5.4.23) are also satisfied.

We are now in a position to verify the noise condition required in Theorem 2.2.1 for $\{\epsilon_{k+1}\}$ given by (5.4.20), i.e., we want to show that for any convergent subsequence $Z_{n_k} \longrightarrow \bar{Z}$

$$\begin{aligned}
&\lim_{T\to 0}\limsup_{k\to\infty} \frac{1}{T}\Big\| \sum_{i=n_k}^{m(n_k,T_k)} \frac{1}{i}[(\Phi C^+ + Z_iP)(Y_{i+1}Y_{i+1}^T - R_y)P \\
&\quad -(S_{i+1}Y_{i+1}^T - R_{sy})P]\Big\| = 0, \quad \forall T_k \in [0,T]. \qquad (5.4.32)
\end{aligned}$$

By (5.4.24)

$$\lim_{k\longrightarrow\infty}\sum_{i=n_k}^{m(n_k,T_k)}\frac{1}{i}[(\Phi C^+(Y_{i+1}Y_{i+1}^T-R_y)P-(S_{i+1}Y_{i+1}^T-R_{sy})P]=0,$$

so for (5.4.32) it suffices to show

$$\lim_{T\longrightarrow 0}\limsup_{k\longrightarrow\infty}\frac{1}{T}\|\sum_{i=n_k}^{m(n_k,T_k)}\frac{1}{i}Z_iP(Y_{i+1}Y_{i+1}^T-R_y)\|=0,\quad \forall T_k\in[0,T]. \tag{5.4.33}$$

Again, by (5.4.24) and also by (5.4.23)

$$\begin{aligned}
&\|\sum_{i=n_k}^{m(n_k,T_k)}\frac{1}{i}Z_iP(Y_{i+1}Y_{i+1}^T-R_y)\|\\
&\quad\le\|\bar{Z}\sum_{i=n_k}^{m(n_k,T_k)}\frac{1}{i}P(Y_{i+1}Y_{i+1}^T-R_y)\|\\
&\qquad+\|\sum_{i=n_k}^{m(n_k,T_k)}\frac{1}{i}(Z_i-\bar{Z})P(Y_{i+1}Y_{i+1}^T-R_y)\|\\
&\quad\le\|\bar{Z}\sum_{i=n_k}^{m(n_k,T_k)}\frac{1}{i}P(Y_{i+1}Y_{i+1}^T-R_y)\|\\
&\qquad+2c_0T\sum_{i=n_k}^{m(n_k,T)}\frac{1}{i}\|Y_{i+1}Y_{i+1}^T-R_y\|\xrightarrow[k\longrightarrow\infty]{}2c_0T^2E\|Y_1Y_1^T-R_y\|,
\end{aligned}$$

which implies (5.4.33).

By Theorem 2.2.1, there is k_0 such that for $k\ge k_0$ Z_k is defined by (5.4.17) and Z_k converges to the root set J for $f(Z)$ given by (5.4.14). This completes the proof for the theorem. □

Remark 5.4.1 For the unconstrained problem $\Phi=0$ and $C=0$, the algorithm (5.4.16) becomes

$$\begin{aligned}
Z_{k+1}&=Z_k-a_k(Z_kY_{k+1}Y_{k+1}^T-S_{k+1}Y_{k+1}^T)\\
&\quad\cdot I_{[\|Z_k-a_k(Z_kY_{k+1}Y_{k+1}^T-S_{k+1}Y_{k+1}^T)\|\le M_{\sigma_k}]},\\
\sigma_k&=\sum_{i=1}^{k-1}I_{[\|Z_i-a_i(Z_iY_{i+1}Y_{i+1}^T-S_{i+1}Y_{i+1}^T)\|>M_{\sigma_i}]},\quad a_k=\frac{1}{k}.
\end{aligned}$$

Further, if $R_y > 0$, then $X_{opt} = R_{sy}R_y^{-1}$, and Theorem 5.4.1 asserts

$$Z_k \longrightarrow R_{sy}R_y^{-1} \quad \text{a.s.},$$

provided $\begin{bmatrix} Y_k \\ S_k \end{bmatrix}$ is stationary, ergodic, and bounded.

5.5. Adaptive Filtering by Sign Algorithms

We now consider the unconstrained problem mentioned in Section 5.4, but we restrict ourselves to discuss the vector case, i.e., instead of matrix signal $\begin{bmatrix} Y_k \\ S_k \end{bmatrix}$ we now consider $\begin{bmatrix} y_k \\ s_k \end{bmatrix}$, where y_k is r-dimensional and s_k is one-dimensional. However, instead of quadratic criterion (5.4.2) we now minimize the L_1 cost

$$L(x) = E|x^T y_1 - s_1|, \tag{5.5.1}$$

where x is an r-dimensional vector.

Note that the gradient of $L(\cdot)$ is given by

$$L_x(x) = E(y_1 \operatorname{sign}(x^T y_1 - s_1)) \triangleq f(x), \tag{5.5.2}$$

where

$$\operatorname{sign}(x) = \begin{cases} 1, & if \quad x > 0 \\ 0, & if \quad x = 0 \\ -1, & if \quad x < 0. \end{cases}$$

The problem is to find x_{opt} which minimizes $L(x)$, or to approach the root set J of $f(x)$

$$J \triangleq \{x : f(x) = 0\}. \tag{5.5.3}$$

As before, let $\{M_k\}$ be an increasing sequence of positive real numbers such that $M_k \longrightarrow \infty$ as $k \longrightarrow \infty$. Define the algorithm as follows

$$x_{k+1} = (x_k - a_k y_k \operatorname{sign}(x_k^T y_k - s_k)) I_{[|x_k - a_k y_k \operatorname{sign}(x_k^T y_k - s_k)| \le M_{\sigma_k}]}, \tag{5.5.4}$$

$$\sigma_k = \sum_{i=1}^{k-1} I_{[|x_i - a_i y_i \operatorname{sign}(x_i^T y_i - s_i)| > M_{\sigma_i}]}, \quad \sigma_0 = 0, \tag{5.5.5}$$

$$a_k = \frac{1}{k}.$$

Theorem 5.5.1 *Assume* $\begin{bmatrix} s_k \\ y_k \end{bmatrix}$ *is stationary and ergodic with*

$$E\begin{bmatrix} s_k \\ y_k \end{bmatrix}[s_k \quad y_k^T] \triangleq R > 0. \tag{5.5.6}$$

Then

$$d(x_k, J) \xrightarrow[k\longrightarrow\infty]{} 0 \quad a.s.,$$

where $\{x_k\}$ *is defined by (5.5.4) and (5.5.5) with an arbitrary initial value. In addition, in a finite number of steps truncations cease to exist in (5.5.4).*

Proof. Define

$$m(k,T) \triangleq \max\{m : \sum_{i=k}^{m} \frac{1}{i} \le T\}, \tag{5.5.7}$$

and

$$f_k(x) \triangleq y_k \operatorname{sign}(x^T y_k - s_k). \tag{5.5.8}$$

Let S be a countable set that is dense in $\mathbb{R}^r$, let $\{T_j\}$ and $\{\lambda_j\}$ be two sequences of positive real numbers such that $T_j \longrightarrow 0$ and $\lambda_j \longrightarrow 0$ as $j \longrightarrow \infty$, and denote

$$\xi_k \triangleq \sum_{i=1}^{k} \|y_i\|, \tag{5.5.9}$$

$$g_k(x) \triangleq \sum_{i=1}^{k} [f(x) - f_i(x)], \quad x \in S, \tag{5.5.10}$$

and

$$h_k(x, l, T_j, \lambda_n) \triangleq \sum_{i=1}^{k} \|y_i\| I_{[((lT_j+\lambda_n)\|y_i\| > |x^T y_i - s_i|]}, \tag{5.5.11}$$

where $x \in S$ and l is an integer.

The summands of (5.5.9)–(5.5.11) are stationary with finite expectations for any $x \in S$, any integer l, any T_j, and any λ_n, and then the ergodic theorem yields that

$$\frac{1}{k}\xi_k \xrightarrow[k\longrightarrow\infty]{} E\|y_1\| \quad \text{a.s.} \tag{5.5.12}$$

$$\frac{1}{k}g_k(x) \xrightarrow[k\longrightarrow\infty]{} 0 \quad \text{a.s.} \tag{5.5.13}$$

and

$$\frac{1}{k}h_k(x,l,T_j,\lambda_n) \xrightarrow[k\longrightarrow\infty]{} E\|y_1\|I_{[(lT_j+\lambda_n)\|y_1\|\geq|x^Ty_1-s_1|]}, \quad \text{a.s.} \tag{5.5.14}$$

Therefore, there is an ω-set Ω_0 such that $P\Omega_0 = 1$ and for each $\omega \in \Omega_0$ the convergence for (5.5.12)–(5.5.14) takes place for any $x \in S$, any integer l, any T_j, and any λ_n, $j = 1, 2, \ldots$, $n = 1, 2, \ldots$.

Let us fix an $\omega \in \Omega_0$.

We first show that for any fixed j

$$x_{k+1} = x_k - a_k y_k \operatorname{sign}(x_k^T y_k - s_k), \quad \forall k : n \leq k \leq m(n, T_j), \tag{5.5.15}$$

if n is large enough (say, for $n \geq n_1$), and in addition,

$$\|x_k - x_n\| \leq cT_j, \quad \forall k : n \leq k \leq m(n, T_j), \tag{5.5.16}$$

where c is a constant which may depend on ω but is independent of n.

In what follows $c_i, i = 1, 2, \ldots$, always denote constants that may depend on ω but are independent of n. By (5.4.24) we have for any $t \geq 0$

$$\|\sum_{i=n}^{m(n,t)} a_i y_i \operatorname{sign}(x_i^T y_i - s_i)\| \leq \sum_{i=n}^{m(n,t)} a_i\|y_i\| \xrightarrow[n\longrightarrow\infty]{} E\|y_1\|t. \tag{5.5.17}$$

There are two cases to be considered. If $\sigma_k \longrightarrow \infty$, then for k large enough, $M_{\sigma_k} > E\|y_1\|T_j$, and (5.5.15) holds. If σ_k is bounded, then the truncations cease to exist after a finite number of steps. So, (5.5.15) also holds if n is sufficiently large. Then (5.5.16) follows immediately from (5.5.15) and (5.5.17).

Let us define

$$\epsilon_{k+1} \triangleq f(x_k) - f_k(x_k) = f(x_k) - y_k \operatorname{sign}(x_k^T y_k - s_k), \tag{5.5.18}$$

where $f(\cdot)$ is given by (5.5.2). Then (5.5.15) can be represented as

$$x_{k+1} = x_k - a_k f(x_k) + a_k \epsilon_{k+1}, \quad \forall k : n_1 \leq k \leq m(n_1, T_j).$$

Let x_{n_k} be a convergent subsequence of $\{x_k\}$: $x_{n_k} \xrightarrow[k\longrightarrow\infty]{} \overline{x}$, and let $x(n) \in S$ be such that $\|x(n) - \overline{x}\| < \frac{\lambda_n}{2}$, $\forall n$. We now show that

$$\limsup_{j\longrightarrow\infty} \limsup_{k\longrightarrow\infty} \frac{1}{T_j}\|\sum_{i=n_k}^{m(n_k,T_j)} a_i \epsilon_{i+1}\| = 0. \tag{5.5.19}$$

$$\sum_{i=n_k}^{m(n_k,T_j)} a_i \epsilon_{i+1} = \sum_{i=n_k}^{m(n_k,T_j)} a_i(f(x_i) - f(\overline{x})) + \sum_{i=n_k}^{m(n_k,T_j)} a_i(f(\overline{x}) - f(x(n)))$$
$$+ \sum_{i=n_k}^{m(n_k,T_j)} a_i(f(x(n)) - f_i(x(n)))$$
$$+ \sum_{i=n_k}^{m(n_k,T_j)} a_i(f_i(x(n)) - f_i(x_i)) \tag{5.5.20}$$

Let $\Delta x_i = x_i - x_{n_k}$. By (5.5.16) $\|\Delta x_i\| \leq c_1 T_j$, or $\|\Delta x_i\| < lT_j$ for some integer l, $\forall i$: $n_k \leq i \leq m(n_k, T_j)$.

We examine that the terms on the right-hand side of (5.5.20) satisfy (5.5.19) .

For the first term on the right-hand side of (5.5.20) we have

$$\limsup_{k \longrightarrow \infty} \|\frac{1}{T_j} \sum_{i=n_k}^{m(n_k,T_j)} a_i(f(x_i) - f(\bar{x}))\|$$
$$= \limsup_{k \longrightarrow \infty} \|\frac{1}{T_j} \sum_{i=n_k}^{m(n_k,T_j)} a_i E y_1(\text{sign}((x_{n_k} - \overline{x} + \overline{x} + \Delta x_i)^T y_1 - s_1)$$
$$- \text{sign}(\bar{x}^T y_1 - s_1))\|$$
$$\leq \limsup_{k \longrightarrow \infty} \frac{2}{T_j} \sum_{i=n_k}^{m(n_k,T_j)} a_i E(\|y_1\| I_{[(lT_j + \|x_{n_k} - \overline{x}\|)\|y_1\| > |\bar{x}^T y_1 - s_1|]})$$
$$= 2E(\|y_1\| I_{[lT_j \|y_1\| \geq |\bar{x}^T y_1 - s_1|]}), \tag{5.5.21}$$

where Δx_i, x_{n_k}, and $\overline{x}$ are deterministic for a fixed ω, and the expectation is taken with respect to y_1 and s_1.

Since (5.5.6), $|\bar{x}^T y_1 - s_1| \neq 0$ a.s., applying the dominated convergence theorem yields

$$\lim_{j \longrightarrow \infty} E\|y_1\| I_{[lT_j \|y_1\| \geq |\bar{x}^T y_1 - s_1|]} = 0. \tag{5.5.22}$$

Then from (5.5.21) it follows that

$$\lim_{j \longrightarrow \infty} \lim_{k \longrightarrow \infty} \|\frac{1}{T_j} \sum_{i=n_k}^{m(n_k,T_j)} a_i(f(x_i) - f(\bar{x}))\| = 0. \tag{5.5.23}$$

Similarly, for the second term on the right-hand side of (5.5.20) we have

$$\begin{aligned}
&\lim_{j\longrightarrow\infty}\limsup_{k\longrightarrow\infty}\|\frac{1}{T_j}\sum_{i=n_k}^{m(n_k,T_j)}a_i(f(\overline{x})-f(x(n)))\| \\
&\le \lim_{j\longrightarrow\infty}\limsup_{k\longrightarrow\infty}\frac{2}{T_j}\sum_{i=n_k}^{m(n_k,T_j)}a_iE(\|y_1\|I_{[\|x(n)-\overline{x}\|\|y_1\|\ge|\overline{x}^Ty_1-s_1|]}) \\
&=2E\|y_1\|I_{[\|x(n)-\overline{x}\|\|y_1\|\ge|\overline{x}^Ty_1-s_1|]}\triangleq\delta_n^{(1)}\xrightarrow[n\longrightarrow\infty]{}0, \qquad (5.5.24)
\end{aligned}$$

since $|\overline{x}^Ty_1-s_1|\neq 0$ a.s.

For the third term on the right-hand side of (5.5.20) by (5.4.24), (5.5.10), and (5.5.13) we have

$$\lim_{k\longrightarrow\infty}\sum_{i=n_k}^{m(n_k,T_j)}a_i(f(x(n))-f_i(x(n)))=0,\quad \forall n, \qquad (5.5.25)$$

since $x(n)\in S$.

Finally, for the last term in (5.5.20), by (5.5.14) and (5.4.24) we have

$$\begin{aligned}
&\limsup_{k\longrightarrow\infty}\|\frac{1}{T_j}\sum_{i=n_k}^{m(n_k,T_j)}a_i(f_i(x(n))-f_i(x_i))\| \\
&\le\limsup_{k\longrightarrow\infty}\frac{2}{T_j}\sum_{i=n_k}^{m(n_k,T_j)}a_i\|y_i\|I_{[\|x_i-x(n)\|\|y_i\|>|x^T(n)y_i-s_i|]} \\
&\le\limsup_{k\longrightarrow\infty}\frac{2}{T_j}\|\sum_{i=n_k}^{m(n_k,T_j)}a_i\|y_i\|I_{[(lT_j+\lambda_n)\|y_i\|\ge|x^T(n)y_i-s_i|]} \\
&=2E(\|y_1\|I_{[(lT_j+\lambda_n)\|y_1\|\ge|x^T(n)y_1-s_1|]})\xrightarrow[j\longrightarrow\infty]{}2E\|y_1\|I_{[\lambda_n\|y_1\|\ge|x^T(n)y_1-s_1|]} \\
&\triangleq\delta_n^{(2)}\xrightarrow[n\longrightarrow\infty]{}0, \qquad (5.5.26)
\end{aligned}$$

where the last convergence follows from the fact that $I_{[\lambda_n\|y_1\|\ge|x^T(n)y_1-s_1|]}\longrightarrow 0$ a.s. as $n\longrightarrow\infty$ since $\lambda_n\longrightarrow 0$ and $x^T(n)y_1-s_1\longrightarrow\overline{x}^Ty_1-s_1\neq 0$ a.s.

Combining (5.5.23)–(5.5.26) yields that

$$\limsup_{j\longrightarrow\infty}\limsup_{k\longrightarrow\infty}\frac{1}{T_j}\|\sum_{i=n_k}^{m(n_k,T_j)}a_i\epsilon_{i+1}\|\le\delta_n^{(1)}+\delta_n^{(2)}. \qquad (5.5.27)$$

Since the left-hand side of (5.5.27) is free of n, tending n to infinity in (5.5.27) leads to (5.5.19). Then the conclusion of the theorem follows from Theorem 2.2.1 by noticing that as $v(x)$ in A2.2.2 one may take $-L(x)$.

5.6. Asynchronous Stochastic Approximation

When dealing with large interconnected systems, it is natural to consider the distributed, asynchronous SA algorithms. For example, in a communication network with d servers, each server has to allocate audio and video bandwidths in an appropriate portion in order to minimize the average time $L(x)$ of queueing delay. Denote by x^i the bandwidth ratio for the ith server, and $x = [x^1, \ldots, x^l]$. Assume the average delay time $L(x)$ depends on x only and $L(x)$ is differentiable, $f(x) \triangleq \nabla L(x)$. Then, to minimize $L(x)$ is equivalent to find the root of $f(x)$. Assume the time, denoted by τ^{ij}, spent on transmitting data from the ith server to the jth server is not negligible. Then at the ith server for the $(k+1)$th iteration we can observe $L(\cdot)$ or $f(\cdot)$ only at $[x^1_{t^i_k - \tau^{1i}}, \ldots, x^l_{t^i_k - \tau^{li}}]^T$, where t^i_k denotes the total time spent until completion of k iterations for the ith server. This is a typical problem solved by asynchronous SA. Similar problem arises also from job-scheduling for computers in a computer network.

We now precisely define the problem and the algorithm.

At time k denote by $x_k \triangleq [x^1_k, \ldots, x^l_k]^T$ the estimate for the unknown root x^0 of $f(\cdot) \triangleq [f^1(\cdot), \ldots, f^l(\cdot)]^T$. Components f^i of f are observed by different processors, and the communication delays τ^{ij}_k from the jth processor to the ith processor at time k are taken into account. The observation of the ith processor is carried only at $(x^1_{k-\tau^{i1}_k}, \ldots, x^l_{k-\tau^{il}_k})$, i.e.,

$$y^i_{k+1} = f^i(x^1_{k-\tau^{i1}_k}, \ldots, x^l_{k-\tau^{il}_k}) + \epsilon^i_{k+1},$$

where ϵ^i_{k+1} is the observation noise.

In contrast to the synchronous case, the update steps now are different for different processors, so it is unreasonable to use the same step size a_k for all processors in an asynchronous environment. At time k the step size used in the ith processor is known and is denoted by $a(k,i)$.

We will still use the expanding truncation technique, but we are unable to simultaneously change estimates in different processors when the estimate exceeds the truncation bound because of the communication delay.

Assume all processors start at the same given initial value $x_0 = [x^1_0, \ldots, x^l_0]^T$ and $\sigma^i_0 = 0$ for all $i = 1, \ldots, l$. The observation y^i_1 at the ith processor is $f^i(x_0)+\epsilon_1^i$ and x_0^i is updated to x_1^i by the rule given below. Because of the communication delay the estimate produced by the jth processor cannot reach the ith processor for the initial steps: $k \leq \tau_k^{ij}$. By agreement we will take x_0^j to serve as $x^j_{k-\tau_k^{ij}}$ whenever $k \leq \tau_k^{ij}$.

At the ith processor, $i = 1, \ldots, d$ there are two sequences $\{x_k^i\}$ and $\{\sigma_k^i\}$ are recursively generated, where x_k^i is the estimate for the ith component of x^0 at time k and σ_k^i is connected with the number of truncations up-to and including time k at the ith processor. For the ith processor at time k the newest information about other processors is $(x^j_{k-\tau_k^{ij}}, \sigma^j_{k-\tau_k^{ij}})$, $j = 1, \ldots, l$, $j \neq i$. In all algorithms discussed until now all components of $f(\cdot)$ are observed at the same point x_k at time k, and this makes updating x_k to x_{k+1} meaningful. In the present case, although we are unable to make all processors to observe $f(\cdot)$ at the same points at each time, it is still desirable to require all processors observe $f(\cdot)$ at points located as close as possible. Presumably, this would make estimate updating reasonable. For this, by noticing that the estimate x_k gradually changes after a truncation, the ideal is to keep all $\{\sigma_k^i, i = 1, \ldots, l\}$ are equal, but for this the best we can do is to equalize σ_k^i with other $\sigma^j_{k-\tau_k^{ij}}$, $j = 1, \ldots, l$, $j \neq i$.

Keeping this idea in mind, we now define the algorithm and the observations for the ith processor, $i = 1, \ldots, l$.

Let $x_* \triangleq [x_*^1, \ldots, x_*^l]^T$ be a fixed point from where the algorithm restarts after a truncation.

i) If there exists j with $\sigma^j_{k-\tau_k^{ij}} > \sigma_k^i$, then reset σ_k^i to equal the biggest one among $\sigma^j_{k-\tau_k^{ij}}$ and pull x_k^i back to the fixed point x_*^i, although x_k^i may not exceed the truncation bound. Precisely, in this case define

$$x_k^i = x_*^i, \tag{5.6.1}$$

$$\sigma_k^i = \max_j \sigma^j_{k-\tau_k^{ij}}, \tag{5.6.2}$$

and observe

$$y_{k+1}^i = f^i(x_*) + \epsilon_{k+1}^i. \tag{5.6.3}$$

ii) If $\sigma^j_{k-\tau_k^{ij}} \leq \sigma_k^i$ for any j, then observe $f^i(\cdot)$ at

$$x_k' = [x^1_{k-\tau_k^{i1}} I_{[\sigma^1_{k-\tau_k^{i1}} = \sigma_k^i]} + x_*^1 I_{[\sigma^1_{k-\tau_k^{i1}} < \sigma_k^i]}, \ldots,$$

$$x^l_{k-\tau^{il}_k} I_{[\sigma^d_{k-\tau^{id}_k}=\sigma^i_k]} + x^l_* I_{[\sigma^l_{k-\tau^{il}_k}<\sigma^i_k]}]^T,$$

i.e.,

$$y^i_{k+1} = f^i(x'_k) + \epsilon^i_{k+1}, \tag{5.6.4}$$

For both cases i) and ii), x^i_k and σ^i_k are updated as follows:

$$\begin{aligned} x^i_{k+1} =&(x^i_k + a(k,i)y^i_{k+1})I_{[|x^i_k+a(k,i)y^i_{k+1}|\le M_{\sigma^i_k}]} \\ &+ x^i_* I_{[|x^i_k+a(k,i)y^i_{k+1}|>M_{\sigma^i_k}]}, \end{aligned} \tag{5.6.5}$$

$$\sigma^i_{k+1} =\sigma^i_k + I_{[|x^i_k+a(k,i)y^i_{k+1}|>M_{\sigma^i_k}]}, \quad \sigma^i_0 = 0, \quad i = 1,\dots,l, \tag{5.6.6}$$

where $a(k,i)$ is the step size at time k and may be random, and $\{M_k, k = 0, 1, \dots\}$ is a sequence of positive numbers increasingly diverging to infinity.

Let us list conditions to be used.

A5.6.1 *$f(\cdot)$ is locally Lipschitz continuous.*

A5.6.2 $a(k,i) > 0$, $a(k,i) \xrightarrow[k\to\infty]{} 0$, $\sum_{k=1}^{\infty} a(k,i) = 0$, $\forall i : 1 \le i \le l$, *and there exist two positive constants a_i, b_i such that*

$$a_i \le \liminf_{k\to\infty} \frac{a(k,i)}{a(k,1)} \le \limsup_{k\to\infty} \frac{a(k,i)}{a(k,1)} \le b_i, \quad a.s. \tag{5.6.7}$$

A5.6.3 *There is a twice continuously differentiable function (not necessarily being nonnegative) $v(\cdot) : \mathbb{R}^l \longrightarrow R$ such that*

$$\max_{u_i\in[a_i,b_i]} \beta(u_2,\dots,u_l) < 0, \quad \forall x \notin J \tag{5.6.8}$$

and $v(J) \triangleq \{v(x) : x \in J\}$ is nowhere dense, where

$$\beta(u_2,\dots,u_l) = f^T(x)\, diag(1,u_2,\dots,u_l) v_x(x),$$

$J \triangleq \{x \in \mathbb{R}^l, f(x) = 0\}$, and $v_x(\cdot)$ denotes the gradient of $v(\cdot)$.

A5.6.4 *For any convergent subsequence $\{x^i_{n_k}\}$, any $i = 1,\dots,l$, and any $T_k \in [0,T]$,*

$$\limsup_{T\to0} \limsup_{k\to\infty} \frac{1}{T} \Big| \sum_{s=n_k}^{m(n_k,T_k)\wedge r(i,\sigma^i_{n_k}+1)} a(s,i)\epsilon^i_{s+1}\Big| = 0, \tag{5.6.9}$$

where $m(n,T) = \inf\{k \geq n, \sum_{s=n}^{k} a(s,1) > T\}$, $r(i,m) = \inf\{k > 0, \quad \sigma_k^i = m\}$, *and* $a \wedge b = \min(a,b)$.

A5.6.5

$$\lim_{k \longrightarrow \infty} \sum_{s=k-\tau_k^{ij}}^{k} a(s,1) = 0, \quad a.s. \tag{5.6.10}$$

Note that (5.6.10) holds if τ_k^{ij} is bounded, since $a(s,1) \xrightarrow[s\longrightarrow\infty]{} 0$. Note also that A5.6.3 holds if $f^i(x)v_x^i(x) < 0$, $\forall i = 1,\ldots,l$, and $\forall x \notin J$.

Theorem 5.6.1 *Let* $x_k \triangleq [x_k^1,\ldots,x_k^l]^T$ *be given by (5.6.1)–(5.6.6) with initial value* $x_0 = [x_0^1,\ldots,x_0^l]^T$. *Assume A5.6.1–A5.6.5 hold, and there is a constant* c_0 *such that* $v(x_*) < \inf_{(x:\max_i |x^i|=c_0)} v(x)$ *and* $\max_i |x_*^i| < c_0$, *where* $v(\cdot)$ *is given in A5.6.3. Then*

$$d(x_k, J) \xrightarrow[k\longrightarrow\infty]{} 0, \quad a.s.,$$

where $d(x,J) = \inf\{\|x-y\| : \forall y \in J\}$.

The proof of the theorem is separated into lemmas. From now on we always assume that A5.6.1–A5.6.5 hold.

We first introduce an auxiliary sequence $\{\tilde{x}_k\}$ and its associated observation noise $\{\tilde{\epsilon}_{k+1}\}$. It will be shown that $\{\tilde{x}_k\}$ differs from $\{x_k\}$ only by a finite number of steps. Therefore, for convergence of $\{x_k\}$ it suffices to prove convergence of $\{\tilde{x}_k\}$.

Let $\{[x_k,\sigma_k] = [(x_k^1,\sigma_k^1),\ldots,(x_k^l,\sigma_k^l)],\ k = 0,1,2,\ldots\}$ be a sample path generated by the algorithm (5.6.1)–(5.6.6), where σ_k^i is the one after resetting according to (5.6.2). Let $r(m) = \inf_i r(i,m)$, where $r(i,m)$ is defined in A5.6.4. Assume $r(j,m) = r(m)$. By the resetting rule given in i), for any k after resetting we have $\sigma_k^i \geq \sigma_{k-\tau_k^{ij}}^j$. For $k = r(i,m)$ we have $\sigma^j_{r(i,m)-\tau^{ij}_{r(i,m)}} \leq \sigma^i_{r(i,m)} = m$, and by the definition of $r(j,m)$

$$\begin{aligned} r(i,m) &\leq r(j,m) + \tau^{ij}_{r(i,m)} \\ &= r(m) + \tau^{ij}_{r(i,m)} \leq r(m) + \max_j \tau^{ij}_{r(i,m)}. \end{aligned} \tag{5.6.11}$$

In the ith processor we take x_*^i and $\tilde{y}_{k+1}^i = 0$ to replace x_k^i and y_{k+1}^i, respectively, and define $\tilde{\epsilon}_{k+1}^i = -f^i(x_*)$ for those k: $r(m) \leq k < r(i,m)$. Further, define $\tilde{x}_k^i = x_k^i$, $\tilde{y}_{k+1}^i = y_{k+1}^i$, and $\tilde{\epsilon}_{k+1}^i = \epsilon_{k+1}^i$ for $k : r(i,m) \leq$

$k < r(m+1)$. Then we obtain new sequences associated with $\{\tilde{x}_k^i\}$. By (5.6.1)–(5.6.6), if $k+1 = r(m)$, then there exists a j with $|\tilde{x}_k^j + a(k,j)\tilde{y}_{k+1}^j| > M_m$, so $x_{k+1}^i = x_*^i$ and

$$\tilde{x}_{k+1}^i = \tilde{x}_k^i + a(k,i)\tilde{y}_{k+1}^i,$$

since $\tilde{x} = x_*^i$ and $y_{k+1}^* = f^i(x_*) + \tilde{\epsilon}_{k+1}^i = 0$ for $r(m) \le k < r(i,m)$. Because during the period $r(i,m) \le k < r(m+1)$ there is no truncation for x_k^i, $\forall i = 1,\dots,l$, the sequences $\{x_k^i\}$, $i = 1,\dots,l$, are recursively updated as follows:

$$\hat{x}_{k+1}^i = \tilde{x}_k^i + a(k,i)\tilde{y}_{k+1}^i, \tag{5.6.12}$$

$$\tilde{x}_{k+1}^i = \hat{x}_{k+1}^i I_{[|\hat{x}_{k+1}^j| < M_{\tilde{\sigma}_k}, \forall j]} + x_*^i I_{[\exists j,\ |\hat{x}_{k+1}^j| \ge M_{\tilde{\sigma}_k}]}, \tag{5.6.13}$$

$$\tilde{\sigma}_{k+1} = \tilde{\sigma}_k + I_{[\exists j,\ |\hat{x}_{k+1}^j| \ge M_{\tilde{\sigma}_k}]}, \quad \tilde{\sigma}_0 = 0, \tag{5.6.14}$$

where $\tilde{\sigma}_k = \max_j \sigma_k^j$.

Define delays $\{\tau_k^{ij}\}$ for $\{\tilde{x}_k\}$ as follows

$$\tilde{\tau}_k^{ij} \triangleq \min\{m : k \ge m \ge 0; \tilde{x}_{k-m}^j$$
$$\text{is available to the } i\text{th processor at time } k\}.$$

Lemma 5.6.1 *For any i, $i = 1,\dots,l$, any convergent subsequence $\{\tilde{x}_{n_k}^i\}$, and any $T_k \in [0,T]$, $\{\tilde{\epsilon}_k\}$ satisfies the following condition*

$$\lim_{T\longrightarrow 0} \limsup_{k\longrightarrow\infty} \frac{1}{T} \Big| \sum_{s=n_k}^{m(n_k,T_k)\wedge r(m_k+1)} a(s,i)\tilde{\epsilon}_{s+1}^i \Big| = 0, \tag{5.6.15}$$

where $m_k = \sup\{m : r(m) \le n_k\}$.

Proof. Since $\tilde{x}_k^i$ equals either x_k^i or x_*^i, which is available at time k, it is seen that

$$\tau_k^{ij} \ge \tilde{\tau}_k^{ij}, \qquad \forall i,j,k. \tag{5.6.16}$$

For $k : r(m) \le k < r(i,m)$ by definition of $\{\tilde{x}_k\}$ we have $\tilde{x}_{r(m)}^j = x_*^j$, which is certainly available to the ith processor. Therefore,

$$\tilde{x}_{k-\tilde{\tau}_k^{ij}}^j = x_*^j. \tag{5.6.17}$$

We rewrite $\tilde{x}_{r(m)}^j = x_*^j$ as $\tilde{x}_{k-(k-r(m))}^j = x_*^j$. By the definition of $\tilde{\tau}_k^{ij}$ and paying attention to (5.6.17) we see

$$k - r(m) \ge \tilde{\tau}_k^{ij}. \tag{5.6.18}$$

We now show that (5.6.18) is true for all $k : r(m) \leq k < r(m+1)$. For $k : r(i,m) \leq k < r(m+1)$, there is no truncation for the ith processor, and hence $\sigma_k^i \geq \sigma^j_{k-\tau_k^{ij}}$, $\forall j$, by the resetting rule i). If $\sigma^j_{k-\tau_k^{ij}} = \sigma_k^i (= m)$ for some $j \neq i$, then by (5.6.16) and the definition of $r(m)$ it follows that

$$k - \widetilde{\tau}_k^{ij} \geq k - \tau_k^{ij} \geq r(m),$$

which implies (5.6.18).

If $\sigma_k^i > \sigma^j_{k-\tau_k^{ij}}$ for some j, then as explained above for the ith processor at time k the latest information about the estimate produced by the jth processor is x_*^j. In other words,

$$\widetilde{x}^j_{k-\widetilde{\tau}_*^{ij}} = x_*^j.$$

However, by definition of $r(j,m)$, $\widetilde{x}^j_{r(j,m)} = x^j_{r(j,m)} = x_*^j$, which yields $k - \widetilde{\tau}_k^{ij} \geq r(j,m) \geq r(m)$. This again implies (5.6.18).

In summary, we have

$$k - \widetilde{\tau}_k^{ij} \geq r(m) \quad \text{for} \quad k : r(m) \leq k < r(m+1). \tag{5.6.19}$$

This means that for $\{\widetilde{x}_k\}$ there is no truncation at any time equal to $k - \widetilde{\tau}_k^{ij}$, and the observation $\widetilde{y}_{k+1}^i$ is carried out at $[\widetilde{x}^1_{k-\widetilde{\tau}_k^{i1}}, \ldots, \widetilde{x}^l_{k-\widetilde{\tau}_k^{il}}]^T$, i.e.,

$$\widetilde{y}_{k+1}^i = f^i(\widetilde{x}^1_{k-\widetilde{\tau}_k^{i1}}, \ldots, \widetilde{x}^l_{k-\widetilde{\tau}_k^{il}}) + \widetilde{\epsilon}_{k+1}^i. \tag{5.6.20}$$

For any i, any convergent subsequence $\{\widetilde{x}_{n_k}^i\}$, and any $T_k \in [0,T]$ we have

$$\begin{aligned}
&\Big|\sum_{s=n_k}^{m(n_k,T_k)\wedge r(m_k+1)} a(s,i)\widetilde{\epsilon}_{m+1}^i\Big| \\
&\leq \Big|\sum_{s=n_k}^{m(n_k,T_k)\wedge r(m_k+1)} a(s,i)\epsilon_{s+1}^i I_{[r(i,m)\leq s<r(m_k+1)]}\Big| \\
&\quad + \Big|f^i(x_*)\sum_{s=n_k}^{m(n_k,T_k)\wedge r(m_k+1)} a(s,i) I_{[r(m)\leq s<r(i,m_k)]}\Big|.
\end{aligned} \tag{5.6.21}$$

By (5.6.11), $r(i,m_k) - r(m_k) \leq \max_j \tau^{ij}_{r(i,m_k)}$. Then from A5.6.2 and A5.6.5 it follows that $\sum\limits_{s=r(m_k)}^{r(i,m_k)} a(s,i) \xrightarrow[k\to\infty]{} 0$, and hence the second term on the right-hand side of (5.6.21) tends to zero as $k \longrightarrow \infty$. Further, from the definition of m_k, there is j such that $\sigma^j_{n_k} = m_k$. Hence the first term on the right-hand side of (5.6.21) is of order $o(T)$ by A5.6.4. Consequently, from A5.6.2, A5.6.4 and A5.6.5 it follows that $\{\widetilde{\epsilon}^i_k\}$ satisfies (5.6.15). □

Lemma 5.6.2 *Let $\{\widetilde{x}_k\}$ be generated by (5.6.12)–(5.6.14). For any convergent subsequence $\{\widetilde{x}_{n_k}\}$ of $\{\widetilde{x}_n\}$, if $\{\widetilde{x}_s, r(\widetilde{\sigma}_{n_k}) \le s \le n_k\}$ is bounded, then there are $c_1 > 0$, $T_1 > 0$, and k_T such that*

$$\|\widetilde{x}_m - \widetilde{x}_{n_k}\| \le c_1 T, \quad \forall T \in [0, T_1], \forall m : n_k \le m \le m(n_k, T), \tag{5.6.22}$$

where $\widetilde{\sigma}_k$ is given in (5.6.14).

Proof. Let $M = \sup_{r(\widetilde{\sigma}_{n_k}) \le s \le n_k} \|\widetilde{x}_s\|$ and $c_1 = 2\bar{b} l H_1$, where

$$H_1 = \max_{\{x : |x^i| < M+1, 1 \le i \le l\}} \|f(x)\|, \quad \bar{b} = 2 \max_{1 \le i \le l} b_i,$$

where b_i is given in A5.6.2.

By (5.6.15) for convergent subsequence $\{\widetilde{x}_{n_k}\}$ there exists $T_0 > 0$ such that for any $T < T_0$ and $T_k \in [0, T]$

$$\limsup_{k \longrightarrow \infty} \frac{1}{T} \Big| \sum_{i=n_k}^{m(n_k, T_k) \wedge r(m_k+1)} a(s,i) \widetilde{\epsilon}^i_{s+1} \Big| < \frac{c_1}{2l}. \tag{5.6.23}$$

Choose $T_1 < T_0$ such that $c_1 T_1 < 1$. For any $T \in [0, T_1]$ let

$$s_k = \max\{s > n_k, \|\bar{x}_s - \widetilde{x}_{n_k}\| \le c_1\}.$$

Then for any $s : n_k \le s \le s_k$,

$$\|\widetilde{x}_s\| \le \|\widetilde{x}_{n_k}\| + c_1 T, \text{ i.e., } \|\widetilde{x}_s\| < M + 1. \tag{5.6.24}$$

If $\sigma \triangleq \lim_{k \longrightarrow \infty} \widetilde{\sigma}_k < \infty$, then $\widetilde{\sigma}_{n_k} = \sigma$, $\forall k > k_1$ if k_1 is sufficiently large, i.e., no truncation occurs after n_{k_1}, and hence for $k > k_1$

$$\widetilde{x}^i_{s+1} = \widetilde{x}^i_s + a(s,i) \widetilde{y}^i_{s+1}, \quad \forall s : n_k \le s \le s_k.$$

If $\lim_{k \longrightarrow \infty} \widetilde{\sigma}_k = \infty$, then there exists k_0 such that $M_{\widetilde{\sigma}_{n_k}} > M + 1$ for any $k > k_0$. From (5.6.24) it follows that

$$\widetilde{x}^i_{s+1} = \widetilde{x}^i_s + a(s,i) \widetilde{y}^i_{s+1}, \quad \forall s : n_k \le s \le s_k. \tag{5.6.25}$$

Therefore, in both cases $s_k < r(m_k + 1)$.

If $s_k < m(n_k, T)$, then for sufficiently large k

$$|\widetilde{x}^i_{s_k+1} - \widetilde{x}^i_{n_k}| < H_1 \sum_{s=n_k}^{s_k} a(s,i) + |\sum_{s=n_k}^{s_k} a(s,i)\widetilde{\epsilon}^i_{s+1}| < \frac{c_1}{l} T,$$

i.e.,

$$\|\widetilde{x}_{s_k+1} - \widetilde{x}_{n_k}\| \leq c_1 T.$$

This contradicts the definition of s_k. Therefore, $s_k \geq m(n_k, T)$. □

Lemma 5.6.3 *Let $\{\tilde{x}_n\}$ be given by (5.6.12)–(5.6.14). For any $[\delta_1, \delta_2]$, $\delta_1 \leq \delta_2$, with $d([\delta_1, \delta_2], v(J)) > 0$, the following assertions take place:*

i) In the case, $\delta_1 < \delta_2$, $v(\widetilde{x}_n)$ cannot cross $[\delta_1, \delta_2]$ infinitely many times keeping $\{\widetilde{x}_s, r(\widetilde{\sigma}_{n_k}) \leq s \leq n_k\}$ bounded, where $\widetilde{x}_{n_k}$ are the starting points of crossing;

ii) In the case $\delta_1 = \delta_2$, $v(\widetilde{x}_n)$ cannot converge to δ_1 keeping $\{\widetilde{x}_s, r(\widetilde{\sigma}_{n_k}) \leq s \leq n_k\}$ bounded.

Proof. i) Since $\{\widetilde{x}_{n_k}\}$ is bounded, there exists a convergent subsequence, which is still denoted by $\{\widetilde{x}_{n_k}\}$ for notational simplicity, $\bar{x} = \lim_{k\longrightarrow\infty} \widetilde{x}_{n_k}$. By the boundedness of $\{\widetilde{x}_n, r(\widetilde{\sigma}_{n_k}) \leq s \leq n_k\}$ and (5.6.22) for sufficiently large k there is no truncation between n_k and $m(n_k, T)$, and hence

$$v(\widetilde{x}_{m(n_k,T)+1}) - v(\widetilde{x}_{n_k}) = \sum_{s=n_k}^{m(n_k,T)} \widetilde{y}^T_{s+1} A(s) v_x(\widetilde{x}_{n_k}) + o(T), \qquad (5.6.26)$$

where $A(s) = \text{diag}(a(s,1), a(s,2), \ldots, a(s,l))$. By (5.6.20), (5.6.22) and $\widetilde{x}_{n_k} \xrightarrow[k\longrightarrow\infty]{} \bar{x}$, it follows that

$$\begin{aligned}
&\sum_{s=n_k}^{m(n_k,T)} \widetilde{y}^T_{s+1} A(s) v_x(\widetilde{x}_{n_k}) \\
&= \sum_{s=n_k}^{m(n_k,T)} \widetilde{y}^T_{s+1} A(s) v_x(\bar{x}) + (\widetilde{x}_{m(n_k,T)+1} - \widetilde{x}_{n_k})(v_x(\widetilde{x}_{n_k} - v_x(\bar{x})) \\
&= \sum_{s=n_k}^{m(n_k,T)} \sum_{i=1}^{l} a(s,i) f^i(\widetilde{x}^1_{s-\tau^{i1}_s}, \ldots, \widetilde{x}^l_{s-\widetilde{\tau}^{il}_s}) v^i_x(\bar{x}) \\
&= \sum_{s=n_k}^{m(n_k,T)} \widetilde{\epsilon}^T_{s+1} A(s) v_x(\bar{x}) + o(T). \qquad (5.6.27)
\end{aligned}$$

By A5.6.2 and A5.6.3 we have

$$\sum_{s=n_k}^{m(n_k,T)} \widetilde{\epsilon}_{s+1}^T A(s) v_x(\bar{x}) + o(T). \tag{5.6.28}$$

Then by A5.6.1

$$|f^i(\widetilde{x}^1_{s-\widetilde{\tau}_s^{i1}}, \dots, \widetilde{x}^d_{s-\widetilde{\tau}_s^{id}}) - f^i(\widetilde{x}_s)| \le L_M \max_j |\widetilde{x}^j_{s-\widetilde{\tau}_s^{ij}} - \widetilde{x}^j_s|,$$

where L_M is the Lipschitz coefficient of f in $\{x \in I\!R^l, |x^i| \le M, \forall i\}$ and $M = \max_{r(\widetilde{\sigma}_{n_k}) \le s \le n_k} \|\widetilde{x}_s\|$. By the boundedness of $\{\widetilde{x}_s, \tilde{r}(\sigma_{n_k}) \le s \le n_k\}$ and the fact that there is no truncation between $r(\widetilde{\sigma}_{n_k})$ and n_k, it follows that

$$\begin{aligned}\widetilde{x}^j_s - \widetilde{x}^j_{s-\widetilde{\tau}_s^{ij}} = &\sum_{k=s-\widetilde{\tau}_s^{ij}}^{s-1} a(k,j) f^j(\widetilde{x}^1_{k-\widetilde{\tau}_k^{j1}}, \dots, \widetilde{x}^l_{k-\widetilde{\tau}_k^{il}}) \\ &+ \sum_{k=s-\widetilde{\tau}_s^{ij}}^{s-1} a(k,j) \widetilde{\epsilon}^j_{k+1}.\end{aligned}$$

Without loss of generality, we may assume $\{\widetilde{x}_{s-\widetilde{\tau}_s^{ij}}\}$ is a convergent sequence. Then by A5.6.3 and A5.6.5

$$\max_j |x^j_{s-\widetilde{\tau}_s^{ij}} - \widetilde{x}^j_s| = o(T). \tag{5.6.29}$$

Therefore,

$$\begin{aligned}&\sum_{i=1}^{l} a(s,i) f^i(\widetilde{x}^1_{s-\widetilde{\tau}_s^{i1}}, \dots, \widetilde{x}^l_{s-\widetilde{\tau}_s^{il}}) v_x^i(\bar{x}) \\ &\quad = f^T(\bar{x}) A(s) v_x(\bar{x}) + (f^T(\widetilde{x}_s) - f^T(\bar{x}) A(s) v_x(\bar{x}) \\ &\quad\quad + \sum_{i=1}^{l} a(s,i)(f^i(\widetilde{x}^1_{s-\widetilde{\tau}_s^{i1}}, \dots, \widetilde{x}^l_{s-\widetilde{\tau}_s^{il}}) - f^i(\widetilde{x}_s)) v_x^i(\bar{x}) \\ &\quad = a(s,1) f^T(\bar{x}) B(s) v_x(\bar{x}) + o(T),\end{aligned} \tag{5.6.30}$$

where

$$B(s) = \mathrm{diag}(1, \frac{a(s,2)}{a(s,1)}, \dots, \frac{a(s,l)}{a(s,1)}).$$

Since $\beta(u_2, \dots, u_l)$ is continuous for fixed x, by A5.6.4 there exists a $\mu > 0$ for $\bar{x}$ such that

$$\beta \triangleq \max_{u_i \in [a_i - \mu, b_i + \mu]} \beta(u_2, \dots, u_l) < 0.$$

Thus, for sufficiently small T and sufficiently large k we have

$$v(\widetilde{x}_{m(n_k,T)+1}) - v(\widetilde{x}_{n_k}) \leq -\beta T/3. \qquad (5.6.31)$$

On the other hand, by Lemma 5.6.2

$$\max_{n_k \leq m \leq m(n_k,T)} |v(\widetilde{x}_m) - v(\widetilde{x}_{n_k})| \xrightarrow[T \longrightarrow 0]{} 0.$$

Thus, $v(\widetilde{x}_{m(n_k,T)+1}) \in [\delta_1, \delta_2]$ for sufficiently small T, and

$$\liminf_{k \longrightarrow \infty} \|\widetilde{x}_{m(n_k,T)+1}\| > \delta_1.$$

This contradicts (5.6.31), and i) is proved.

ii) If $\{\widetilde{x}_n\}$ is bounded, then there is a convergent subsequence $\widetilde{x}_{n_k} \longrightarrow \bar{x}$. Then the assertion can be deduced by a similar way as that for i). □

Lemma 5.6.4 *Under the conditions of Theorem 5.6.1*

$$\lim_{k \longrightarrow \infty} \widetilde{\sigma}_k = \sigma < \infty,$$

where $\widetilde{\sigma}_k$ is given by (5.6.14).

Proof. If $\widetilde{\sigma}_k \xrightarrow[k \longrightarrow \infty]{} \infty$, then there exists a sequence $\{n_k\}$ such that $\widetilde{\sigma}_{n_k} = \widetilde{\sigma}_{n_k - 1} + 1$. From (5.6.12)–(5.6.14) we have $\widetilde{x}_{n_k} = x_*$.

Choose a small positive constant δ such that $\delta < (\inf_{(x: \max_i |x^i| = c_0)} - v(x_*))/3$ and $[v(x_*) + \delta, v(x_*) + 2\delta] \cap v(J) = \emptyset$. Let A_1 be a connected set containing x_* and included in the set $\{x \in \mathbb{R}^l,\ v(x) - v(x_*) < \delta\}$ and let A_2 be a connected set containing x_* and included in the set $\{x \in \mathbb{R}^l,\ v(x) - v(x_*) \leq 2\delta\}$. Clearly, $A_1 \subset A_2$, and A_1 and A_2 are bounded.

Since $\{M_k\}$ diverges to infinity, there exists k_0 such that $M_k > 2c_0$ for $k > k_0$. Noting that there exists i such that

$$|\widetilde{x}^i_{n_{k+1}-1} + a(n_{k+1} - 1, i)\widetilde{y}^i_{n_{k+1}-1}| > M_{\sigma_{n_k}}$$

and $v(\widetilde{x}_{n_k}) = v(x_*)$, we can define $m_k \triangleq \inf\{s > n_k,\ v(\widetilde{x}_s) \notin A_2\}$ and $s_k = \sup\{s < m_k,\ v(\widetilde{x}_s) \in A_1\}$ for $k > k_0$.

Since $\widetilde{x}_{s_k} \in A_1$, there is a convergent subsequence in $\{\widetilde{x}_{s_k}\}$, also denoted by $\{\widetilde{x}_{s_k}\}$. Let $\bar{x}$ be a limit point of $\{\widetilde{x}_{s_k}\}$.

By the definition of m_k, $\{\widetilde{x}_s, n_k \leq s \leq s_k\}$ is bounded. But $\{\widetilde{x}_s, s_k \leq s \leq m_k + 1\}$ crosses $[v(x_*) + \delta, v(x_*) + 2\delta]$ infinitely many times, and it is impossible by Lemma 5.6.3. Thus, $\sigma < \infty$. □

Proof of Theorem 5.6.1

By Lemma 5.6.4 $\{\widetilde{x}_n\}$ is bounded. Let

$$v_1 \triangleq \liminf_{n\longrightarrow\infty} v(\widetilde{x}_n) \leq \limsup_{n\longrightarrow\infty} v(\widetilde{x}_n) \triangleq v_2.$$

If $v_1 = v_2$, then by Lemma 5.6.3, we have $d(v_1, v(J)) = 0$, i.e., $\lim_{n\longrightarrow\infty} d(v(\widetilde{x}_n), v(J)) = 0$. If $v_1 < v_2$, then there are δ_1 and δ_2 such that $[\delta_1, \delta_2] \subset [v_1, v_2]$ and $[\delta_1, \delta_2] \cap v(J) = \emptyset$, since $v(J)$ is nowhere dense. But by Lemma 5.6.3 this is impossible. Therefore, $v_1 = v_2$.

We now show $d(\widetilde{x}_n, J) \xrightarrow[n\longrightarrow\infty]{} 0$. If there is a convergent subsequence $\widetilde{x}_{n_k} \xrightarrow[k\longrightarrow\infty]{} \bar{x}$ and $d(\bar{x}, J) > 0$, then (5.6.26)–(5.6.30) still hold. Hence, $v(\widetilde{x}_{m(n_k,T)}) \xrightarrow[k\longrightarrow\infty]{} v_1 - \beta T/2$. This is a contradiction to $\lim_{n\longrightarrow\infty} v(\widetilde{x}_n) = v_1$. Consequently, $d(\bar{x}, J) = 0$, i.e., $d(\widetilde{x}_n, J) \xrightarrow[n\longrightarrow\infty]{} 0$.

Since $\max_i \sigma_k^i = \widetilde{\sigma}_k$ and $\lim_{k\longrightarrow\infty} \widetilde{\sigma}_k < \infty$, the truncations occur only for finitely many times. Therefore, x_k and $\widetilde{x}_k$ differ from each other only for a finite number of k. So, $d(x_k, J) \xrightarrow[k\longrightarrow\infty]{} 0$. □

5.7. Notes and References

For blind identification with "block" algorithms we refer to [71, 96]. Recursive blind channel identification algorithms appear to be new. Section 5.1 is written on the basis of the joint work "H. F. Chen, X. R. Cao, and J. Zhu, Convergence of stochastic approximation based algorithms for blind channel identification". Principal component analysis is applied in different areas (see, e.g., [36, 79]). The results presented in Section 5.2 are the improved version of those given in [101]. The principal component analysis is applied to solve the blind identification problem in Section 5.3, which is based on the recent work "H. T. Fang and H. F. Chen, Blind channel identification based on noisy observation by stochastic approximation method". The proof of Lemma 5.3.2 is given in [9].

For adaptive filter we refer to [57]. The results presented in Section 5.4 are stronger than those given in [11, 28]. The sign algorithms are dealt with in [42], but conditions used in Section 5.5 are considerably weaker than those in [42]. Section 5.5 is based on the recent work "H. F. Chen and G. Yin, Asymptotic properties of sign algorithms for adaptive filtering".

Asynchronous stochastic approximation was considered in [9, 88, 89, 99]. Section 5.6 is written on the basis of [50].

Chapter 6

APPLICATION TO SYSTEMS AND CONTROL

Assume a control system depends on a parameter θ and the system operation reaches its ideal status when the parameter equals some θ^0. Since θ^0 is unknown, we have to estimate it during the operation of the system, which, therefore, can work only on the estimate θ_k of θ^0. In other words, the real system is not under the ideal parameter θ^0 and the problem is to on-line estimate θ^0 and to make the system asymptotically operating in the ideal status. It is clear that this kind of system parameter identification can be dealt with by SA methods.

Adaptive control for linear stochastic systems is a typical example for the situation described above. If the system coefficients are known, then the optimal stochastic control may be a feedback control of the system state. The corresponding feedback gain can be viewed as the ideal parameter θ^0 which depends on the system coefficients. In the setup of adaptive control, system coefficients are unknown, and hence θ^0 is unknown. The problem is to estimate θ^0 and to prove that the resulting adaptive control system by using the estimate θ_k as the feedback gain is asymptotically optimal as k tends to infinity.

In Section 6.1 the ideal parameter is identified by SA methods for systems in a general setting, and the results are applied to solving the adaptive quadratic control problem. The adaptive stabilization problem is solved for stochastic systems in Section 6.2, while the adaptive exact pole assignment is discussed in Section 6.3. An adaptive regulation problem for nonlinear and nonparametric systems is considered is Section 6.4.

6.1. Application to Identification and Adaptive Control

Consider the following linear stochastic system depending on parameter $\theta \in \mathbb{R}^l$,

$$x_{k+1}(\theta) = A(\theta)x_k(\theta) + D(\theta)e_{k+1}, \quad x_0(\theta) = 0, \tag{6.1.1}$$

where $x_k(\theta) \in \mathbb{R}^m$, $e_k \in \mathbb{R}^e$, $m \times m$-matrix $A(\theta)$ and $m \times e$-matrix $D(\theta)$ are unknown.

The ideal parameter θ^0 for System (6.1.1) is a root of an unknown function $f(\cdot)$:

$$\theta^0 \in J \triangleq \{\theta : f(\theta) = 0\}. \tag{6.1.2}$$

The system actually operates with θ equal to some estimate θ_k for θ^0, i.e., the real system is as follows:

$$x_{k+1}(\theta_k) = A(\theta_k)x_k(\theta_k) + D(\theta_k)e_{k+1}. \tag{6.1.3}$$

For the notational simplicity, we suppress the dependence $\{\theta_k\}$ on the state $\{x_k\}$ and rewrite (6.1.3) as

$$x_{k+1} = A(\theta_k)x_k + D(\theta_k)e_{k+1}. \tag{6.1.4}$$

The observation at time $k+1$ is

$$y_{k+1} = Q(k, \theta_k, x_{k+1}) + \nu_{k+1}, \tag{6.1.5}$$

where $\{\nu_k\}$ is a noise process.

From (6.1.5) it is seen that the function $f(\cdot)$ is not directly observed, but it is connected with $Q(\cdot,\cdot,\cdot)$ as follows:

$$\lim_{k \longrightarrow \infty} EQ(k, \theta, x_{k+1}(\theta)) = f(\theta), \tag{6.1.6}$$

where $x_k(\theta)$ is generated by (6.1.1).

Let $\{M_k\}$ be a sequence of positive numbers increasingly diverging to infinity and let $\theta^* \in \mathbb{R}^l$ be a fixed point. Fixing an initial value θ_0 we recursively estimate θ^0 by the SA algorithm with expanding truncations:

$$\theta_{k+1} = (\theta_k + a_k y_{k+1}) I_{[\|\theta_k + a_k y_{k+1}\| \le M_{\sigma_k}]} + \theta^* I_{[\|\theta_k + a_k y_{k+1}\| > M_{\sigma_k}]}, \tag{6.1.7}$$

$$\sigma_k = \sum_{i=1}^{k-1} I_{[\|\theta_i + a_i y_{i+1}\| > M_{\sigma_i}]}, \quad \sigma_0 = 0. \tag{6.1.8}$$

We list conditions that will be used.

A6.1.1 $a_k > 0,\ \sum_{k=1}^{\infty} a_k = \infty$ *and* $\sum_{k=1}^{\infty} a_k^2 < \infty.$

A6.1.2 *There is a continuously differentiable function* $v(\cdot) : I\!\!R^l \longrightarrow R$ *such that*

$$\sup_{\delta \le d(\theta,J) \le \Delta} f^T(\theta) v_x(\theta) < 0 \qquad (6.1.9)$$

for any $\Delta > \delta > 0$ *and* $v(J) \triangleq \{v(\theta) : \theta \in J\}$ *is nowhere dense, where* J *is given by (6.1.2). Further,* θ^* *used in (6.1.8) is such that* $v(\theta^*) < \inf_{\|\theta\|=c_0} v(\theta)$ *for some* c_0 *and* $\|\theta^0\| < c_0$.

A6.1.3 *The random sequence* $\{e_k\}$ *in (6.1.1) satisfies a mixing condition characterized by*

$$\varphi_n(k) \triangleq \sup_{B \in \mathcal{F}^{\infty}_{n=k}} \operatorname{esssup}_{\omega} |P(B|\mathcal{F}_1^n) - P(B)| \xrightarrow[k\longrightarrow\infty]{} 0 \qquad (6.1.10)$$

uniformly in n, *where* $\mathcal{F}_i^j \triangleq \sigma\{e_i, i \le k \le j\}$. *Further,* $\{e_k\}$ *is such that* $\sup_k \|e_k\| < \xi$ *a.s., where*

$$E\xi^2 < \infty. \qquad (6.1.11)$$

A6.1.4 *For sufficiently large integer* $N(\ge N_0)$

$$\lim_{T\longrightarrow 0} \limsup_{k\longrightarrow\infty} \frac{1}{T} \Big\| \sum_{i=n_k}^{m(n_k,T_k)} a_i \nu_{i+1} I_{[\|\theta_i\| \le N]} \Big\| = 0 \ a.s.\ \forall T_k \in [0,T] \quad (6.1.12)$$

for any $\{n_k\}$ *such that* θ_{n_k} *converges, where* $m(k,T)$ *is given by (1.3.2).*

Let $D_s \triangleq \{\theta : A(\theta) \text{ is stable}\}$, and let D_c be an open, connected subset of D_s with $\overline{D}_c \subset D_s$.

A6.1.5 Q *and* f *are connected by (6.1.6) and (6.1.1) for each* $\theta \in D_R$. Q *satisfies a local Lipschitz condition on* $D_s \times I\!\!R^m$:

$$\begin{aligned} &\|Q(k,\theta_1,x_1) - Q(k,\theta_2,x_2)\| I_{[\|\theta_1\|+\|\theta_2\|+\|x_1\|+\|x_2\| \le N]} \\ &\qquad \le K(N)(\|\theta_1 - \theta_2\| + \|x_1 - x_2\|) \end{aligned} \qquad (6.1.13)$$

with $EK^2(c_1 + c_2\xi)(1+\xi^2) < \infty$ *for any constants* c_1 *and* c_2, *where* ξ *is given in A6.1.3.*

A6.1.6 *$A(\cdot)$ and $D(\cdot)$ in (6.1.1) are globally Lipschitz continuous:*

$$\|A(\theta_1) - A(\theta_2)\| + \|D(\theta_1) - D(\theta_2)\| \leq L\|\theta_1 - \theta_2\|,$$

where L is a constant.

A6.1.7 *$\{\theta_k\}$ given by (6.1.7) is D_c-valued. If $\{\theta_{n_k}\}$ converges for some ω, then $\|x_{n_k}\| < c$ where c may depend on ω.*

Theorem 6.1.1 *Assume A6.1.1–A6.1.7 hold. Then*

$$\lim_{k \longrightarrow \infty} d(\theta_k, J^*) = 0 \quad a.s.,$$

where J^* is a connected subset of $\overline{J}$.

Proof. By (6.1.5) we rewrite the observation y_{k+1} in the standard from

$$y_{k+1} = f(\theta_k) + \varepsilon_{k+1} \tag{6.1.14}$$

where

$$\varepsilon_{k+1} = Q(k, \theta_k, x_{k+1}) - f(\theta_k) + \nu_{k+1}. \tag{6.1.15}$$

By Theorem 2.2.2 and Condition A6.1.4, the assertion of the theorem will immediately follow if we can show that for almost all ω condition (2.2.2) is satisfied with $\varepsilon_{i+1}(x_i(\omega), \omega)$ replaced by $Q(i, \theta_i, x_{i+1}) - f(\theta_i)$.

Let $Q(i, \theta_i, x_{i+1}) - f(\theta_i)$ be expressed as a sum of seven terms:

$$Q(i, \theta_i, x_{i+1}) - f(\theta_i) = \sum_{j=1}^{7} \nu_{i+1}^{(j)} \tag{6.1.16}$$

where

$$\begin{aligned}
\nu_{i+1}^{(1)} =& Q(i, \theta_i, x_{i+1}) - Q(i, \theta, x_{i+1}(\theta)),\\
\nu_{i+1}^{(2)} =& Q(i, \theta, x_{i+1}(\theta)) - E[Q(i, \theta, x_{i+1}(\theta))|\mathcal{F}_1^{i-k}],\\
\nu_{i+1}^{(3)} =& E[Q(i, \theta, x_{i+1}(\theta))|\mathcal{F}_1^{i-k}]\\
& - \int Q(i, \theta, x(i, k, \theta)) dF(y_{i-\frac{k}{2}+1}, \cdots, y_{i+1}|\mathcal{F}_1^{i-k}),\\
\nu_{i+1}^{(4)} =& \int Q(i, \theta, x(i, k, \theta)) dF(y_{i-\frac{k}{2}+1}, \cdots, y_{i+1}|\mathcal{F}_1^{i-k})\\
& - \int Q(i, \theta, x(i, k, \theta)) dF(y_{i-\frac{k}{2}+1}, \cdots, y_{i+1}),\\
\nu_{i+1}^{(5)} =& \int Q(i, \theta, x(i, k, \theta)) dF(y_{i-\frac{k}{2}+1}, \cdots, y_{i+1}) - EQ(i, \theta, x_{i+1}(\theta)),\\
\nu_{i+1}^{(6)} =& EQ(i, \theta, x_{i+1}(\theta)) - f(\theta),\\
\nu_{i+1}^{(7)} =& f(\theta) - f(\theta_i),
\end{aligned}$$

where

$$x(i,k,\theta) = A^{k-1}(\theta)x_{i-k}(\theta) + \sum_{j=i-\frac{k}{2}+1}^{i+1} A^{i+1-j}(\theta)D(\theta)y_j, \qquad (6.1.17)$$

and $F(y_s,\ldots,y_{i+1})$ and $F(y_s,\ldots,y_{i+1}|\mathcal{F}_1^k)$ denote the distribution and conditional distribution of $(e_s,\ldots,e_{i+1})$ given $\mathcal{F}_1^k$, respectively.

To prove the theorem it suffices to show that there exists Ω_0 with $P\Omega_0 = 1$ such that for each $\omega \in \Omega_0$, all $\{\nu_i^{(j)}\}$, $j = 1,\ldots,7$ satisfy (2.2.2) with $\varepsilon_{i+1}(x_i(\omega),\omega)$ respectively identified to $\nu_{i+1}^{(j)}$, $j = 1,\ldots,7$.

By definition, for any $\theta \in \overline{D}_c \subset D_s$ there is $\lambda(\theta) \in (0,1)$ such that

$$\|A^n(\theta)\| \le c_1(\theta)\lambda^n(\theta), \quad \forall n \ge 1, \qquad (6.1.18)$$

where $c_1(\theta)$ is independent of ω.

Let us first show that $\{\nu_i^{(j)}\}, j = 2,3,4,5$ satisfy (2.2.2).

Solving (6.1.1) yields

$$x_{k+n}(\theta) = A^n(\theta)x_k(\theta) + \sum_{j=k+1}^{k+n} A^{k+n-j}(\theta)D(\theta)e_j, \quad x_0 = 0. \qquad (6.1.19)$$

By A6.1.3 $\{e_k\}$ is bounded. Hence, by (6.1.18) $\{x_k(\theta)\}$ is bounded and by A6.1.5 $\{Q(i,\theta,x_{k+1}(\theta)\}$ is also bounded:

$$\|x_i(\theta)\| \le c(\theta), \quad \|Q(i,\theta,x_{i+1}(\theta))\| \le c'(\theta), \quad \forall i, \qquad (6.1.20)$$

where

$$c(\theta) = \frac{c_1(\theta)\|B(\theta)\|\xi}{1-\lambda(\theta)}, \; c'(\theta) = \|Q(0,0,0)\| + K(c(\theta)+\|\theta\|)(c(\theta)+\|\theta\|),$$

where $K(\cdot)$ is given in A6.1.5.

Since $E\xi^2 < \infty$, we have $E(c'(\theta))^2 < \infty$.

We now show that $x_i(\theta)$ and $Q(i,\theta,x_{i+1}(\theta))$ are continuous in θ uniformly with respect to i.

By (6.1.18) and (6.1.20), from (6.1.19) it follows that

$$\begin{aligned}\|x_{i+1}(\theta_1) - x_{i+1}(\theta_2)\| =& \Big\|\sum_{j=1}^{i+1} A^{i+1-j}(\theta_1)\Big[(A(\theta_1)-A(\theta_2))x_j(\theta_2) \\ &+ (D(\theta_1)-D(\theta_2))e_{j+1}\Big]\Big\| \\ \le& c_1(\theta_1)L\sum_{j=1}^{i+1}\lambda^{i+1-j}(\theta_1)\|\theta_1-\theta_2\|(c(\theta_2)+\xi) \\ \le& \frac{c_1(\theta_1)L(c(\theta_2)+\xi)}{1-\lambda(\theta_1)}\|\theta_1-\theta_2\|, \end{aligned} \qquad (6.1.21)$$

which implies the uniform continuity of $x_i(\theta)$. This together with (6.1.13) yield that $Q(i,\theta,x_{i+1}(\theta))$ is also uniformly continuous.

Let D_R be a countable dense subset of D_c.

Noticing that $Q(i,\theta,x_{k+1}(\theta))$ is $\mathcal{F}_1^{i+1}$-measurable, and expressing $\nu_{i+1}^{(2)}$ as a sum of k martingale difference sequences

$$\nu_{i+1}^{(2)} = \sum_{s=0}^{k} E(Q(i,\theta,x_{k+1}(\theta))|\mathcal{F}_1^{i+1-s}) - E(Q(i,\theta,x_{i+1}(\theta))|\mathcal{F}_i^{i-s}),$$

by (6.1.20) and $E(c'(\theta))^2 < \infty$, we find that there is Ω_1 with $P\Omega_1 = 1$ such that for each $\omega \in \Omega_1$

$$\sum_{i=1}^{\infty} a_i \nu_{i+1}^{(2)} < \infty$$

for any integer k and any $\theta \in D_R$. From here by uniform continuity of $Q(i,\theta,x_{i+1}(\theta))$, it follows that for $\omega \in \Omega_1$ and for any integer k

$$\limsup_{n\longrightarrow\infty} \frac{1}{T}\|\sum_{j=n}^{m(n,t)} a_j \nu_{j+1}^{(2)}\| = 0, \quad \forall t \in [0,T],\ \forall T > 0,\ \forall \theta \in \overline{D}_c. \quad (6.1.22)$$

Note that

$$\begin{aligned}
&\int Q(i,\theta,x(i,k,\theta))dF(y_{i-\frac{k}{2}+1},\dots,y_{i+1}|\mathcal{F}_1^{i-k})\\
&= \int Q(i,\theta,A^{k+1}(\theta)x_{i-k}(\theta) + \sum_{j=i-\frac{k}{2}+1}^{i+1} A^{i+1-j}(\theta)D(\theta)y_j))\cdot\\
&\quad \cdot dF(y_{i-\frac{k}{2}+1},\dots,y_{i+1}|\mathcal{F}_1^{i-k})\\
&= E[Q\big(i,\theta,A^{k+1}(\theta)x_{i-k}(\theta) + \sum_{j=i-\frac{k}{2}+1}^{i+1} A^{i+1-j}(\theta)D(\theta)e_j\big)|\mathcal{F}_1^{i-k}]
\end{aligned}$$

and

$$x_{i+1}(\theta) = A^{k+1}(\theta)x_{i-k}(\theta) + \sum_{j=i-k+1}^{i+1} A^{i+1-j}(\theta)D(\theta)e_j.$$

By (6.1.18) (6.1.20) and the Lipschitz condition A6.1.5 for Q it follows that

$$\|\nu_{i+1}^{(3)}\| \le E[K(\|\theta\| + c(\theta))\|\sum_{j=i-k+1}^{i-\frac{k}{2}} A^{i+1-j}(\theta)D(\theta)e_j\||\mathcal{F}_1^{i-k}]$$

$$\leq \frac{c_1(\theta)\|D(\theta)\|(\lambda(\theta))^{\frac{k}{2}+1}}{1-\lambda(\theta)} E[\xi K(\|\theta\|+c(\theta))|\mathcal{F}_1^{i-k}] \xrightarrow[(k\leq i)]{k\longrightarrow\infty} 0, \quad (6.1.23)$$

since $E[\xi K(\|\theta\|+c(\theta))|\mathcal{F}_1^i]$ is bounded by the martingale convergence theorem. It is worth noting that (6.1.23) holds a.s. for any $\theta \in \bar{D}_c$, but without loss of generality (6.1.23) may be assumed to hold for all $\theta \in \overline{D}_c$ on $\Omega_2 \subset \Omega_1$ with $P\Omega_2 = 1$. To see this, we first select $\Omega_2 \subset \Omega_1$ with $P\Omega_2 = 1$ such that (6.1.23) holds for any $\theta \in D_R$. This is possible because D_R is a countable set. Then, we notice that $Q(i, \theta, x_{i+1}(\theta))$ is continuous in θ uniformly with respect to i. Thus, we have

$$\|\nu_{i+1}^{(3)}\| \xrightarrow[i\longrightarrow\infty]{} 0, \quad \forall\, \theta \in \overline{D}_c, \quad \forall \omega \in \Omega_2. \quad (6.1.24)$$

Similarly, we can find $\Omega_3 \subset \Omega_2$ with $P\Omega_3 = 1$ such that for $\forall \omega \in \Omega_3$ and $\forall \theta \in \overline{D}_c$

$$\|\nu_{i+1}^{(5)}\| \xrightarrow[i\longrightarrow\infty]{} 0. \quad (6.1.25)$$

This is because by (6.1.18) and (6.1.20) we have the following estimate:

$$\begin{aligned}
\|\nu_{i+1}^{(5)}\| =& \|EQ(i,\theta, A^{k+1}(\theta)y + \sum_{j=i-\frac{k}{2}+1}^{i+1} A^{i+1-j}(\theta)D(\theta)e_j)|_{y=x_{i-k}(\theta)} - \\
& - EQ(i,\theta, \sum_{j=1}^{i-\frac{k}{2}} A^{i+1-j}(\theta)D(\theta)e_j + \sum_{j=i-\frac{k}{2}+1}^{i+1} A^{i+1-j}(\theta)D(\theta)e_j)\| \\
\leq & E[K(\|\theta\|+c(\theta)) \cdot \|A^{k+1}(\theta)y - \sum_{j=1}^{i-\frac{k}{2}} A^{i+1-j}(\theta)D(\theta)e_j\|]|_{y=x_{i-k}(\theta)} \\
\leq & c_1(\theta)\lambda^{\frac{k}{2}+1}(\theta)E[K(\|\theta\|+c(\theta))(\lambda^{\frac{k}{2}}(\theta)\|y\| + \|x_{i-\frac{k}{2}}(\theta)\|)]|_{y=x_{i-k}(\theta)} \\
\leq & c_1(\theta)\lambda^{\frac{k}{2}+1}(\theta)[c(\theta)\lambda^{\frac{k}{2}}(\theta)EK(\|\theta\|+c(\theta)) + Ec(\theta)K(\|\theta\|+c(\theta))] \\
& \xrightarrow[k\leq i]{k\longrightarrow\infty} 0 \qquad (6.1.26)
\end{aligned}$$

We now estimate $\nu_{i+1}^{(4)}$ by the treatment used in Lemma 2.5.2. By applying the Jordan-Hahn decomposition to the signed measure

$$\begin{aligned}
& dG(y_{i-\frac{k}{2}+1}, \cdots, y_{i+1}; \omega) \\
& \triangleq dF(y_{i-\frac{k}{2}+1}, \cdots, y_{i+1}|\mathcal{F}_1^{i-k}) - dF(y_{i-\frac{k}{2}+1}, \cdots, y_{i+1})
\end{aligned}$$

it is seen that there is a Borel set D in the sampling space $(y_{i-\frac{k}{2}}, \dots, y_{i+1})$ such that for any A in the sampling space

$$\begin{aligned}&\int_A dG^+(y_{i-\frac{k}{2}+1}, \dots, y_{i+1}; \omega)\\&= \int_{A\cap D^c} dG(y_{i-\frac{k}{2}+1}, \dots, y_{i+1}; \omega) \le \varphi_{i-k}(\frac{k}{2}+1)\end{aligned} \tag{6.1.27}$$

$$\begin{aligned}&\int_A dG^-(y_{i-\frac{k}{2}+1}, \dots, y_{i+1}; \omega)\\&= \int_{A\cap D} dG(y_{i-\frac{k}{2}+1}, \dots, y_{i+1}; \omega) \le \varphi_{i-k}(\frac{k}{2}+1)\end{aligned} \tag{6.1.28}$$

and

$$\begin{aligned}&dG(y_{i-\frac{k}{2}+1}, \dots, y_{i+1}; \omega)\\&= dG^+(y_{i-\frac{k}{2}+1}, \dots, y_{i+1}; \omega) - dG^-(y_{i-\frac{k}{2}+1}, \dots, y_{i+1}; \omega)\end{aligned}$$

$$\begin{aligned}&dG^+(y_{i-\frac{k}{2}+1}, \dots, y_{i+1}; \omega) + dG^-(y_{i-\frac{k}{2}+1}, \dots, y_{i+1}; \omega)\\&= dF(y_{i-\frac{k}{2}+1}, \dots, y_{i+1}|\mathcal{F}_1^{i-k}) + dF(y_{i-\frac{k}{2}+1}, \dots, y_{i+1}),\end{aligned} \tag{6.1.29}$$

where $\varphi_{i-k}(\frac{k}{2}+1)$ is the mixing coefficient given in A6.1.3. Thus, by (6.1.27)–(6.1.29) we have

$$\begin{aligned}\|\nu_{i+1}^{(4)}\| \le &\int \|Q(i,\theta, A^{k+1}(\theta)x_{i-k}(\theta) + \sum_{j=i-\frac{k}{2}+1}^{i+1} A^{i+1-j}(\theta)D(\theta)y_j)\|\\&\cdot (dG^+(y_{i-\frac{k}{2}+1}, \dots, y_{i+1}; \omega) + dG^-(y_{i-\frac{k}{2}+1}, \dots, y_{i+1}; \omega))\\ \le&\varphi_{i-k}^{\frac{1}{2}}(\frac{k}{2}+1)[(\int \|Q(i,\theta, A^{k+1}(\theta)x_{i-k}(\theta)\\&+ \sum_{j=i-\frac{k}{2}+1}^{i+1} A^{i+1-j}(\theta)D(\theta)y_j)\|^2 dG^+(y_{i-\frac{k}{2}+1}, \dots, y_{i+1}; \omega)]^{\frac{1}{2}}\\&+ \varphi_{i-k}^{\frac{1}{2}}(\frac{k}{2}+1)[(\int \|Q(i,\theta, A^{k+1}(\theta)x_{i-k}(\theta)\\&+ \sum_{j=i-\frac{k}{2}+1}^{i+1} A^{i+1-j}(\theta)D(\theta)y_j)\|^2 dG^-(y_{i-\frac{k}{2}+1}, \dots, y_{i+1}; \omega)]^{\frac{1}{2}}\\ \le&\sqrt{2}\varphi_{i-k}^{\frac{1}{2}}(\frac{k}{2}+1)[\int \|Q(i,\theta, A^{k+1}(\theta)x_{i-k}(\theta)\end{aligned}$$

$$
\begin{aligned}
&+ \sum_{j=i-\frac{k}{2}+1}^{i+1} A^{i+1-j}(\theta)D(\theta)y_j)\|^2 \cdot (dF(y_{i-\frac{k}{2}+1}, \dots, y_{i+1}|\mathcal{F}_1^{i-k}) \\
&+ dF(y_{i-\frac{k}{2}+1}, \dots, y_{i+1}))]^{\frac{1}{2}} \\
=&\sqrt{2}\varphi_{i-k}^{\frac{1}{2}}(\frac{k}{2}+1)[E(\|Q(i,\theta,A^{k+1}(\theta)x_{i-k}(\theta) \\
&+ \sum_{j=i-\frac{k}{2}+1}^{i+1} A^{i+1-j}(\theta)D(\theta)e_j)\|^2|\mathcal{F}_1^{i-k}) \\
&+ (E\|Q(i,\theta,A^{k+1}(\theta)y + \sum_{j=i-\frac{k}{2}+1}^{i+1} A^{i+1-j}(\theta)D(\theta)e_j)\|^2)|_{y=x_{i-k}(\theta)}]^{\frac{1}{2}}.
\end{aligned}
\tag{6.1.30}
$$

By A6.1.5, (6.1.18), (6.1.20), and noticing $\|e_i\| < \xi$, we find

$$
\begin{aligned}
&\|Q(i,\theta,A^{k+1}(\theta)x_{i-k}(\theta) + \sum_{j=i-\frac{k}{2}+1}^{i+1} A^{i+1-j}(\theta)D(\theta)e_j)\|^2 \\
&\leq \left[\|Q(0,0,0) + K(\|\theta\| + c(\theta))(\|\theta\| + c(\theta))\right]^2 = (c'(\theta))^2
\end{aligned}
$$

whose expectation is finite as explained for (6.1.20). Therefore, on the right-hand side of (6.1.30) the conditional expectation is bounded with respect to i by the martingale convergence theorem, and the last term is also bounded with respect to i. Thus, by (6.1.10) from (6.1.30) it follows that there is $\Omega_4 \subset \Omega_3$ with $P\Omega_4 = 1$ such that

$$
\|\nu_{i+1}^{(4)}\| \xrightarrow[k \leq i]{k\longrightarrow\infty} 0, \quad \forall\theta \in \overline{D}_c, \quad \forall\omega \in \Omega_4. \tag{6.1.31}
$$

Let $\omega \in \Omega_4$ be fixed. Assume $\{\theta_{n_k}\}$ is a convergent subsequence $\theta_{n_k} \xrightarrow[k\longrightarrow\infty]{} \theta \in \overline{D}_c \subset D_s$.

Define

$$
\Delta_{m,n_k}(\theta) \triangleq \max_{n_k \leq i \leq m} \|\theta_i - \theta\|, \quad m \geq n_k. \tag{6.1.32}
$$

Write (6.1.4) as

$$
x_i = A(\theta)x_{i-1} + (A(\theta_{i-1}) - A(\theta))x_{i-1} + D(\theta_{i-1})e_i.
$$

Since $\|x_{n_k}\| \leq c$ by A6.1.7, by A6.1.5 and A6.1.6 it follows that for $i \geq n_k$

$$\begin{aligned}\|x_i\| \leq & c_1(\theta)\lambda^{i-n_k}(\theta)\|x_{n_k}\| + c_1(\theta) \sum_{j=n_k+1}^{i} \lambda^{i-j}(\theta)[\|A(\theta_{j-1}) - A(\theta)\|\|x_{j-1}\| \\ & + (\|D(\theta_{j-1}) - D(\theta)\| + \|D(\theta)\|)\|e_j\|] \\ \leq & c_1(\theta)c\lambda^{i-n_k}(\theta) + c_1 L \sum_{j=n_k+1}^{i} \lambda^{i-j}(\theta)\Delta_{j-1,n_k}(\theta)(\|x_{j-1}\| + 1) \\ & + c_1(\theta) \sum_{j=n_k+1}^{i} \lambda^{i-j}(\theta)\|D(\theta)\|\xi).\end{aligned}$$

Consequently, we have

$$1 + \|x_i\| \leq c_2(\theta) + c_1(\theta)L \sum_{j=n_k+1}^{i} \lambda^{i-j}(\theta)\Delta_{j-1,n_k}(\theta)(1 + \|x_{j-1}\|),$$

where and hereafter $c_i(\theta)$ always denotes a constant for fixed ω and, without loss of generality, we assume $c_2(\theta) \geq 1$.

Define $h_i = \lambda^{-i}(\theta)(1 + \|x_i\|)$ so that

$$h_i \leq c_2(\theta)\lambda^{-i}(\theta) + \frac{c_1(\theta)L}{\lambda(\theta)} \sum_{j=n_k+1}^{i} \Delta_{j-1,n_k}(\theta)h_{j-1}. \tag{6.1.33}$$

Applying the Gronwall inequality to (6.1.33) we obtain the inequality

$$\begin{aligned}h_i \leq & c_2(\theta)\lambda^{-i}(\theta) \\ & + \sum_{s=n_k}^{i-1} c_2(\theta)\lambda^{-s}(\theta)(1 + \frac{c_1(\theta)L\Delta_{i-1,n_k}(\theta)}{\lambda(\theta)})^{i-s-1},\end{aligned}$$

and hence

$$\begin{aligned}1 + \|x_i\| \leq & c_2(\theta) + \lambda(\theta) \sum_{s=n_k}^{i-1} (\lambda(\theta) + c_1(\theta)L\Delta_{i-1,n_k}(\theta))^{i-s-1} \\ \leq & c_2(\theta) + \sum_{s=n_k}^{i-1} (\lambda(\theta) + c_1(\theta)\Delta_{i-1,n_k}(\theta))^{i-s}.\end{aligned} \tag{6.1.34}$$

Now choose $T(\theta) \in (0, 1]$ sufficiently small so that

$$\lambda_1(\theta) \triangleq \lambda(\theta) + c_1(\theta)Lc_3(\theta)T(\theta) < 1, \tag{6.1.35}$$

where $c_3(\theta) \triangleq 1 + \|f(\theta)\|$.

By induction we now show that

$$\Delta_{m,n_k}(\theta) < c_3(\theta)T(\theta), \quad \forall m : n_k \le m \le m(n_k, T(\theta)) \tag{6.1.36}$$

for all suitable large k.

For any fixed $T(\theta)$, $\Delta_{n_k,n_k} \le \frac{1}{2}T(\theta)$ if k is large enough, since $\theta_{n_k} \longrightarrow \theta$.

Therefore (6.1.36) holds for $m = n_k$, since $c_3(\theta) = 1 + \|f(\theta)\| > \frac{1}{2}$.

Assume (6.1.36) holds for some $m \in \{n_k, \dots, m(n_k, T(\theta))\}$. By noticing $c_2 \ge 1$, from (6.1.34) and (6.1.35) it follows that

$$\|x_{i+1}\| \le \frac{c_2(\theta)}{1 - \lambda_1(\theta)}, \quad i = n_k, \dots, m. \tag{6.1.37}$$

By using (6.1.20) (6.1.37) and the inductive assumption and applying (6.1.19) to $\nu_{i+1}^{(1)}$ it follows that

$$\|\nu_{i+1}^{(1)}\| \le K(\Delta_{i,n_k}(\theta) + \|\theta\| + \frac{c_2(\theta)}{1 - \lambda_1(\theta)} + c(\theta))(\Delta_{i,n_k}(\theta) + \tilde{x}_{i+1}) \tag{6.1.38}$$

for $i = n_k, \dots, m$, where $\tilde{x}_{i+1} = x_{i+1}(\theta_i) - x_{i+1}(\theta)$ and $\tilde{x}_{i+1}$ satisfies the following equation

$$\tilde{x}_{i+1} = A(\theta)\tilde{x}_i + [A(\theta_i) - A(\theta)]x_i + [D(\theta_i) - D(\theta)]e_{i+1}.$$

By A6.1.7 and (6.1.20) we have

$$\|\tilde{x}_{n_k}\| \le C + c_4(\theta),$$

and using (6.1.18), (6.1.37), and the inductive assumption we derive

$$\begin{aligned}
\|\tilde{x}_{i+1}\| \le & c_1(\theta)c_4(\theta)\lambda^{i+1-n_k}(\theta) + c_1(\theta) \sum_{j=n_k+1}^{i+1} \lambda^{i+1-j}(\theta)[\|A(\theta_{j-1}) - A(\theta)\| \\
& \cdot \|x_{j-1}\| + \|D(\theta_{j-1}) - D(\theta)\|\xi] \\
\le & c_1(\theta)c_4(\theta)\lambda^{i+1-n_k}(\theta) + c_1(\theta)L \sum_{j=n_k+1}^{i+1} \lambda^{i+1-j}(\theta)\Delta_{j-1,n_k}(\theta) \\
& \cdot (\|x_{j-1}\| + \xi) \\
\le & c_1(\theta)c_4(\theta)\lambda^{i+1-n_k}(\theta) + c_1(\theta)Lc_3(\theta)T(\theta)\Big(\frac{c_2(\theta)}{1 - \lambda_1(\theta)} + \xi\Big)\frac{1}{1 - \lambda(\theta)}.
\end{aligned}$$

This combining with (6.1.38) leads to that there are real numbers $c_5(\theta)$ and $c_6(\theta)$ such that

$$\|\nu_{i+1}^{(1)}\| \leq c_5(\theta)\lambda^{i+1-n_k}(\theta) + c_6(\theta)T(\theta)$$

for $i = n_k, \ldots, m$. From here it follows that

$$\|\sum_{i=n_k}^{m} a_i \nu_{i+1}^{(1)}\| \leq \max_{n_k \leq i \leq m} a_i c_5(\theta) \frac{\lambda(\theta)}{1-\lambda(\theta)} + c_6(\theta)T^2(\theta). \tag{6.1.39}$$

From the inductive assumption it follows that for $i = n_k, \ldots, m$

$$\|\theta_i\| \leq \|\theta\| + \Delta_{m,n_k}(\theta) \leq \|\theta\| + c_3(\theta)T(\theta) \leq N$$

for some large enough integer N. Then by (6.1.12)

$$\|\sum_{i=n_k}^{j} a_i \nu_{i+1}\| = o(T(\theta)), \quad \text{as} \quad k \longrightarrow \infty, \quad \forall j = n_k, \ldots, m. \tag{6.1.40}$$

Setting

$$g_{n_k,j}(\theta) = \max_{n_k \leq i \leq j} a_i c_5(\theta) \frac{\lambda(\theta)}{1-\lambda(\theta)} + \max_{n_k \leq i \leq j} \|EQ(i, \theta, x_{i+1}(\theta)) - f(\theta)\| T(\theta),$$

we derive

$$\begin{aligned}
&\|\sum_{i=n_k}^{j} a_i (Q(i, \theta_i, x_{i+1}) + \nu_{i+1})\| \\
&\leq \|\sum_{i=n_k}^{j} a_i \nu_{i+1}^{(1)}\| + \|\sum_{i=n_k}^{j} a_i \nu_{i+1}^{(2)}\| + \|\sum_{i=n_k}^{j} a_i \nu_{i+1}^{(3)}\| \\
&\quad + \|\sum_{i=n_k}^{j} a_i \nu_{i+1}^{(4)}\| + \|\sum_{i=n_k}^{j} a_i \nu_{i+1}^{(5)}\| + \|\sum_{i=n_k}^{j} a_i \nu_{i+1}^{(6)}\| \\
&\quad + \|\sum_{i=n_k}^{j} a_i \nu_{i+1}\| + \|f(\theta)\| T(\theta) \\
&\leq g_{n_k,j}(\theta) + (c_6(\theta)T(\theta) + \|f(\theta)\| + o(1))T(\theta),
\end{aligned} \tag{6.1.41}$$

where (6.1.22), (6.1.24), (6.1.25), (6.1.31), (6.1.39), and (6.1.40) are used.

Choose $T(\theta)$ sufficiently small so that $T(\theta) < 1$, (6.1.35) holds, and $c_6(\theta)T(\theta) < \frac{1}{4}$.

Since $a_i \longrightarrow 0$, by A6.1.5 there is k_1 such that

$$\frac{g_{n_k,j}(\theta)}{T(\theta)} < \frac{1}{4} \quad \text{and} \quad c_6(\theta)T(\theta) + o(1) < \frac{1}{4}$$

for all $k > k_1$. From (6.1.41) it then follows that

$$\|\sum_{i=n_k}^{j} a_i(Q(i,\theta_i,x_{i+1}) + \nu_{i+1})\| \le (\frac{1}{2} + \|f(\theta)\|)T(\theta), \tag{6.1.42}$$

$$\forall k \ge k_1, \quad \forall j : n_k \le j \le m.$$

It can be assumed that k_1 is sufficiently large so that

$$1 + \|f(\theta)\| + \|\theta\| < M_{\sigma_{n_k}}, \quad \forall k \ge k_1.$$

Since $\|\theta_{n_k} - \theta\| = \Delta_{n_k,n_k} \le \frac{1}{2}T(\theta)$, by (6.1.42) it follows that

$$\|\theta_{n_k} + \sum_{i=n_k}^{j} a_i(Q(i,\theta_i,x_{i+1}) + \nu_{i+1})\| \le \|\theta\| + (1 + \|f(\theta)\|)T(\theta) < M_{\sigma_{n_k}},$$

$\forall j = n_k, n_k+1, \ldots, m, \forall k \ge k_1$, and hence there is no truncation at times $j = n_k+1, \ldots, m$, and

$$\theta_{m+1} = \theta_m + a_m(Q(m,\theta_m,x_{m+1}) + \nu_{m+1}).$$

Thus, we have

$$\begin{aligned}\|\theta_{m+1} - \theta\| &\le \|\theta_{n_k} - \theta\| + \|\sum_{i=n_k}^{m} a_i(Q(i,\theta_i,x_{i+1}) + \nu_{i+1}\| \\ &< c_3(\theta)T(\theta),\end{aligned}$$

or equivalently,

$$\Delta_{m+1,n_k}(\theta) < c_3(\theta)T(\theta),$$

which proves (6.1.36).

Consequently, (6.1.39) is valid for $m : n_k \le m \le m(n_k, T(\theta))$, and hence

$$\lim_{T \longrightarrow 0} \limsup_{k \longrightarrow \infty} \frac{1}{T}\|\sum_{i=n_k}^{m(n_k,t)} a_i \nu_{i+1}^{(1)}\| = 0, \quad \forall t \in [0,T]. \tag{6.1.43}$$

From (6.1.21) and (6.1.13) it is seen that $EQ(i,\theta,x_{i+1}(\theta))$ is continuous in θ uniformly with respect to i. Therefore, its limit $f(\theta)$ is a continuous function. Then by (6.1.36) it follows that

$$\lim_{T\longrightarrow 0}\limsup_{k\longrightarrow\infty}\frac{1}{T}\|\sum_{i=n_k}^{m(n_k,t)} a_i\nu_{i+1}^{(7)}\| \le \lim_{T\longrightarrow 0} o(T)=0.$$

Finally, noticing that A6.1.5 assumes (6.1.6), we conclude that for each $\omega\in\Omega_4\triangleq\Omega_0$, all $\{\nu_i^{(j)}\}, j=1,\cdots,7$, satisfy (2.2.2) with $\varepsilon_{i+1}(x_i(\omega),\omega)$ respectively replaced by $\nu_{i+1}^{(j)}, j=1,\cdots,7$. The proof of the theorem is completed. □

We now apply the obtained result to an adaptive control problem.

Assume that $\theta^0\in\mathbb{R}^l$ is the ideal parameter for the system, being the unique zero of an unknown function $f(\cdot)$. The system in the ideal condition is described by the equation

$$x_{k+1}(\theta^0)=A(\theta^0)x_k(\theta^0)+B(\theta^0)u_k+D(\theta^0)e_{k+1}, \tag{6.1.44}$$

where u_k is the feedback control which is required to minimize

$$J(u)=\limsup_{n\longrightarrow\infty}\frac{1}{n}\sum_{k=1}^{n}(x_k^T(\theta^0)Q_1x_k(\theta^0)+u_k^TQ_2u_k), \tag{6.1.45}$$

where Q_1 and Q_2 are symmetric such that $Q_1\ge 0$ and $Q_2>0$.

Let $\mathcal{F}_k=\sigma\{e_i,1\le i\le k\}(=\mathcal{F}_1^k$ given by A6.1.3). The control $\{u_k\}$ should be selected in the family U of admissible controls:

$$\begin{aligned}U=\Big\{u:u=\{u_k\},u_k\in\mathcal{F}_k,\ &\text{and}\\ &\limsup_{n\longrightarrow\infty}\frac{1}{n}\sum_{i=0}^{n-1}(\|u_i\|^2+\|x_i(\theta^0)\|^2)<\infty\quad a.s.,\\ &\limsup_{n\longrightarrow\infty}\frac{\|x_n(\theta^0)\|^2}{n}=0\quad a.s.\Big\},\end{aligned} \tag{6.1.46}$$

where $x_i(\theta^0)$ and u_i are related by (6.1.44).

However, since the ideal θ^0 is unknown, the real system satisfies the equation

$$x_{k+1}=A(\theta_k)x_k+B(\theta_k)u_k+D(\theta_k)e_{k+1}, \tag{6.1.47}$$

where θ_k is the estimate of θ^0.

Let θ_k be given by (6.1.7) and (6.1.8) with y_{k+1} given by (6.1.5).

In order to give adaptive control we need the expression of the optimal control when θ^0 is known.

Lemma 6.1.1 *Suppose that*

i) $\{e_k, \mathcal{F}_k\}$ *is a martingale difference sequence with*

$$\sup_{k\geq 0} E[\|\varepsilon_{k+1}\|^2|\mathcal{F}_k] < \infty \quad a.s. \tag{6.1.48}$$

$$\lim_{n\longrightarrow\infty} \frac{1}{n}\sum_{i=1}^{n} \varepsilon_i\varepsilon_i^T = R > 0 \quad \text{a.s.} \tag{6.1.49}$$

ii) $(A(\theta^0)), B(\theta^0), C)$ *where* $C^TC = Q_1$ *is controllable and observable, i.e.,* $[B(\theta^0), A(\theta^0)B(\theta^0), \cdots, A^{m-1}(\theta^0)B(\theta^0)]$ *and* $[C^T, A^T(\theta^0)C^T, \cdots, (A^{m-1}(\theta^0))^TC^T]$ *are of full rank. Then in the class of nonnegative definite matrices there is an unique* $S(\theta^0) > 0$ *satisfying*

$$\begin{aligned} S(\theta^0) =& A^T(\theta^0)S(\theta^0)A(\theta^0) - A^T(\theta^0)S(\theta^0)B(\theta^0)(Q_2 \\ &+ B^T(\theta^0)S(\theta^0)B(\theta^0))^{-1}B^T(\theta^0)S(\theta^0)A(\theta^0) + Q_1 \end{aligned} \tag{6.1.50}$$

and

$$F(\theta^0) \triangleq A(\theta^0) - B(\theta^0)(Q_2 + B^T(\theta^0)S(\theta^0)B(\theta^0))^{-1}B^T(\theta^0)S(\theta^0)A(\theta^0) \tag{6.1.51}$$

is stable. The optimal control minimizing (6.1.45) is

$$u_k^* = L(\theta^0)x_k(\theta^0), \tag{6.1.52}$$

where

$$L(\theta^0) = -(Q_2 + B^T(\theta^0)S(\theta^0)B(\theta^0))^{-1}B^T(\theta^0)S(\theta^0)A(\theta^0), \tag{6.1.53}$$

and

$$\min_{u\in U} J(u) = J(u^*) = \operatorname{tr} S(\theta^0)D(\theta^0)RD^T(\theta^0). \tag{6.1.54}$$

Proof. The existence of an unique solution to (6.1.50) and stability of F given by (6.1.51) are well-known facts in control theory. We show the optimality of control given by (6.1.52).

For notational simplicity , we temporarily suppress the dependence of $x_k(\theta^0), A(\theta^0), B(\theta^0)$, and $D(\theta^0)$ on θ^0 and write them as x_k, A, B, and D, respectively.

Noticing

$$x_n^TSx_n - x_0^TSx_0 = \sum_{i=0}^{n-1}(x_{i+1}^TSx_{i+1} - x_i^TSx_i)$$

$$
\begin{aligned}
&= \sum_{i=0}^{n-1} [(Ax_i + Bu_i + De_{i+1})^T S (Ax_i + Bu_i + De_{i+1}) \\
&\quad - x_i^T (A^T SA - A^T SB(Q_2 + B^T SB)^{-1} B^T SA + Q_1) x_i],
\end{aligned}
$$

we then have

$$
\begin{aligned}
&\sum_{i=0}^{n-1} (x_i^T Q_1 x_i + u_i^T Q_2 u_i) \\
&= x_0^T S x_0 - x_n^T S x_n + \sum_{i=0}^{n-1} [x_i^T A^T SB(Q_2 + B^T SB)^{-1} B^T SAx_i \\
&+ 2(Bu_i + De_{i+1})^T SAx_i + (Bu_i + De_{i+1})^T S(Bu_i + De_{i+1}) + u_i^T Q_2 u_i] \\
&= x_0^T S x_0 - x_n^T S x_n + \sum_{i=0}^{n-1} [e_{i+1}^T D^T SDe_{i+1} + 2(Ax_i + Bu_i)^T SDe_{i+1}] \\
&\quad + \sum_{i=0}^{n-1} [u_i + (Q_2 + B^T SB)^{-1} B^T SAx_i]^T (Q_2 + B^T SB) \\
&\cdot [u_i + (Q_2 + B^T SB)^{-1} B^T SAx_i]. \qquad (6.1.55)
\end{aligned}
$$

Since $\{u_i\} \in U$, by the estimate for the weighted sum of martingale difference sequence from (6.1.55) it follows that

$$
\begin{aligned}
&\limsup_{n \longrightarrow \infty} \frac{1}{n} \sum_{i=0}^{n-1} (x_i^T Q_1 x_i + u_i^T Q_2 u_i) \\
&\quad = \operatorname{tr} RD^T SD + \limsup_{n \longrightarrow \infty} \frac{1}{n} \sum_{i=0}^{n-1} [u_i + (Q_2 + (B^T SB)^{-1} B^T SAx_i]^T \\
&\qquad \cdot (Q_2 + B^T SB)[u_i + (Q_2 + B^T SB)^{-1} B^T SAx_i]. \qquad (6.1.56)
\end{aligned}
$$

Notice that the last term of (6.1.56) is nonnegative. The conclusions of the lemma follow from (6.1.56). □

According to (6.1.52), by the certainty-equivalence-principle, we form the adaptive control

$$
u_k = L(\theta_k) x_k, \qquad (6.1.57)
$$

where x_k is the state in (6.1.47).

Thus the closed system becomes

$$
x_{k+1} = F(\theta_k) x_k + D(\theta_k) e_{k+1}, \qquad (6.1.58)
$$

which has the same structure as (6.1.4). Therefore, under the assumptions A6.1.1–A6.1.7 with $A(\theta)$ replaced by $F(\theta)$ and with J being a singleton θ^0, by Theorem 6.1.1 it is concluded that

$$\theta_k \longrightarrow \theta^0 \quad a.s. \tag{6.1.59}$$

By continuity and stability of $F(\cdot)$, it is seen that there are $\alpha > 0$ and $\mu \in (0,1)$ possibly depending on ω such that

$$\|F(\theta_n)F(\theta_{n-1})\cdots F(\theta_k)\| \leq \alpha\mu^{n-k}, \quad \forall k, \forall\, n \geq k.$$

This yields the boundedness of $\{x_k\}$ and

$$x_k - x_k(\theta^0) \xrightarrow[k\longrightarrow\infty]{} 0 \quad a.s., \tag{6.1.60}$$

because

$$\tilde{x}_{k+1} = F(\theta^0)\tilde{x}_k + (F(\theta_k) - F(\theta^0))x_k + (D(\theta_k) - D(\theta^0))e_{k+1}.$$

By (6.1.60) it follows that

$$u_k - u_k^* \xrightarrow[k\longrightarrow\infty]{} 0 \quad \text{a.s.}$$

Therefore, the closed system (6.1.58) asymptotically operates under the ideal parameter θ^0 and makes the performance index (6.1.45) minimized.

6.2. Application to Adaptive Stabilization

Consider the single-input single-output system

$$A(z)y_n = B(z)u_n + w_n, \tag{6.2.1}$$

where $u_n, y_n,$ and w_n are the system input, output, and noise, respectively, and

$$A(z) = 1 + a_1 z + \cdots + a_p z^p, \tag{6.2.2}$$
$$B(z) = b_1 z + \cdots + b_p z^p, \tag{6.2.3}$$

where z is the backward shift operator, $zy_n = y_{n-1}$.

The system coefficient

$$\theta = [a_1, \ldots, a_p, b_1, \ldots, b_p]^T$$

is unknown. The purpose of adaptive stabilization is to design control so that

$$\limsup_{n\longrightarrow\infty} \frac{1}{n}\sum_{i=1}^{n}(u_i^2 + y_i^2) < \infty. \tag{6.2.4}$$

If θ is known and if $A(z)$ and $B(z)$ are coprime, then for an arbitrary stable polynomial $A^*(z)$ of degree $2p-1$ there are unique polynomials $L(z)$ and $R(z)$, both of order $(p-1)$ with $L(0)=1$, such that

$$A(z)L(z) + B(z)R(z) = A^*(z). \tag{6.2.5}$$

Then the feedback control generated by

$$L(z)u_k = -R(z)y_k \tag{6.2.6}$$

leads the system (6.2.1) to

$$A^*(z)y_n = L(z)w_n.$$

Then, by stability of $A^*(z)$ (6.2.4) holds if assume

$$\limsup_{n\longrightarrow\infty} \frac{1}{n}\sum_{i=1}^{n} |w_i|^2 < \infty.$$

Considering coefficients of $L(z)$ and $R(z)$ as unknowns, and identifying coefficients of $z^k, k = 0, 1, \cdots, 2p-1$ for both sides of (6.2.5), we derive a system of linear algebraic equations with matrix $M(\theta)$ for unknowns:

$$M(\theta) = \begin{bmatrix} 1 & & 0 & 0 & & 0 \\ & \ddots & & & \ddots & \\ a_1 & & 1 & b_1 & & 0 \\ \vdots & \ddots & \vdots & & \ddots & \\ a_p & & a_1 & b_p & & b_1 \\ & \ddots & \vdots & & \ddots & \vdots \\ 0 & & a_p & 0 & & b_p \end{bmatrix}. \tag{6.2.7}$$

The fact that $L(z)$ and $R(z)$ can be solved from (6.2.5) for any $A^*(z)$ means that

$$h(\theta) \triangleq \det M(\theta) \tag{6.2.8}$$

is nonzero. In other words, the coprimeness of $A(z)$and $B(z)$ is equivalent to $h(\theta) \neq 0$.

In the case θ is unknown the certainly-equivalency-principle suggests replacing θ by its estimate θ_k to derive the adaptive control law. However, for $\theta_k, h(\theta_k)$ may be zero and (6.2.5) may not be solvable with $A(z)$ and $B(z)$ replaced by their estimates.

Let us estimate θ by the following algorithm called the weighted least squares (WLS) estimate, which is convergent for any feedback control

$\{u_k\}$:

$$\theta_{n+1} = \theta_n + \frac{P_n \varphi_n}{g(r_n) + \varphi_n^T P_n \varphi_n}(y_{n+1} - \theta_n^T \varphi_n), \tag{6.2.9}$$

$$P_{n+1} = P_n - \frac{P_n \varphi_n \varphi_n^T P_n}{g(r_n) + \varphi_n^T P_n \varphi_n}, \tag{6.2.10}$$

$$r_n = \|P_0^{-1}\| + \sum_{i=0}^{n} \|\varphi_i\|^2, \tag{6.2.11}$$

where

$$\varphi_n^T = [-y_n, \ldots, -y_{n-p+1}, u_n, \ldots, u_{n-p+1}],$$

$g(x) > 0$ and $\displaystyle\int_M^\infty \frac{dx}{xg(x)} < \infty$ for some $M > 0$.

Though θ_n converges a.s., its limit may not be the true θ. If a bounded sequence $\{\beta_n\}$ can be found such that the modified estimate

$$\overline{\theta}_n = \theta_n + F_n \beta_n, \quad F_n \triangleq P_n^{\frac{1}{2}}, \tag{6.2.12}$$

is convergent and

$$\liminf_{n \longrightarrow \infty} |h(\overline{\theta}_n)| > 0, \tag{6.2.13}$$

then the control obtained from (6.2.6) with θ replaced by $\overline{\theta}_n$ solves the adaptive stabilization problem, i.e., makes (6.2.4) to hold.

Therefore, the central issue in adaptive stabilization is to find a bound -ed sequence $\{\beta_n\}$ such that $\{\overline{\theta}_n\}$ given by (6.2.12) is convergent and (6.2.13) is fulfilled. This gives rise to the following definition.

Definition. *System (6.2.1) is called adaptively stabilizable by the use of parameter estimate θ_n if there is a bounded sequence $\{\beta_n\}$ such that (6.2.13) holds and $\|\overline{\theta}_n\|$ given by (6.2.12) is convergent.*

It can be shown that if system (6.2.1) is controllable, i.e., $A(z)$ and $B(z)$ are coprime, then it is adaptively stabilizable by the use of the WLS estimate. It can also be shown that the system is adaptively stabilizable by use of $\{\theta_n\}$ if and only if $f(\theta' + F\beta) \not\equiv 0$, where θ' and F denote the limits of θ_n and $P_n^{\frac{1}{2}}$, respectively, which are generated by (6.2.9)–(6.2.11).

We now use an SA algorithm to recursively produce $\{\beta_n\}$ such that $\{\beta_n\}$ is convergent and the resulting estimate $\{\overline{\theta}_n\}$ by (6.2.12) satisfies (6.2.13).

Let $l = 2p$ and β be l–dimensional, and let

$$f(\beta) \triangleq -(\frac{\partial h(\theta' + F\beta)}{\partial \beta} - l\|\beta\|^{l-2}\beta)(h(\theta' + F\beta) - \|\beta\|^l - b), \quad b > 0. \tag{6.2.14}$$

As a matter of fact,

$$f(\beta) = -\frac{1}{2}\frac{\partial v(\beta)}{\partial \beta}, \tag{6.2.15}$$

where

$$v(\beta) = (h(\theta' + F\beta) - \|\beta\|^2 - b)^2. \tag{6.2.16}$$

The root set of $f(\cdot)$ is denoted by $J = \{\beta : f(\beta) = 0\} = J_1 \cup J_2$, where

$$J_1 = \{\beta : \frac{\partial h(\theta' + F\beta)}{\partial \beta} - l\|\beta\|^{l-2}\beta = 0\}, \tag{6.2.17}$$

$$J_2 = \{\beta : h(\theta' + F\beta) - \|\beta\|^l - b = 0\}. \tag{6.2.18}$$

From algebraic geometry it is known that $v(J) \triangleq \{v(\beta) : \beta \in J\}$ is a finite set.

However, $f(\beta)$ is not directly observed; the real observation is

$$\begin{aligned} y_{k+1} = & -(\frac{\partial h(\theta_k + F_k\hat{\beta}_k)}{\partial \beta} - l\|\hat{\beta}_k\|^{l-2}\hat{\beta}_k) \\ & \cdot (h(\theta_k + F_k\hat{\beta}_k) - \|\hat{\beta}_k\|^l - b), \end{aligned} \tag{6.2.19}$$

which can be written as

$$y_{k+1} = f(\hat{\beta}_k) + \varepsilon_{k+1}, \tag{6.2.20}$$

where

$$\varepsilon_{k+1} = y_{k+1} - f(\hat{\beta}_k), \tag{6.2.21}$$

θ_k is generated by (6.2.9)–(6.2.11), F_k is defined by (6.2.11), and $\hat{\beta}_k$ is recursively defined by an SA given below.

Let us take a few real sequences defined as follows:

$$M_k > 0, \quad M_k \uparrow \infty, \quad \nu_k > 0, \quad \nu_1 < \frac{1}{2}, \quad \nu_k \downarrow 0,$$

$$\delta_k > 0, \quad \delta_k \downarrow 0, \quad \delta_1 < \frac{1}{2}.$$

Let $\{e_i^{(1)}\}$ be l-dimensional with only one nonzero element equal to either $+1$ or -1, $i = 1, \ldots, 2l$. Similarly, let $\{e_i^{(j)}\}$ be l-dimensional with only j nonzero elements, each of which equals either 1 or -1, $i = 1, 2, \ldots, \dfrac{l(l-1)\cdots(l-j+1)2^j}{j!}$.

The total number of such vectors is L

$$L = \sum_{j=1}^{l} C_j^l 2^j.$$

Normalize these vectors and denote the resulting vectors by $e_1, \ldots, e_L$ in the nondecreasing order of the number of nonzero elements in e_i.

Define $e_i^0 = e_i, i = 1, \ldots, L$ and $e_i^0 = e_i - [\frac{i}{L}]L$ for $i > L$. Introduce

$$v_k(\beta) = [h(\theta_k + F_k\beta) - \|\beta\|^l - b]^2. \tag{6.2.22}$$

Define the recursive algorithm for $\{\hat{\beta}_k\}$ as follows:

$$\overline{\beta}_{k+1} = \hat{\beta}_k + a_k y_{k+1}, \tag{6.2.23}$$

$$\begin{aligned}\hat{\beta}_{k+1} = &\overline{\beta}_{k+1} I_{[\|\overline{\beta}_{k+1}\| \le M_{\sigma_k}] \cap [\|\overline{\beta}_{k+1}\| \ge \nu_{\tau_k}]} \\ &+ \beta^* I_{[\|\overline{\beta}_{k+1}\| > M_{\sigma_k}]} + \delta_{\mu_k} e^0_{\gamma_k} I_{[\|\overline{\beta}_{k+1}\| < \nu_{\tau_k}]},\end{aligned} \tag{6.2.24}$$

$$\beta_0 >> \nu_1, \quad \text{say,} \quad \hat{\beta}_0 = 10,$$

$$\sigma_n = \sum_{k=1}^{n-1} I_{[\|\overline{\beta}_{k+1}\| \le M_{\sigma_k}]}, \quad \tau_n = \sum_{k=1}^{n-1} I_{[\|\overline{\beta}_{k+1}\| < \nu_{\tau_k}]}, \tag{6.2.25}$$

$$\gamma_n = \sum_{k=1}^{n-1} I_{[v_{k+1}(\delta_{\mu_k} e^0_{\gamma_k}) \ge b^2]}, \tag{6.2.26}$$

$$\mu_n = \sum_{k=L+1}^{n-1} I_{[\tau_k - \tau_{k-L} = \gamma_k - \gamma_{k-L}]}, \tag{6.2.27}$$

and β^* is a fixed vector, $\|\beta^*\| > 1$.

The algorithm (6.2.23)–(6.2.27) is the RM algorithm with expanding truncations, but it differs from the algorithm given by (2.1.1)–(2.1.3) as follows. The algorithm (2.1.1)–(2.1.3) is truncated at the upper side only, but the present algorithm is truncated not only at the upper side but also at the lower side: $\hat{\beta}_k$ is allowed neither to diverge to infinity nor to tend to zero; whenever it reaches the truncation bounds the estimate $\hat{\beta}_k$ is pulled back to β^* and M_{σ_k} is enlarged to $M_{\sigma_{k+1}}$ at the upper side, while at the lower side $\hat{\beta}_k$ is pulled back to $e^0_{\gamma_k}$, which will change to the next $e^0_{\gamma_{k+1}}$ whenever $[v_{k+1}(\delta_{\mu_k} e^0_{\gamma_k}) \geq b^2]$ is satisfied. If for L successive resettings of $e^0_{\gamma_k}$ we have to change to the next one, then we reduce δ_{μ_k} to $\delta_{\mu_{k+1}}$.

Lemma 6.2.1 *Assume the following conditions hold:*

A6.2.1 $a_k > 0, a_k \longrightarrow 0,$ *and* $\sum_{k=1}^{\infty} a_k = \infty;$

A6.2.2 *System (6.2.1) is adaptively stabilizable by use of* $\{\theta_k\}$ *generated by (6.2.9)–(6.2.11), i.e.,* $h(\theta' + F\beta) \not\equiv 0$.

If $h(\theta') = 0$, *then after a finite number of steps the algorithm (6.2.23)–(6.2.27) becomes the RM algorithm*

$$\hat{\beta}_{k+1} = \hat{\beta}_k + a_k y_{k+1}, \tag{6.2.28}$$

$v(\hat{\beta}_n)$ *converges and* $\lim_{n\longrightarrow\infty} d(\hat{\beta}_n, J\backslash\{0\}) = 0.$

Proof. The basic steps of the proof are essentially the same as those for proving Theorem 2.2.1, but some modifications should be made because of truncations at the lower side.

Step 1. Let $\{\hat{\beta}_{n_k}\}$ be a convergent subsequence of $\{\hat{\beta}_n\}, \hat{\beta}_{n_k} \xrightarrow[k\longrightarrow\infty]{} \beta'$. For any n_k define the RM algorithm

$$\beta'_{m+1} = \beta'_m + a_m y'_{m+1},$$

$$y'_{m+1} = -(h(\theta_m + F\beta'_m) - \|\beta'_m\|^l - b)(\frac{\partial h(\theta_m + F_m\beta'_m)}{\partial\beta} - l\|\beta'_m\|^{l-2}\beta'_m)$$

with $\beta'_{n_k} = \hat{\beta}_{n_k}$ or for some $i : n_k \leq i < m(n_k, T),\ \beta'_i = \delta_{\mu_j} e^0_{\gamma_j}$ for some j.

We show that there are $M > 0,\ T > 0$ such that $\|\beta'_m\| \leq M,\ \forall m \in \{n_k, \ldots, m(n_k, T)\}$ when $\beta'_{n_k} = \hat{\beta}_{n_k}$, and $\forall m \in \{i, i+1, \ldots, m(n_k, T)\}$ when $\beta'_i = \delta_{\mu_j} e^0_{\gamma_j}, \forall k \geq k_0$, if k_0 is large enough, where $m(k, T)$ is given by (1.3.2).

Let $c > 1$ be a constant such that

$$\|\beta'_{n_k}\| < c, \quad \forall k \geq 1. \tag{6.2.29}$$

It is clear that

$$\|\beta'_i\| = \delta_{\mu_j} \|e^0_{\gamma_j}\| = \delta_{\mu_j} < \frac{1}{2} < c. \tag{6.2.30}$$

Since θ_k and F_k are convergent, there is $\alpha < \infty$ such that

$$\max_{\substack{\|\beta\|\leq 2c \\ 1\leq k<\infty}} |h(\theta_k + F_k\beta) - \|\beta\|^l - b\Big|\|\frac{\partial h(\theta_k + F_k\beta)}{\partial\beta} - l\|\beta\|^{l-2}\beta\| = \alpha < \infty.$$

Let $T < (c/2\alpha)$. By (6.2.29) and (6.2.30), we have

$$\|\beta'_j\| \leq M \tag{6.2.31}$$

for $j = n_k$ if $\beta'_{n_k} = \hat{\beta}_{n_k}$, and for $j = i$ if $\beta'_{n_k} = \delta_{\mu_j} e^0_{\gamma_j}$, where $M \triangleq \frac{3c}{2}$.

Let (6.2.31) hold for $j = n_k, n_k+1, \dots, m < m(n_k, T)$ or $j = i, i+1, \dots, m < m(n_k, T)$. It then follows that

$$\begin{aligned}\|\beta'_{m+1}\| \leq& \|\beta'_j\| + \sum_{s=j}^{m} a_s |h(\theta_s + F\beta'_s) - \|\beta'_s\|^l - b| \\ &\cdot \|\frac{\partial h(\theta_s + F_s\beta'_s)}{\partial\beta} - l\|\beta'_s\|^l\beta'_s\| \\ <& c + \sum_{s=j}^{m} a_s\alpha \leq c + T\alpha < \frac{3c}{2},\end{aligned}$$

where $j = n_k$ or $j = i$.

Thus, (6.2.31) has been inductively proved for $j \in \{n_k, \dots, m(n_k, T)\}$ or $j = \{i, i+1, \dots, m(n_k, T)\}, \forall k \geq k_0$.

Step 2. Let $\{\hat{\beta}_{n_k}\}$ be a convergent subsequence. We show that there are $M > 0$ and $T > 0$ such that

$$\|\hat{\beta}_m\| \leq M, \quad \forall m \in \{n_k, \dots, m(n_k, T)\}, \quad \forall k \geq k_0 \tag{6.2.32}$$

if k_0 is large enough.

If σ_n defined by (6.2.25) is bounded, then (6.2.32) directly follows.

Again take $c > 0$ such that $\|\hat{\beta}_{n_k}\| < c, \forall k \geq 1$ and set $M = \frac{3c}{2}$.

Assume $\sigma_n \longrightarrow \infty$. Then there is a k_0 such that

$$M_{\sigma_{k_0}} > M.$$

By the result proved in Step 1, starting from $\hat{\beta}_{n_k}, k \geq k_0$, the algorithm for $\hat{\beta}_s$ cannot directly hit the sphere with radius $M_{\sigma_{k_0}}$ without a truncation for $s \in \{n_k, n_k+1, \dots, m(n_k, T)\}$. So it may first hit some lower bound at time $i : n_k < i < m(n_k, T)$ and switch to some $\delta_{\mu_j} e^0_{\gamma_j}$, from which again by Step 1 $\hat{\beta}_s$ cannot directly reach $M_{\sigma_{k_0}}$ without a truncation. The only possibility is to be truncated again at a lower bound. Therefore, (6.2.32) takes place.

Step 3. Since θ_k and F_k are convergent, by (6.2.32) it follows that from any convergent subsequence $\{\hat{\beta}_{n_k}\}$ there are constants $c_1 > 0$ and $T > 0$ such that

$$\|y_m\| < c_1, \quad \forall m \in \{n_k, \ldots, m(n_k, T)\}, \quad \forall k \geq k_0,$$

if k_0 is large enough.

Consequently, there is $c_2 > 0$ such that

$$\|\hat{\beta}_{m+1} - \hat{\beta}_{n_k}\| \leq c_2 T, \quad \forall m \in \{n_k, \ldots, m(n_k, T)\}. \tag{6.2.33}$$

By (6.2.32) and the convergence of θ_k and F_k it also follows that

$$\varepsilon_{m+1} = y_{m+1} - f(\hat{\beta}_m) \longrightarrow 0, \quad \forall m \in \{n_k, \ldots, m(n_k, T)\}$$

as $k \longrightarrow \infty$.

Therefore,

$$\lim_{T \longrightarrow 0} \lim_{k \longrightarrow \infty} \frac{1}{T} \| \sum_{i=n_k}^{m(n_k,T)} a_i \varepsilon_{i+1} \| = 0. \tag{6.2.34}$$

Using (6.2.33) and (6.2.34) by the same argument as that given in Step 3 of the proof for Theorem 2.2.1, we arrive at the following conclusion. If starting from $\hat{\beta}_{n_k}$, $k = 1, 2, \ldots$, the algorithm (6.2.24) is calculated as an RM algorithm and $\{\|\hat{\beta}_{n_k}\|, \ k = 1, 2, \ldots\}$ is bounded, then for any $[\alpha_1, \alpha_2]$ with $d([\alpha_1, \alpha_2], v(J)) > 0$ and $\alpha_2 > \alpha_1, v(\hat{\beta}_n)$ cannot cross $[\alpha_1, \alpha_2]$ infinitely often.

Step 4. We now show that $\{\hat{\beta}_n\}$ is bounded.

If $\{\hat{\beta}_n\}$ is unbounded, then $\sigma_n \longrightarrow \infty$ as $n \longrightarrow \infty$. Therefore, $\{v(\hat{\beta}_n)\}$ is unbounded and comes back to the fixed point $v(\beta^*)$ infinitely many times.

Notice that $v(J)$ is a finite set and

$$v(\delta_{\mu_k} e^0_{\gamma_k}) \leq \max_{\|\beta\| \leq \frac{1}{2}} v(\beta) \triangleq \xi < \infty.$$

We see that there is an interval $[\alpha_1, \alpha_2]$ with $\alpha_1 > \xi$ and $d([\alpha_1, \alpha_2], v(J)) > 0$ such that $v(\hat{\beta}_n)$ crosses $[\alpha_1, \alpha_2]$ infinitely often, and during each crossing the algorithm (6.2.24) behaves like an RM algorithm with staring point $\hat{\beta}_{n_k}$. It is clear that $\{\hat{\beta}_{n_k}\}$ is bounded because $v(\beta) \longrightarrow \infty$ as $\|\beta\| \longrightarrow \infty$. But by Step 3, this is impossible. Thus, we conclude that $\{\hat{\beta}_n\}$ is bounded, and after a finite number of steps (6.2.24) becomes

$$\hat{\beta}_{k+1} = \overline{\beta}_{k+1} I_{[\|\overline{\beta}_{k+1}\| \geq \nu_{\tau_k}]} + \delta_{\mu_k} e^0_{\gamma_k} I_{[\|\overline{\beta}_{k+1}\| < \nu_{\tau_k}]}. \tag{6.2.35}$$

Step 5. We now show (6.2.28), i.e., after a finite number of steps the algorithm (6.2.35) ceases to truncate at the lower side.

Since $h(\theta') = 0$ and $h(\theta' + F\beta) \not\equiv 0$ by A6.2.2, it follows that there is at least one nonzero coefficient in the polynomial $h(\theta' + F\beta)$ for some β^i with $i > 0$ $(i \leq l-1)$. Therefore, for some e_i^0 and a small $\varepsilon > 0$

$$h(\theta' + F\varepsilon e_j^0) > 0.$$

From (6.2.16) it is seen that for sufficiently small $\varepsilon > 0$, we have

$$v(\varepsilon e_j^0) < b^2 = v(0).$$

This combining with convergence of θ_k and F_k leads to

$$v_k(\varepsilon e_j^0) < b^2 \tag{6.2.36}$$

for sufficiently large k.

From (6.2.26) and (6.2.36) it follows that $\{\gamma_k\}$ must be bounded, and hence $\{\mu_k\}$ is bounded. This means that there is a k_0 such that

$$\delta_{\mu_k} = \delta_{\mu_{k_0}}, \quad e^0_{\gamma_k} = e^0_{\gamma_{k_0}}, \quad \forall k \geq k_0.$$

We now show that $\{\tau_n\}$ is bounded.

Since $v_k(\delta_{\mu_{k_0}} e^0_{\gamma_{k_0}}) < b^2$ for all sufficiently large k, it follows that

$$v(\delta_{\mu_{k_0}} e^0_{\gamma_{k_0}}) \triangleq a < b^2 = v(0). \tag{6.2.37}$$

If $\{\tau_n\}$ were unbounded, then by (6.2.37) the algorithm, starting from $\delta_{\mu_{k_0}} e^0_{\gamma_{k_0}}$, would infinitely many times enter the sphere with radius ε : $\{\beta : \|\beta\| < \varepsilon\}$, where ε is small enough such that

$$\min_{\|\beta\| \leq \varepsilon} v(p) = a + \delta \leq b^2, \quad \delta > 0.$$

Then $v(\hat{\beta}_n)$ would cross infinitely often an interval $[\alpha_1, \alpha_2] \in (a, a+\delta)$. Since $v(J)$ is a finite set, we may assume $d([\alpha_1, \alpha_2], v(J)) > 0$. It is clear that during the crossing the algorithm behaves like an RM algorithm. By Step 4, this is impossible.

Therefore, there is a k_0 such that

$$\hat{\beta}_{k+1} = \hat{\beta}_k + a_k y_{k+1}, \quad \forall k \geq k_0. \tag{6.2.38}$$

Noticing (6.2.20), (6.2.34), and that $v(\beta)$ serves as the Lyapunov function for $f(\beta)$, from Theorem 2.2.1 we conclude the remaining assertions of the lemma. □

Using $\{\hat{\beta}_k\}$ we now define $\{\beta_k\}$ in (6.2.12) satisfying (6.2.13) and thus solving the adaptive stabilization problem.

Let $\eta_k > 0, \eta_k \downarrow 0$.

1) If $|h(\theta_k)| > \eta_{\lambda_k}$, then set $\beta_k = 0$, i.e., $\overline{\theta}_k = \theta_k$. Using θ_k we produce the adaptive control u_k from (6.2.6) with $L(z)$ and $R(z)$ defined from (6.2.5) with θ replaced by θ_k and go back to 1) for $k+1$.

2) If $|h(\theta_k)| \le \eta_{\lambda_k}$, then define

a) $\beta_k = \beta_{k-1}$ for the case where $|h(\theta_k - F_k\beta_{k-1})| > \eta_{\lambda_k}$;

b) $\beta_k = \hat{\beta}_k$ defined by (6.2.24) for the case where $|h(\theta_k + F_k\beta_{k-1})| \le \eta_{\lambda_k}$, but $|h(\theta_k + F_k\hat{\beta}_k)| > \eta_{\lambda_k}$;

c) $\beta_k = (1+\sqrt{\eta_{\lambda_k}})\hat{\beta}_k$ for the case, where $|h(\theta_k + F_k\beta_{k-1})| \le \eta_{\lambda_k}$, $|h(\theta_k + F_k\hat{\beta}_k)| \le \eta_{\lambda_k}$, but

$$|h(\theta_k + F_k(1+\sqrt{\eta_{\lambda_k}}\hat{\beta}_k)| > \eta_{\lambda_k}.$$

Define

$$\overline{\theta}_k = \theta_k + F_k\beta_k \tag{6.2.39}$$

and use $\overline{\theta}_k$ to produce the adaptive control u_k as in 1), and go back to 1) for $k+1$.

3) If $|h(\theta_k)| \le \eta_{\lambda_k}$ and none of a)-c) of 2) is the case, then set $\beta_k = 0$ and $u_k = 0$, go back to 1) for $k+1$, and at the same time change η_{λ_k} to η_{λ_k+1}, i.e.,

$$\lambda_k = \sum_{i=1}^{k-1} I_{[|h(\theta_i)|\le\eta_{\lambda_i};|h(\theta_i+F_i\beta_{i-1})|\le\eta_{\lambda_i};|h(\theta_i+F_i\hat{\beta}_i)|\le\eta_{\lambda_i};|h(\theta_i+(1+\sqrt{\eta_{\lambda_i}}F_i\hat{\beta}_i)|\le\eta_{\lambda_i}]}. \tag{6.2.40}$$

Theorem 6.2.1 *Assume conditions A6.2.1 and A6.2.2 hold. Then there is k_0 such that $\beta_k \equiv \beta_{k_0}, \forall k \ge k_0$, and $\{\overline{\theta}_k\}$ converges and*

$$\liminf_{k\longrightarrow\infty} |h(\overline{\theta}_k)| > 0,$$

where $\{\beta_k\}$ and $\{\overline{\theta}_k\}$ are defined by 1)-3) described above.

Proof. The key step is to show that $\lim_{k\longrightarrow\infty} \lambda_k < \infty$.

Assume the converse: $\lambda_k \xrightarrow[k\longrightarrow\infty]{} \infty$.

Case i) ($|h(\theta')| \ne 0$): The assumption $\lambda_k \xrightarrow[k\longrightarrow\infty]{} \infty$ implies that $\eta_{\lambda_k} \xrightarrow[k\longrightarrow\infty]{} 0$ and $|h(\theta_k)| \le \eta_{\lambda_k}$ occurs infinitely many times. However, this is impossible, since $\theta_k \longrightarrow \theta'$ and $|h(\theta')| \ne 0$. The contradiction shows $\lim_{k\longrightarrow\infty} \lambda_k < \infty$.

Case ii) ($|h(\theta')|=0$): The assumption $\lambda_k \xrightarrow[k\to\infty]{} \infty$ implies that there is a sequence of integers $\{n_k\}$ such that $n_k \longrightarrow \infty$, and $\lambda_{n_k+1} = \lambda_{n_k}+1$, i.e., for all $k=1,2,\dots,$ the following indicator equals one

$$I_{[|h(\theta_{n_k})|\le \eta_{\lambda_{n_k}};|h(\theta_{n_k}+F_{n_k}\beta_{n_k-1})|<\eta_{\lambda_{n_k}};|h(\theta_{n_k}+F_{n_k}\hat\beta_{n_k})|\le\eta_{\lambda_{n_k}}; \atop |h(\theta_{n_k}+(1+\sqrt{\eta_{\lambda_{n_k}}})F_{n_k}\hat\beta_{n_k})|\le\eta_{\lambda_{n_k}}]} = 1. \tag{6.2.41}$$

Take a convergent subsequence of $\{\hat\beta_{n_k}\}$. For notational simplicity denote by $\{\hat\beta_{n_k}\}$ itself its convergent subsequence. Thus $\hat\beta_{n_k} \xrightarrow[k\to\infty]{} \beta^0$.

By Lemma 6.2.1, $\beta^0 \in J\backslash\{0\}$.

1) If $\beta^0 \in J_2$, then

$$h(\theta_{n_k}+F_{n_k}\hat\beta_{n_k}) \longrightarrow h(\theta'+F\beta^0) = \|\beta^0\|^l + b > \eta_{\lambda_{n_k}}$$

for all sufficiently large k. Thus, (6.2.41) may take place at most a finite number of times. The contradiction shows that $\lim\limits_{k\to\infty}\lambda_k<\infty$.

2) If $\beta^0\in J_1\backslash\{0\}$, then as $\delta\longrightarrow 0$ we have

$$\begin{aligned} h(\theta'+(1+\sqrt{\delta})F\beta^0) &= h(\theta'+F\beta^0) + (\frac{\partial h(\theta'+F\beta^0)}{\partial\beta})^T\sqrt{\delta}\beta^0 + O(\delta) \\ &= h(\theta'+F\beta^0)+\sqrt{\delta}l\|\beta^0\|^{l-2}\|\beta\|^2+O(\delta). \end{aligned} \tag{6.2.42}$$

Since $\theta_k\longrightarrow\theta'$, $F_k\longrightarrow F$, and $\hat\beta_{n_k}\longrightarrow\beta^0$, for sufficiently large k from (6.2.42) it follows that

$$\begin{aligned} &h(\theta_k+(1+\sqrt{\eta_{\lambda_{n_k}}})F_{n_k}\hat\beta_{n_k}) \\ &= h(\theta_{n_k}+F_{n_k}\hat\beta_{n_k})+\sqrt{\eta_{\lambda_{n_k}}}l\|\beta^0\|^l+O(\eta_{\lambda_{n_k}}). \end{aligned} \tag{6.2.43}$$

If $|h(\theta_{n_k}+F_{n_k}\hat\beta_{n_k})|\le\eta_{\lambda_{n_k}}$, then from (6.2.43) it follows that

$$|h(\theta_k+(1+\sqrt{\eta_{\lambda_{n_k}}})F_{n_k}\hat\beta_{n_k})| > \frac{1}{2}\sqrt{\eta_{\lambda_{n_k}}}l\|\beta^0\|^l>\eta_{\lambda_{n_k}}$$

for all sufficiently large k. Again, this means that (6.2.41) may take place at most a finite number of times, and we conclude that $\lim\limits_{k\to\infty}\lambda_k<\infty$.

Thus, there is k_0 such that

$$\lambda_k\equiv\lambda_{k_0},\quad \forall k\ge k_0,$$

and the algorithm defining $\{\beta_k\}$ will run over the following cases: 1) and 2a)-2c). Since θ_k and F_k are convergent, the inequality

$$|h(\theta_k+F_k\hat\beta_k)|>\eta_{\lambda_{k_0}} \quad \text{implies} \quad |h(\theta_{k+i}+F_{k+i}\hat\beta_k)|>\eta_{\lambda_{k_0}},\quad \forall i\le 1$$

for sufficiently large k. This means that the algorithm can be at 2b) only for finitely many times. By the same reason it cannot be at 2c) for infinitely many times. Therefore, the algorithm will stick on 1) if $|h(\theta')| > \eta_{\lambda_{k_0}}$ and on 2a) if $|h(\theta')| \leq \eta_{\lambda_{k_0}}$, and in both cases there is a k_0 such that $\beta_k = \beta_{k_0}, \forall k \geq k_0$ and

$$\liminf_{k \longrightarrow \infty} |h(\overline{\theta}_k)| > \eta_{\lambda_{k_0}} > 0.$$

The convergence of $\overline{\theta}_k$ follows from the convergence of θ_k and F_k. □

Remark 6.2.1 For the case $h(\theta') = 0$, the origin $\beta = 0$ is not a stable equilibrium for the equation

$$\frac{\partial \beta_t}{\partial t} = f(\beta_t).$$

Consequently, the truncation at the lower bound in (6.2.24) should be very rare. The computation will be simplified if there is no lower bound truncation.

6.3. Application to Pole Assignment for Systems with Unknown Coefficients

Consider the linear stochastic system

$$x_{k+1} = Ax_k + Bu_k + w_{k+1}, \tag{6.3.1}$$

where x_k is the n-dimensional state, u_k is the one-dimensional control, and w_{k+1} is the n-dimensional system noise.

The task of pole assignment is to define the feedback control

$$u_k = Fx_k$$

in order that the characteristic polynomial

$$\det(\lambda I - (A + BF)) \tag{6.3.2}$$

of the closed-loop system coincides with a given polynomial

$$\beta(\lambda) = \lambda^n + \beta_1 \lambda^{n-1} + \cdots + \beta_{n-1}\lambda + \beta_n. \tag{6.3.3}$$

The pair (A_1, B_1) is called similar to (A, B) if there exists a nonsingular matrix $T \triangleq [t_1, \ldots, t_n]$ such that

$$TA_1T^{-1} = A, \quad TB_1 = B, \tag{6.3.4}$$

where t_i denotes the $i$the column of T.

Define

$$A_1 = \begin{bmatrix} -a_1 & -a_2 & \cdots & -a_n \\ 1 & 0 & & 0 \\ 0 & \ddots & & \vdots \\ \cdots & & \ddots & \vdots \\ 0 & 0 & 1 & 0 \end{bmatrix}, \quad B_1 = \begin{bmatrix} 1 \\ 0 \\ \vdots \\ 0 \end{bmatrix}, \tag{6.3.5}$$

where a_i, $i = 1, \dots, n$, are coefficients of

$$\det(\lambda I - A) = \lambda^n + a_1\lambda^{n-1} + \cdots + a_{n-1}\lambda + a_n. \tag{6.3.6}$$

The pair (A_1, B_1) is called the controller form associated to the pair (A, B).

If (A, B) is controllable, i.e., $Q \triangleq [B, AB, \cdots, A^{n-1}B]$ is of full rank, then (A, B) is similar to its controller form. To see this, we note that (6.3.4) implies $t_1 = B$, and from $[t_1, \dots, t_n]A_1 = A[t_1, \dots, t_n]$ it follows that

$$\begin{aligned} -a_1t_1 + t_2 &= At_1, \\ -a_2t_1 + t_3 &= At_2, \\ &\cdots \\ -a_nt_1 + t_n &= At_{n-1}, \\ -a_nt_1 &= At_n, \end{aligned}$$

which imply

$$T = QR, \tag{6.3.7}$$

where

$$R = \begin{bmatrix} 1 & a_1 & \cdots & \cdots & \cdots & a_{n-1} \\ 0 & 1 & a_1 & \cdots & \cdots & a_{n-2} \\ \vdots & & & \ddots & & \\ & & & & & a_1 \\ 0 & \cdots & \cdots & & 0 & 1 \end{bmatrix}. \tag{6.3.8}$$

So, T is nonsingular if and only if Q is nonsingular.

Assume that (A, B) is controllable and (A, B) is already in its controller form (6.3.5). For notational simplicity, we will write (A, B) rather than (A_1, B_1).

With feedback control $u_k = Fx_k$ the closed-loop system takes the form

$$x_{k+1} = (A + BF)x_k + w_{k+1}. \tag{6.3.9}$$

Since (A, B) is in controller form,

$$\det(\lambda I - (A + BF)) \\ = \lambda^n + (a_1 - f_1)\lambda^{n-1} + \cdots + (a_{n-1} - f_{n-1})\lambda + (a_n - f_n), \tag{6.3.10}$$

where f_i, $i = 1, \dots, n$ are elements of the row vector F:

$$F = [f_1, \dots, f_n].$$

Therefore, if (A, B) is known, then comparing (6.3.10) with (6.3.3) gives the solution $F^0 = [f_1^0, \dots, f_n^0]$ to the pole assignment problem, where

$$f_i^0 = a_i - \beta_i, \qquad i = 1, \dots, n. \tag{6.3.11}$$

We now solve the pole assignment problem by learning for the case where (A, B) is unknown.

Let us combine the vector equation (6.3.9) for n initial values to form the matrix equation

$$X_{k+1} = (A + BF)X_k + W_{k+1}. \tag{6.3.12}$$

Let $X_0 = I$. In learning control, X_1 can be observed at any fixed F

$$X_1 = (A + BF)X_k + W_1.$$

For any $F = F_k$, the observation of X_1 is denoted by

$$Y_{k+1} = (A + BF_k) + Z_{k+1}, \tag{6.3.13}$$

where Z_{k+1} is the system noise at time "1" for the system with feedback gain F_k applied.

Having observed Y_k, we compute its characteristic polynomial $\det(\lambda I - Y_k)$, which is a noise-corrupted characteristic polynomial of $A + BF_{k-1}$.

Let F_k be the kth estimate for F^0. By observing $\det(\lambda I - Y_k)$ we actually learn the difference $\det(\lambda I - Y_k) - \beta(\lambda)$, which in a certain sense reflects how far $\det(\lambda I - (A + BF_k))$ differs from the ideal polynomial $\beta(\lambda)$.

For any $F_k \triangleq [f_{1k}, \dots, f_{nk}]$, let

$$\beta(F_k) \triangleq [\beta_1(F_k), \dots, \beta_n(F_k)] \tag{6.3.14}$$

be the row vector composed of coefficients of

$$\det(\lambda I-(A+BF_k))\triangleq\lambda^n+\beta_1(F_k)\lambda^{n-1}+\cdots+\beta_{n-1}(F_k)\lambda+\beta_n(F_k). \tag{6.3.15}$$

By (6.3.10)

$$\beta_i(F_k)=a_i-f_{ik},\quad i=1,\ldots,n. \tag{6.3.16}$$

Similarly, define row vectors

$$y_{k+1}=[y_{k+1,1},\ldots,y_{k+1,n}]\quad\text{and}\quad\beta=[\beta_1,\ldots,\beta_n] \tag{6.3.17}$$

composed of coefficients of

$$\det(\lambda I-Y_{k+1})=\lambda^n+y_{k+1,1}\lambda^{n-1}+\cdots+y_{k+1,n} \tag{6.3.18}$$

and $\beta(\lambda)$, respectively.

Take a sequence of positive real numbers $\{M_k\}$, $M_k<M_{k+1}$, $\forall k\geq 0$, and $M_k\xrightarrow[k\to\infty]{}\infty$.

Calculate the estimate F_k for F^0 by the following RM algorithm with expanding truncations:

$$\begin{aligned}F_{k+1}=&[F_k+a_k(y_{k+1}-\beta)]I_{[\|F_k+a_k(y_{k+1}-\beta)\|\leq M_{\sigma_k}]}\\&+F^*I_{[\|F_k+a_k(y_{k+1}-\beta)\|>M_{\sigma_k}]},\end{aligned} \tag{6.3.19}$$

$$\sigma_k=\sum_{i=1}^{k-1}I_{[\|F_i+a_i(y_{i+1}-\beta)\|>M_{\sigma_i}]},\quad\sigma_0=0 \tag{6.3.20}$$

with fixed $F^*:\|F^*\|<M_1$.

Theorem 6.3.1 *Assume that (A,B) is controllable and is in the controller form. Further, assume the following conditions A6.3.1 and A6.3.2 hold:*

A6.3.1 *The components of $\{z_{k,ij},i=1,\ldots,n,j=1,\ldots,n,k=1,2,\ldots\}$ of $\{Z_k\}$ in (6.3.13) are mutually independent with*

$$Ez_{k,ij}=0,\quad\sup_{i,j,k}E|z_{k,ij}|^p<\infty\ \textit{for some}\ p\in(1,2];$$

A6.3.2 $a_k>0$, $\sum_{k=1}^{\infty}a_k=\infty$, *and* $\sum_{k=1}^{\infty}a_k^p<\infty$,
where p is the same as that in A6.3.1.
Then there is Ω' with $P\Omega'=1$ such that for each $\omega\in\Omega'$, $F_k\longrightarrow F^0$ as

$k \longrightarrow \infty$, where F^0 is the desired feedback gain realizing the exact pole assignment.

Proof. Define

$$\xi_{k+1} \stackrel{\Delta}{=} y_{k+1} - \beta(F_k), \quad \xi_{k+1} = [\xi_{k+1,1}, \dots, \xi_{k+1,n}], \tag{6.3.21}$$

where $\beta(F_k)$ and y_{k+1} are given by (6.3.14) and (6.3.17), respectively.

By (6.3.11) and (6.3.16) it follows that

$$\begin{aligned} y_{k+1} - \beta &= \beta(F_k) + \xi_{k+1} - \beta \\ &= -(F_k - F^0) + \xi_{k+1}. \end{aligned}$$

Thus, (6.3.19) and (6.3.20) become

$$\begin{aligned} F_{k+1} =& [F_k - a_k(F_k - F^0) + a_k\xi_{k+1}]I_{[\|F_k - a_k(F_k-F^0)+a_k\xi_{k+1}\| \le M_{\sigma_k}]} \\ &+ F^* I_{[\|F_k - a_k(F_k-F^0)+a_k\xi_{k+1}\| > M_{\sigma_k}]}, \end{aligned} \tag{6.3.22}$$

$$\sigma_k = \sum_{i=1}^{k-1} I_{[\|F_i - a_i(F_i - F^0)+a_i\xi_{i+1}\| > M_{\sigma_i}]}. \tag{6.3.23}$$

It is clear that the recursive algorithm for F_k^T has the same structure as (2.1.1)–(2.1.3). For the present case, as function $v(\cdot)$ required in A2.2.2 we may take

$$v(F) = (F - F^0)(F - F_0)^T.$$

Therefore, the conclusion of the theorem will follow from Theorem 2.2.1, if we can show that for any integer N

$$\sum_{k=1}^{\infty} a_k \xi_{k+1} I_{[\|F_k\| \le N]} < \infty \quad \text{a.s.} \tag{6.3.24}$$

From (6.3.21) by (6.3.18), (6.3.15), and (6.3.13) it follows that

$$\begin{aligned} &\xi_{k+1,1}\lambda^{n-1} + \cdots + \xi_{k+1,n-1}\lambda + \xi_{k+1,n} \\ &\quad = \det(\lambda I - (A + BF_k) - Z_{k+1}) - \det(\lambda I - (A + BF_k)). \end{aligned}$$

From here it is seen that $\xi_{k+1,i}$, $i = 1, \dots, n$ is a sum of products of i elements from $\{A_{ij}, f_{jk}, z_{k+1,ij}\}$, $i, j = 1, \dots, n$, with $+1$ and -1 as multiple for each product, where A_{ij} and f_{jk} denote elements of A and F_k, respectively. It is important to note that each product in $\xi_{k+1,i}$ includes at least one of $\{z_{k+1,ij}\}$ as its factor. Thus, the product is of the form

$$(\pm) A_{i_1 j_1} \cdots A_{i_{la} j_{la}} f_{p_1,k} \cdots f_{p_{l_f} k} z_{k+1,s_1 t_1} \cdots z_{k+1,s_{l_z} t_{l_z}}, \tag{6.3.25}$$

where $i_1 \neq i_2 \neq \cdots \neq i_{la}, j_1 \neq \cdots \neq j_{l_a}, p_1 \neq \cdots \neq p_{l_f}$,

$$s_1 \neq \cdots \neq s_{l_z}, \quad t_1 \neq \cdots \neq t_{l_z}, \quad l_z \geq 1.$$

By A6.3.1 we have

$$\sup_k E\{|A_{i_1 j_1} \cdots A_{i_{la} j_{la}} f_{p_1,k} \cdots f_{p_{l_f} k} z_{k+1,s_1 t_1} \cdots z_{k+1,s_{l_z} t_{l_z}} \cdot I_{[\|F_k\| \leq N]}|^p | \mathcal{F}_k\} < \infty, \qquad (6.3.26)$$

where $\mathcal{F}_k = \sigma\{Z_i, i \leq k\}$.

By A6.3.2 and the convergence theorem for martingale difference sequences it follows that

$$\sum_{k=1}^{\infty} a_k A_{i_1 j_1} \cdots A_{i_{la} j_{la}} f_{p_1,k} \cdots f_{p_{l_f} k} z_{k+1,s_1 t_1} \cdots z_{k+1,s_{l_z} t_{l_z}} \cdot I_{[\|F_k\| \leq N]} < \infty, \quad \text{a.s.}$$

for any integer N, which implies (6.3.24). □

6.4. Application to Adaptive Regulation

We now apply the SA method to solve the adaptive regulation problem for a nonlinear nonparametric system.

Consider the following system

$$x_{k+1} = f(x_k, u_k), \qquad (6.4.1)$$

where $x_k \in \mathbb{R}^n$ is the system state, $u_k \in \mathbb{R}^n$ is the control, and $f(\cdot,\cdot) : \mathbb{R}^n \times \mathbb{R}^n \longrightarrow \mathbb{R}^n$ is an unknown nonlinear function with $(0, u^0)$ being the unknown equilibrium for the system (6.4.1).

Assume the state x_k is observed, but the observations are corrupted by noise:

$$y_k = x_k + \epsilon_k, \qquad (6.4.2)$$

where $\epsilon_k \in \mathbb{R}^n$ is the observation noise, which may depend on u_{k-1}.

The purpose of adaptive regulation is to define adaptive control based on measurements in order the system state to reach the desired value, which, without loss of generality, may be assumed to be equal to zero.

We need the following conditions.

A6.4.1 *$a_i > 0$, $a_i \longrightarrow 0$, and* $\sum_{i=1}^{\infty} a_i = \infty$*;*

A6.4.2 *The upper bound b for u^0 is known, i.e., $\|u^0\| < b$, and u^0 is a robust stabilizing control in the sense that for any $d_k \xrightarrow[k\longrightarrow\infty]{} 0$ the state x_k tends to zero for the following system*

$$x_{k+1} = f(x_k, u^0) + d_k;$$

A6.4.3 *The system (6.4.1) is BIBS stable, i.e., for any bounded input, the system state is also bounded;*

A6.4.4 *$f(x, \cdot)$ is continuous for bounded x, i.e., for any $a > 0$*

$$\sup_{\|x\|\leq a} \|f(x, u + \Delta u) - f(x, u)\| \xrightarrow[\|\Delta u\|\longrightarrow 0]{} 0;$$

A6.4.5 *The system (6.4.1) is strictly input passive, i.e., there are β and $\epsilon > 0$ such that for any input $\{u_i\}$*

$$\sum_{i=1}^{n} u_i^T x_{i+1} \geq \epsilon \sum_{i=1}^{n} \|u_i\|^2 + \beta, \quad \forall n; \tag{6.4.3}$$

A6.4.6 *For any convergent subsequence $\{u_{n_k}\}$*

$$\lim_{T\longrightarrow 0} \limsup_{k\longrightarrow\infty} \frac{1}{T} \| \sum_{i=n_k}^{m(n_k,t)} a_i \epsilon_{i+1} \| = 0, \quad \forall t \in [0, T]$$

where $m(n, t)$ is defined by (1.3.2).

It is worth noting that A6.4.6 becomes

$$\lim_{T\longrightarrow 0} \limsup_{n\longrightarrow\infty} \frac{1}{T} \| \sum_{i=n}^{m(n,t)} a_i \epsilon_{i+1} \| = 0, \quad \forall t \in [0, T]$$

if $\{\epsilon_i\}$ is independent of $\{u_i\}$.

The adaptive control u_k is given according to the following recursive algorithm:

$$u_{k+1} = (u_k - a_k y_{k+1}) I_{[\|u_k - a_k y_{k+1}\| < 2b]}, \tag{6.4.4}$$

where b is specified in A6.4.2.

Theorem 6.4.1 *Assume A6.4.1–A6.4.6. Then the system (6.4.1), (6.4.2), and (6.4.4) has the desired properties:*

$$u_k \xrightarrow[k\longrightarrow\infty]{} u^0, \quad x_k \xrightarrow[k\longrightarrow\infty]{} 0$$

at sample paths where A6.4.6 holds.

Proof. Let u_{n_i} be a convergent subsequence of $\{u_k\}$ such that $u_{n_i} \xrightarrow[i\to\infty]{} \overline{u}$ and $\|\overline{u}\| < 2b$.

We have

$$u_{m+1} = u_m - a_m y_{m+1}, \tag{6.4.5}$$
$$\|u_{m+1} - u_{n_i}\| \le ct, \quad \forall m : n_i \le m \le m(n_i, t), \forall t \in [0, T], \tag{6.4.6}$$

for sufficiently large i and small enough T, where c is a constant to be specified later on. The relationships (6.4.5) and (6.4.6) can be proved along the lines of the proof for Theorem 2.2.1, but here u_i is known to be bounded, and (6.4.5) and (6.4.6) can be proved more straightforwardly. We show this.

Since the system (6.4.1) is BIBS, from $\|u_k\| < 2b$ it follows that there is $a > 0$ such that $\|x_k\| \le a$, $\forall k$.

By A6.4.6 for large i and small $T > 0$,

$$\sum_{j=n_i}^{m} a_j \epsilon_{j+1}\| \le at, \quad \forall m : n_i \le m \le m(m_i, t), \forall t \in [0, T].$$

This implies that

$$\begin{aligned} \|\sum_{j=n_i}^{m} a_j y_{j+1}\| &= \|\sum_{j=n_i}^{m} a_j (x_{j+1} + \epsilon_{j+1})\| \\ &\le a \sum_{j=n_i}^{m} a_j + \sum_{j=n_i}^{m} a_j \epsilon_{j+1} \le 2at, \quad \forall m : n_i \le m \le m(n_i, t). \end{aligned}$$

Let i be large enough such that

$$\|u_{n_i} - \overline{u}\| < \frac{1}{2}(2b - \|\overline{u}\|),$$

and let T be small enough such that

$$aT < \frac{1}{4}(2b - \|\overline{u}\|).$$

Then we have

$$\|u_{n_i} - a_{n_i} y_{n_i+1}\| \le \|u_{n_i} - \overline{u}\| + \|\overline{u}\| + \|a_{n_i} y_{n_i+1}\| < 2b,$$

and hence there is no truncation in (6.4.4) for $k = n_i$, i.e., (6.4.5) holds for $m = n_i$. Therefore,

$$\|u_{n_i+1} - u_{n_i}\| = \|a_{n_i} y_{n_i+1}\| \le 2at \triangleq ct.$$

Thus, (6.4.5) and (6.4.6) hold for $m = n_i$. Assume they are true for all $m : m \le k$, $n_i \le k < m(n_i, t)$. We now show that they are true for $m = k+1$ too.

Since

$$\begin{aligned}\|u_{k+1} - a_{k+1}y_{k+1}\| =&\|u_{n_i} - \sum_{j=n_i}^{k+1} a_j y_{j+1}\| \\ &\le \|u_{n_i} - \overline{u}\| + \|\overline{u}\| + \|\sum_{j=n_i}^{k+1} a_j y_{j+1}\| < 2b,\end{aligned}$$

from (6.4.4) it follows that (6.4.5) holds for $m = k+1$. Hence, $\|u_{k+2} - u_{n_i}\| = \|\sum_{j=n_i}^{k+1} a_j y_{j+1}\| \le 2at \triangleq ct$, and (6.4.6) is true for $m = k+1$ indeed.

By induction, the assertions (6.4.5) and (6.4.6) have been proved.

We now show that for any convergent subsequence $\{u_{n_k}\}$, $u_{n_k} \longrightarrow \overline{u} \ne u^0$ there is a $\delta > 0$ such that

$$\liminf_{k\longrightarrow\infty} \frac{1}{T} \sum_{i=n_k}^{m(n_k,T)} a_i (u_i - u^0)^T x_{i+1} \ge \delta \tag{6.4.7}$$

for small enough $T > 0$.

By A6.4.5, we have

$$\begin{aligned}\sum_{j=n_k}^{i} (u_j - u^0)^T x_{j+1} &\ge \epsilon \sum_{j=n_k}^{i} \|u_j - u^0\|^2 + \beta \\ &= \epsilon \sum_{j=n_k}^{i} \|\overline{u} - u^0 + (u_j - \overline{u})\|^2 + \beta \\ &= \epsilon \sum_{j=n_k}^{i} \|\overline{u} - u^0\|^2 + 2\epsilon \sum_{j=n_k}^{i} (\overline{u} - u^0)^T (u_j - \overline{u}) \\ &\quad + \epsilon \sum_{j=n_k}^{i} \|u_j - \overline{u}\|^2 + \beta\end{aligned} \tag{6.4.8}$$

Let us restrict i in (6.4.8) to $\{n_k, n_k+1, \ldots, m(n_k, T)\}$. Then for small T and large k, from (6.4.6) and (6.4.8) it follows that

$$\frac{1}{i - n_k + 1} \sum_{j=n_k}^{i} (u_j - u^0)^T x_{j+1}$$

$$\geq \epsilon\|\overline{u}-u^0\|^2 - 4\epsilon\|\overline{u}-u^0\|cT - 4\epsilon T^2 + \frac{\beta}{i-n_k+1}$$

for $i \in [n_k, \dots, m(n_k, T)]$.

This implies that there exist a $\delta > 0$ and a sufficiently large i_0, which may depend on $\overline{u}$ but is independent of k, such that

$$\frac{1}{i-n_k+1}\sum_{j=n_k}^{i}(u_j-u^0)^T x_{j+1} > \delta, \qquad (6.4.9)$$
$$\forall i \in \{n_k+i_0, n_k+i_0+1, \dots, m(n_k,T)\}$$

for all sufficiently large k and small enough $T > 0$.

Set

$$S_{n_k,i} = \sum_{j=n_k}^{i}(u_j-u^0)^T x_{j+1}, \quad S_{n_k,n_k-1} = 0.$$

Using a partial summation, by (6.4.9) we have

$$\begin{aligned}
&\sum_{i=n_k}^{m(n_k,t)} a_i(u_i-u^0)^T x_{i+1} \\
&= \sum_{i=n_k}^{m(n_k,T)} a_i(S_{n_k,i}-S_{n_k,i-1}) \\
&= a_{m(n_k,T)}S_{n_k,m(n_k,T)} + \sum_{i=n_k}^{m(n_k,T)-1}(a_i-a_{i+1})S_{n_k,i} \\
&> a_{m(n_k,T)}\delta(m(n_k,T)-n_k+1) + \sum_{i=n_k}^{n_k+i_0-1}(a_i-a_{i+1})S_{n_k,i} \\
&\quad + \sum_{i=n_k+i_0}^{m(n_k,T)-1}(a_i-a_{i+1})\delta(i-n_k+1). \qquad (6.4.10)
\end{aligned}$$

Since $\|u_i\| < 2b$ and $\|x_i\| < a$, it is seen that

$$\begin{aligned}
&\|\sum_{i=n_k}^{n_k+i_0-1}(a_i-a_{i+1})S_{n_k,i}\| \leq \sum_{i=n_k}^{n_k+i_0-1}(a_i-a_{i+1})i_0(2b+\|u^0\|)a \\
&= i_0(2b+\|u^0\|)a(a_{n_k}-a_{n_k+i_0}) \xrightarrow[k\longrightarrow\infty]{} 0.
\end{aligned}$$

Then (6.4.10) implies that

$$\begin{aligned}
&\| \sum_{i=n_k}^{m(n_k,T)} a_i(u_i - u^0)^T x_{i+1}\| \\
&\quad > a_{m(n_k,T)}\delta(m(n_k,T) - n_k + 1) + o(1) \\
&\quad - a_{m(n_k,T)}\delta(m(n_k,T) - n_k) + a_{n_k+i_0}\delta(i_0+1) + \delta \sum_{i=n_k+i_0+1}^{m(n_k,T)-1} a_i \\
&\quad = \delta a_{m(n_k,T)} + o(1) + a_{n_k+i_0}\delta(i_0+1) + \delta \sum_{i=n_k+i_0+1}^{m(n_k,T)-1} a_i \xrightarrow[k\longrightarrow\infty]{} \delta T.
\end{aligned}$$

This proves (6.4.7).
Define

$$v(u) = \|u - u^0\|^2.$$

From (6.4.7) it follows that

$$\liminf_{k\longrightarrow\infty} \frac{1}{T} \sum_{i=n_k}^{m(n_k,T)} a_i v_x^T(u_i) f(x_i, u_i) \geq 2\delta \tag{6.4.11}$$

for convergent subsequence $u_{n_k} \longrightarrow \overline{u} \neq u^0$.

Using A6.4.6 and (6.4.11), by completely the same argument as that used in the proof (Steps 3– 6) of Theorem 2.2.1, we conclude that $u_k \longrightarrow u^0$.

Finally, write (6.4.1) as

$$x_{k+1} = f(x_k, u^0) + f(x_k, u_k) - f(x_k, u^0).$$

By A6.4.4 and the boundedness of $\{x_k\}$ we have $d_k \triangleq f(x_k, u_k) - f(x_k, u^0) \xrightarrow[k\longrightarrow\infty]{} 0$, and by A6.4.2 we conclude $x_k \xrightarrow[k\longrightarrow\infty]{} 0$. □

Remark 6.4.1 It is easy to see that A6.4.6 is also necessary if A6.4.1–A6.4.5 hold and $x_k \longrightarrow x^0$ and $u_k \longrightarrow u^0$. This is because for large k the observation noise can be expressed as

$$\epsilon_{k+1} = \frac{u_{k+1} - u_k}{a_k} + f(x_k, u_k),$$

and hence

$$\sum_{i=n_k}^{m(n_k,T)} a_i \epsilon_{i+1} = \sum_{i=n_k}^{m(n_k,T)} (u_{i+1} - u_i) + \sum_{i=n_k}^{m(n_k,T)} a_i f(x_i, u_i),$$

which tends to zero as $k \longrightarrow \infty$ since $x_k \longrightarrow x^0$ and $u_k \longrightarrow u^0$.

Remark 6.4.2 In the formulation of Theorem 6.4.1 the condition A6.4.5 can be replaced either by (6.4.7) or by (6.4.11), which are the consequences of A6.4.5. Further, the quadratic $v(u)$ can be replaced by a continuously differentiable function $v(\cdot) : \mathbb{R}^n \longrightarrow R$ such that $v(u^0) = 0$ and $v(0) < \inf_{\|u\|=2b} v(u)$. In this case, $(u_i - u^0)$ in (6.4.7) should be correspondingly replaced by $v_x(u_i)$.

Example 6.4.1 Let the nonlinear system be affine:

$$x_{k+1} = g(x_k)(u_k - u^0),$$

where the scalar nonlinear function $g(\cdot)$ is bounded from above and from below by positive constants:

$$0 < \alpha \leq g(x) \leq \beta < \infty, \quad \forall x \in \mathbb{R}^n.$$

Note that $(u_j - u^0)^T x_{j+1} = g(x_j)\|u_j - u^0\|^2 \geq \alpha\|u_j - u^0\|^2$, and hence (6.4.7) holds, if $u_{n_k} \longrightarrow \overline{u} \neq u^0$. Assume b is known: $\|u^0\| < b$. Then A6.4.2, A6.4.3, and A6.4.4 are satisfied. Therefore, if $\{\epsilon_i\}$ satisfies A6.4.6, then $\{u_k\}$ given by (6.4.4) leads to $u_k \longrightarrow u^0$ and $x_k \longrightarrow 0$.

In the area of system and control, the SA methods also are successfully applied in discrete event dynamic systems, especially, to the perturbation analysis based parameter optimization.

6.5. Notes and References

For system identification and adaptive control we refer to [10, 23, 54, 62, 75, 90]. The identification problem stated in Section 6.1 was solved in [72] by ODE method. In comparison with [72], conditions used here have considerably been weakened, and the convergence is proved by the TS method rather than the ODE method. Section 6.1 is based on the joint work by H. F. Chen, T. Duncan and B. Pasik-Duncan. The existence and uniqueness of the solution to (6.1.50) can be found, e.g., in [23]. For stochastic quadratic control refer to [2, 10, 12, 33].

Adaptive stabilization for stochastic systems is dealt with in [5, 55, 77]. The convergence of WLS and adaptive stabilization using WLS are given in [55]. The problem is solved by the SA method in [19]. This approach is presented in Section 6.2.

The pole assignment problem for stochastic system with unknown coefficients is solved by SA with the help of learning in Section 6.3, which is based on [20]. For concept of linear control systems we refer to

[1, 46, 60]. The connection between the feedback gain and coefficients of the desired characteristic polynomial is called the Ackermann's formula, which can be found in [46].

Application of SA to adaptive regulation is based on [26].

For perturbation analysis of discrete event dynamic systems we refer to [58]. The perturbation analysis based parameter optimization is dealt with in [29, 86, 87].

Appendix A

In Appendix A we introduce the basic concept of probability theory. Results are presented without proof. For details we refer to [31, 32, 70, 76, 84].

A.1. Probability Space

The basic space is denoted by Ω. The point $\omega \in \Omega$ is called elementary event or sample. The point set in Ω is denoted by A, $A \subset \Omega$.

Let $\mathcal{F}$ be a family of sets in Ω satisfying the following conditions:

1. $\Omega \in \mathcal{F}$;
2. If $A \in \mathcal{F}$, then A^c, the complement of A, also belongs to $\mathcal{F}$, $A^c \in \mathcal{F}$;
3. If $A_i \in \mathcal{F}$, $i = 1, 2, \ldots$, then $\cup_{i=1}^{\infty} A_i \in \mathcal{F}$.

Then, $\mathcal{F}$ is called the σ-algebra or σ-field. The element A of $\mathcal{F}$ is called the measurable set, or random event, or event.

As a consequence of Properties 2 and 3, $\cap_{i=1}^{\infty} A_i \in \mathcal{F}$, if $A_i \in \mathcal{F}$, $i = 1, 2, \ldots$.

A set function ϕ defined on $\mathcal{F}$ is called σ-additive if $\phi(\sum_{i=1}^{\infty} A_i) = \sum_{i=1}^{\infty} \phi(A_i)$ for any sequence of disjoint events $A_i \in \mathcal{F}$. By definition, one of the values $+\infty$ or $-\infty$ is not allowed to be taken by ϕ.

A nonnegative σ-additive set function is called a measure.

Define

$$\phi^+(A) = \sup_{B \in A} \phi(B), \quad \phi^-(A) = -\inf_{B \in A} \phi(B), \quad A, B \in \mathcal{F}.$$

The set functions ϕ^+, ϕ^-, and $\tilde{\phi} \triangleq \phi^+ + \phi^-$ are called the upper, lower, and total variation of ϕ on $\mathcal{F}$, respectively.

Jordan-Hahn Decomposition Theorem If ϕ is σ-additive on $\mathcal{F}$, then there exists a set D such that, for any $A \in \mathcal{F}$,

$$\phi^-(A) = -\phi(AD), \quad \phi^+(A) = \phi(AD^c),$$

ϕ^+ and ϕ^- are measures and $\phi = \phi^+ - \phi^-$.

Let P be a set function defined on $\mathcal{F}$ with the following properties.

1. $PA \geq 0$, $\forall A \in \mathcal{F}$;
2. $P\Omega = 1$;

3. $P\bigcup_{i=1}^{\infty} A_i = \sum_{i=1}^{\infty} PA_i$, if A_i, $i = 1, 2, \ldots$, are disjoint. Then, P is called a probability measure on $\mathcal{F}$. The triple $(\Omega, \mathcal{F}, P)$ is called a probability space.

PA is called the probability of random event A.

It is assumed that any subset of a measurable set of probability zero is measurable and its probability is zero. After such a completion of measurable sets the resulting probability space is called completed.

If a relationship between random variables holds for any ω with possible exception of a set with probability zero, then we say this relationship holds a.s. (almost surely) or with probability one.

A.2. Random Variable and Distribution Function

In R, the real line, the smallest σ-algebra containing all intervals is called the Borel σ-algebra and is denoted by $\mathcal{B}$. The "smallest" means that if there is a σ-algebra $\mathcal{B}$, containing all intervals, then there must be $\mathcal{B} \subset \mathcal{B}_1$ in the sense that $B \in \mathcal{B}_1$ for any $B \in \mathcal{B}$. The Borel σ-algebra $\mathcal{B}^r$ can also be defined in $\mathbb{R}^r$. Any set in $\mathcal{B}$ or $\mathcal{B}^r$ is called the Borel set.

Any interval can be endowed with a measure equal to its length. This measure can be extended to each $B \in \mathcal{B}$, i.e., to each Borel set. Any subset of a set with measure zero is also assumed to be a measurable set with measure zero. After such a completion, the measurable set is called Lebesgue measurable, and the measure the Lebesgue measure. In what follows $\mathcal{B}$ always means the completed Borel σ-algebra.

A real function $\xi = \xi(\omega)$ defined on $(\Omega, \mathcal{F}, P)$ is called measurable, if

$$\{\omega : \xi(\omega) < x\} \in \mathcal{F}, \quad \forall x \in R.$$

If ξ is a real measurable function defined on $(\Omega, \mathcal{F}, P)$ and $P(\omega : |\xi(\omega)| < \infty) = 1$, then ξ is called a random variable. Therefore, if $g(\cdot)$ is a measurable function, then $g(\xi)$ is also a random variable if $|g(\xi)| < \infty$ a.s.

Let ξ be a random variable. The distribution function of ξ is defined as

$$F_\xi(x) \triangleq P(\omega : \xi(\omega) < x), \quad \forall x \in R.$$

By a random vector $\xi = [\xi^1, \ldots, \xi^n]^T$ we mean that each component ξ^i, $i = 1, \ldots, n$, of ξ is a random variable.

The distribution function of a random vector ξ is defined as

$$F_\xi(x^1, \ldots, x^n) = P(\omega : \xi^1 < x^1, \ldots, \xi^n < x^n).$$

If $F_\xi(x)$ is differentiable, then its derivative $f_\xi(x) = \frac{dF_\xi(x)}{dx}$ is called the density of ξ. The density of a random vector is defined by a similar way. The density of l-dimensional normal distribution $v(\mu, R)$ is defined by

$$\frac{1}{(2\pi)^{\frac{l}{2}}(\det R)^{\frac{1}{2}}} \exp(-\frac{1}{2}(x-\mu)^T R^{-1}(x-\mu)).$$

A.3. Expectation

Let ξ be a random variable and let $\xi \geq 0$ a.s.

Define

$$E\xi = \int_\Omega \xi dP \triangleq \lim_{n \longrightarrow \infty} [\sum_{i=0}^{n2^n-1} i2^{-n} P(A_{n_i}) + nP(\xi > n)]$$

where $A_{n_i} = \{\omega : i2^{-n} < \xi \leq (i+1)2^{-n}\}$.

$E\xi$ is called the expectation of ξ.

For an arbitrary random variable define

$$\xi^+ \triangleq \max(\xi, 0), \quad \xi^- \triangleq \max(-\xi, 0).$$

The expectation of ξ is defined as

$$E\xi = E\xi^+ - E\xi^-,$$

if at least one of $E\xi^+$ and $E\xi^-$ is finite .

If $E|\xi| = E\xi^+ + E\xi^- < \infty$, then ξ is called integrable.

The expectation of ξ can be expressed by a Lebesgue-Stieltjes integral with respect to its distribution function $F_\xi(x)$:

$$E\xi = \int_\Omega \xi dP = \int_{-\infty}^{\infty} x dF_\xi(x).$$

If $g(\cdot)$ is a measurable function, then

$$Eg(\xi) = \int_{-\infty}^{\infty} g(x) dF_\xi(x).$$

In the density of l-dimensional random vector ξ with normal distribution,

$$\mu = E\xi, \quad R = E(\xi - E\xi)(\xi - E\xi)^T.$$

A.4. Convergence Theorems and Inequalities

Let $\{\xi_k\}$ be a sequence of random variables and ξ be a random variable.

If $P(\xi_k \xrightarrow[k\to\infty]{} \xi) = 1$, then we say that ξ_k converges to ξ a.s. and write $\xi_k \xrightarrow[k\to\infty]{} \xi$ a.s.

If for any $\epsilon > 0$, $P(|\xi_k - \xi| > \epsilon) \xrightarrow[k\to\infty]{} 0$, then we say that ξ_k converges to ξ in probability and write $\xi_k \xrightarrow{p} \xi$.

If the distribution functions $F_{\xi_k}(x)$ of ξ_k converge to $F_\xi(x)$ at any x where $F_\xi(x)$ is continuous, then we say ξ_k weakly (or in distribution) converges to ξ and write $\xi_k \xrightarrow{w} \xi$ or $\xi_k \xrightarrow{d} \xi$.

If $E|\xi_k - \xi|^2 \xrightarrow[k\to\infty]{} 0$, then we say ξ_k converges to ξ in the mean square sense and write l.i.m. $\xi_k = \xi$.

"$\xi_k \longrightarrow \xi$ a.s." implies "$\xi_k \xrightarrow{P} \xi$", which in turn implies "$\xi_k \xrightarrow[k\to\infty]{w} \xi$".

Monotone Convergence Theorem If random variables ξ_k nondecreasingly (nonincreasingly) converge to ξ a.s., $\xi_k \geq \eta$ ($\xi_k \leq \eta$) a.s. and $E\eta^- < \infty$ ($E\eta^+ < \infty$), then $E\xi_k \uparrow E\xi$ ($E\xi_k \downarrow E\xi$).

Dominated Convergence Theorem If $\xi_k \xrightarrow[k\to\infty]{P} \xi$ and there exists an integrable random variable η such that $|\xi_k| \leq \eta$, then $E|\xi| < \infty$, $E\xi_k \xrightarrow[k\to\infty]{} E\xi$, and $E|\xi_k - \xi| \xrightarrow[k\to\infty]{} 0$.

Fatou Lemma If $\xi_k \geq \eta$ ($\xi_k \leq \eta$), $k = 1, 2, \ldots$, for some random variable η with $E\eta^- < \infty$ ($E\eta^+ < \infty$), then

$$E \liminf_{k\to\infty} \xi_k \leq \liminf_{k\to\infty} E\xi_k, \quad (\limsup_{k\to\infty} E\xi_k \leq E \limsup_{k\to\infty} \xi_k).$$

Chebyshev Inequality

$$P(|\xi| > \epsilon) \leq \frac{E|\xi|}{\epsilon}, \quad \text{for any } \epsilon > 0.$$

Lyapunov Inequality

$$(E|\xi|^s)^{1/s} \leq (E|\xi|^t)^{1/t}, \quad \forall 0 < s \leq t.$$

Hölder Inequality
Let $1 < p < \infty$, $1 < q < \infty$ and $\frac{1}{p} + \frac{1}{q} = 1$. Then

$$E|\xi\eta| \leq (E|\xi|^p)^{1/p}(E|\eta|^q)^{1/q}.$$

In the special case where $p = q = 2$, the Hölder inequality is called the Schwarz inequality.

A.5. Conditional Expectation

Let $(\omega, \mathcal{F}, P)$ be a probability space. $\mathcal{F}_1$ is called a sub-σ-algebra of $\mathcal{F}$ if $\mathcal{F}_1$ is a σ-algebra and $\mathcal{F}_1 \subset \mathcal{F}$, by which it is meant that any $A \in \mathcal{F}_1$ implies $A \in \mathcal{F}$.

Radon-Nikodym Theorem Let $\mathcal{F}_1$ be a sub-σ-algebra of $\mathcal{F}$. For any random variable ξ with at least one of $E\xi^+$ and $E\xi^-$ being finite, there is an unique $\mathcal{F}_1$-measurable random variable denoted by $E(\xi|\mathcal{F}_1)$ such that for any $A \in \mathcal{F}_1$

$$\int_A \xi dP = \int_A E(\xi|\mathcal{F}_1)dP \quad \text{a.s.}$$

The $\mathcal{F}_1$-measurable random variable satisfying the above equality is called conditional expectation of ξ given $\mathcal{F}_1$.

Let $\mathcal{F}^\eta$ be the smallest (see A.2) σ-algebra containing all sets $\{\omega : \eta(\omega) < x\}$, $\forall x \in R$. $\mathcal{F}^\eta$ is called the σ-algebra generated by η.

The conditional expectation $E(\xi|\eta)$ of ξ given η is defined as

$$E(\xi|\eta) = E(\xi|\mathcal{F}^\eta) \quad \text{a.s.}$$

Let A be an event. Conditional probability $P(A|\mathcal{F}_1)$ of A given $\mathcal{F}_1$ is defined by

$$P(A|\mathcal{F}_1) = E(I_A|\mathcal{F}_1) \quad \text{a.s.}$$

Properties of the conditional expectation are listed below.
1) $E(a\xi + b\eta|\mathcal{F}_1) = aE(\xi|\mathcal{F}_1) + bE(\eta|\mathcal{F}_1)$ for constants a and b;
2) $E(E(\xi|\mathcal{F}_1)) = E\xi$;
3) $E(\eta\xi|\mathcal{F}_1) = \eta F(\xi|\mathcal{F}_1)$, if η is $\mathcal{F}_1$-measurable and $E(|\xi||\mathcal{F}_1) < \infty$ a.s.;
4) $E(E(\xi|\mathcal{F}_1)|\mathcal{F}_2) = E(\xi|\mathcal{F}_2)$, if $\mathcal{F}_2 \subset \mathcal{F}_1 \subset \mathcal{F}$;
5) $E(\xi|\mathcal{F}_1) = E\xi$, if $\mathcal{F}_1 = (\Omega, \phi)$.

Convergence theorems and inequalities stated in A.4 remain true with expectation $E(\cdot)$ replaced by the conditional expectation $E(\cdot|\mathcal{F}_1)$. For example, the conditional Hölder inequality

$$E(|\xi\eta||\mathcal{F}_1) \leq (E(|\xi|^p|\mathcal{F}_1))^{1/p}(E(|\eta|^q|\mathcal{F}_1))^{1/q}$$

for $p > 0$, $q > 0$, $\frac{1}{p} + \frac{1}{q} = 1$.

For a sequence $\{\xi_k\}$ of random variables and a sub-σ-algebra $\mathcal{F}_1$, the consistent conditional distribution functions $F_{\xi_1,\dots,\xi_{k_m}}(x_{k_1},\dots,x_{k_m};\omega)$ of $(\xi_{k_1},\dots,\xi_{k_m})$ given

$\mathcal{F}_1$ can be defined such that i) they are $\mathcal{F}_1$-measurable for any $k_1, \ldots, k_m$ and any fixed $x_{k_1}, \ldots, x_{k_m}$; ii) they are distribution functions for any fixed ω; and iii) for any measurable function $g(\cdot)$

$$E(g(\xi_{k_1}, \ldots, \xi_{k_m})|\mathcal{F}_1)$$
$$= \int_{-\infty}^{\infty} \cdots \int_{-\infty}^{\infty} g(x_{k_1}, \ldots, x_{k_m}) dF_{\xi_{k_1},\ldots,\xi_{k_m}}(x_{k_1}, \ldots, x_{k_m}; \omega) \quad \text{a.s.}$$

A.6. Independence

Let A_i $i = 1, 2, \ldots$, be a sequence of events.

If for any set of indices $\{i_1, \ldots, i_k\}$

$$P(\bigcap_{j=1}^{k} A_{i_j}) = \prod_{j=1}^{k} PA_{i_j},$$

then $\{A_i\}$ is called mutually independent.

Let $\mathcal{F}_i$, $i = 1, 2, \ldots$, be a sequence of σ-algebras. If events $A_1, \ldots, A_k$ are mutually independent whenever $A_i \in \mathcal{F}_i$, $i = 1, \ldots, k$, $\forall k \geq 1$, then the family $\{\mathcal{F}_i\}$ of σ-algebras is called mutually independent.

Let $\{\xi_i\}$ be a sequence of random variables and let $\mathcal{F}^{\xi_i}$ be the σ-algebra generated by ξ_i. If $\{\mathcal{F}^{\xi_i}\}$ is mutually independent, then the sequence $\{\xi_i\}$ of random variables is called mutually independent.

Law of iterated logarithm Let $\{\xi_i\}$ be a sequence of independent and identically distributed (iid) random variables, $E\xi_1 = 0$, $E\xi_1^2 = \sigma^2 < \infty$. Then

$$\limsup_{n \longrightarrow \infty} \frac{|\sum_{i=1}^{n} \xi_i|^2}{2\sigma^2 n \log \log n} = 1 \quad \text{a.s.}$$

Proposition A.6.1 *Let $f(x, y)$ be a measurable function defined on $(\mathbb{R}^l \times \mathbb{R}^m, \mathcal{B}^l \times \mathcal{B}^m)$. If the l-dimensional random vector ξ is independent of the m-dimensional random vector η, then*

$$E(f(\xi, \eta)|\xi) = g(\xi)$$

where

$$g(x) = Ef(x, \eta),$$

provided $Ef(x, \eta)$ exists for all $x \in \mathbb{R}^l$ in the range of ξ.

From this proposition it follows that

$$E(\eta|\xi) = E\eta$$

if ξ is independent of η.

A.7. Ergodicity

Let $\{\xi_k\}$ be a sequence of random variables and let $F_{k_1,\ldots,k_n}(x_{k_1}, \ldots, x_{k_n})$ be the distribution function of $(\xi_{k_1}, \ldots, \xi_{k_n})$.

If $F_{k_1+i,\ldots,k_n+i}(x_{k_1}, \ldots, x_{k_n}) = F_{k_1,\ldots,k_n}(x_{k_1}, \ldots, x_{k_n})$ for any integer i, then $\{\xi_k\}$ is called stationary, or $\{\xi_k\}$ is a stationary process.

Proposition A.7.1 *Let $\{\xi_k\}$ be stationary.*

If $E\xi_1$ exists, then

$$\frac{1}{n}\sum_{i=1}^{n}\xi_i \xrightarrow[n\longrightarrow\infty]{} E(\xi_1|\mathcal{C}) \quad a.s.,$$

where $\mathcal{C}$ is a sub-σ-algebra of $\mathcal{F}$ and is called invariant σ-algebra.

If $E(\xi_1|\mathcal{C}) = E\xi_1$, then the stationary process $\{\xi_k\}$ is called ergodic. Thus, for stationary and ergodic process $\{\xi_k\}$ we have

$$\frac{1}{n}\sum_{i=1}^{n}\xi_i \xrightarrow[n\longrightarrow\infty]{} E\xi_1 \quad \text{a.s.}$$

If $\{\xi_k\}$ is a sequence of mutually independent and identically distributed (and hence stationary) random variables, then $\mathcal{C} = (\Omega, \phi)$ and the sequence is ergodic.

Appendix B

In Appendix B we present the detailed proof of convergence theorems for martingales and martingale difference sequences.

Let $\{\xi_k\}$ be a sequence of random variables, and let $\{\mathcal{F}_k\}$ be a family of nondecreasing σ-algebras, i.e.,

$$\mathcal{F}_i \subset \mathcal{F}_j \subset \mathcal{F}, \quad \forall i \leq j.$$

If ξ_k is $\mathcal{F}_k$-measurable for any k, then we write $(\xi_k, \mathcal{F}_k)$ and call it as an adapted process.

An adapted process $(\xi_k, \mathcal{F}_k)$ with $E|\xi_k| < \infty, \forall k$ is called a martingale if $E(\xi_j|\mathcal{F}_i) = \xi_i$, a.s. $\forall j \geq i$, a supermartingale if $E(\xi_j|\mathcal{F}_i) \leq \xi_i$ a.s., $\forall j \geq i$, and a submartingale if $E(\xi_j|\mathcal{F}_i) \geq \xi_i$ a.s. $\forall j \geq i$.

An adapted process $(\xi_k, \mathcal{F}_k)$ is called a martingale difference sequence (MDS) if $E(\xi_k|\mathcal{F}_{k-1}) = 0$ a.s. $\forall k \geq 1$.

A sequence of mutually independent random vectors $\{\xi_k\}$ with $E\xi_k = 0$, $\forall k$ is an obvious example of MDS.

An integer-valued measurable function τ is called a Markov time with respect to $\{\mathcal{F}_k\}$ if

$$\{\omega : \tau(\omega) = n\} \in \mathcal{F}_n \quad \text{for any } n \geq 0.$$

If, in addition, $P(\tau < \infty) = 1$, then τ is called a stopping time.

B.1. Convergence Theorems for Martingale

Lemma B.1.1 Let $\{\xi_k, \mathcal{F}_k\}$ be adapted, τ a Markov time, and B a Borel set. Let τ_B be the first time at which the process $\{\xi_k\}$ hits the set B after time τ, i.e.,

$$\tau_B = \begin{cases} inf\{k : \tau < k, \xi_k \in B\} \\ \infty, \text{ if } \xi_k \notin B \text{ for } \forall k > \tau. \end{cases}$$

Then τ_B is a Markov time.

Proof. The conclusion follows from the following expression:

$$\{\tau_B = k\} = \bigcup_{i=0}^{k-1} \{(\tau = i)(\xi_{i+1} \notin B, \cdots, \xi_{k-1} \notin B, \xi_k \in B)\} \in \mathcal{F}_k, \quad \forall k \geq 0.$$

□

For defining the number of up-crossing of an interval (a, b) by a submartingale $\{\xi_k, \mathcal{F}_k\}$, $k = 1, \dots, N$ we first define

$$\begin{aligned}
\tau_0 &= 0,\\
\tau_1 &= \begin{cases} \min\{0 < k < N : \xi_k \le a\},\\ N, \text{ if } \xi_k > a, \forall k = 1, \dots, N,\end{cases}\\
\tau_2 &= \begin{cases} \min\{\tau_1 < k \le N : \xi_k \ge b\},\\ N, \text{ if } \xi_k < b, \forall k : \tau_1 < k \le N,\end{cases}\\
&\dots\\
\tau_{2m-1} &= \begin{cases} \min\{\tau_{2m-2} < k \le N : \xi_k \le a, \},\\ N, \text{ if } \xi_k > a, \forall k : \tau_{2m-2} < k \le N,\end{cases}\\
\tau_{2m} &= \begin{cases} \min\{\tau_{2m-1} < k \le N : \xi_k \ge b\},\\ N, \text{ if } \xi_k < b, \forall k : \tau_{2m-1} < k \le N.\end{cases}
\end{aligned}$$

The largest m for which $\xi_{\tau_{2m}} \ge b$ is called the number of up-crossing of the interval (a, b) by the process $(\xi_k, \mathcal{F}_k)$ and is denoted by $\beta(a, b)$.

By Lemma B.1.1 $\{\tau_1 = k\} \in \mathcal{F}_k$, $\forall k \le N$ and $\{\tau_1 = N\} = (\cup_{i=0}^{N-1}\{\tau_1 = i\})^c \in \mathcal{F}_N$. So, τ_1 is a Markov time.

Assume τ_i is a Markov time. Again, by Lemma B.1.1, $\{\tau_{i+1} = k\} \in \mathcal{F}_k$, $\forall k < N$ and $\{\tau_{i+1} = N\} = (\cup_{i=0}^{N-1}\{\tau_{i+1} = j\})^c \in \mathcal{F}_N$.

Therefore, all τ_i, $i = 0, 1, \dots, 2m$, are Markov times.

Theorem B.1.1 (Doob) For submartingales $\{\xi_k, \mathcal{F}_k\}$ the following inequalities hold

$$E\beta(a, b) \le \frac{E(\xi_N - a)^+}{b - a} \le \frac{E\xi_N^+ + |a|}{b - a},$$

where $\xi^+ = \begin{cases} \xi, \text{ if } \xi > 0,\\ 0, \text{ if } \xi \le 0.\end{cases}$

Proof. Note that $\beta(a, b)$ equals the number $\beta(0, b-a)$ of up-crossing of the interval $(0, b - a)$ by the submartingale $\{\xi_k - a, \mathcal{F}_k\}$ or by $\{(\xi_k - a)^+, \mathcal{F}_k\}$.

Since for $k \ge m$

$$\begin{aligned}
(\xi_m - a)^+ \le& [E(\xi_k - a|\mathcal{F}_m)]^+\\
=&\{E[(\xi_k - a)^+ - (\xi_k - a)^-]|\mathcal{F}_m\}^+ \le E[(\xi_k - a)^+|\mathcal{F}_m],
\end{aligned}$$

$\{(\xi_k - a)^+, \mathcal{F}_k\}$ is a submartingale.

Thus, without loss of generality, it suffices to prove that for a nonnegative submartingale $\{\xi_k, \mathcal{F}_k\}$,

$$E\beta(0, b) \le \frac{E\xi_N}{b}. \tag{B.1.1}$$

Define

$$\eta_i = \begin{cases} 0, \text{ if } \tau_{m-1} < i \le \tau_m \text{ and } m \text{ is odd},\\ 1, \text{ if } \tau_{m-1} < i \le \tau_m \text{ and } m \text{ is even}.\end{cases}$$

Define also $\xi_0 = 0$. Then for even m, ξ_k crosses $(0, b)$ from time τ_{m-1} to τ_m. Therefore,

$$\sum_{i=\tau_{m-1}+1}^{\tau_m} \eta_i(\xi_i - \xi_{i-1}) = \sum_{i=\tau_{m-1}+1}^{\tau_m} (\xi_i - \xi_{i-1})$$
$$= \xi_{\tau_m} - \xi_{\tau_{m-1}} \geq \xi_{\tau_m} \geq b$$

and

$$\sum_{i=1}^{N} \eta_i(\xi_i - \xi_{i-1}) \geq b\beta(0, b). \tag{B.1.2}$$

Further, the set $\{\eta_i = 1\}$ is $\mathcal{F}_{i-1}$-measurable since τ_i is a Markov time, $i = 1, 2, \ldots$, and

$$\{\eta_i = 1\} = \cup_{k\geq 1}(\{\tau_{2k-1} < i\} \cap \{\tau_{2k} < i\}^c).$$

Taking expectation of both sides of (B-1-2) yields

$$\begin{aligned} bE\beta(0, b) \leq & E\sum_{i=1}^{N} \eta_i(\xi_i - \xi_{i-1}) \\ = & \sum_{i=1}^{N} \int_{\{\eta_i=1\}} (\xi_i - \xi_{i-1})dP \\ = & \sum_{i=1}^{N} \int_{\{\eta_i=1\}} E[(\xi_i - \xi_{i-1})|\mathcal{F}_{i-1}]dP \\ \leq & \sum_{i=1}^{N} \int_{\{\eta_i=1\}} [E(\xi_i|\mathcal{F}_{i-1}) - \xi_{i-1}]dP \\ \leq & \sum_{i=1}^{N} \int_{\Omega} [E(\xi_i|\mathcal{F}_{i-1}) - \xi_{i-1}]dP = E\xi_N, \end{aligned}$$

where the last inequality holds because $\{\xi_i, \mathcal{F}_i\}$ is a submartingale and hence the integrand is nonnegative.

Thus (B.1.1) and hence the theorem is proved. □

Theorem B.1.2 (Doob) Let $\{\xi_k, \mathcal{F}_k\}$ be a submartingale with $\sup_k E\xi_k^+ < \infty$ a.s.

Then there is a random variable ξ with $E|\xi| < \infty$ such that

$$\lim_{k\longrightarrow\infty} \xi_k = \xi \qquad \text{a.s.}$$

Proof. Set

$$\limsup_{k\longrightarrow\infty} \xi_k \triangleq \xi^*, \quad \liminf_{k\longrightarrow\infty} \xi_k = \xi_*.$$

Assume the converse: $P(\xi^* > \xi_*) > 0$.

Then

$$(\xi^* > \xi_*) = \cup_{a<b}(\xi^* > b > a > \xi_*)$$

where a and b run over all rational numbers.

By the converse assumption there exist rational numbers $a < b$ such that

$$P(\xi^* > b > a > \xi_*) > 0. \tag{B.1.3}$$

Let $\beta_N(a,b)$ be the number of up-crossing of the interval (a,b) by $\{\xi_k, \mathcal{F}_k\}$, $k \leq N$. By Theorem B.1.1

$$E\beta_N(a,b) \leq \frac{E\xi_N^+ + |a|}{b-a}. \tag{B.1.4}$$

By the monotone convergence theorem from (B-1-4) it follows that

$$\begin{aligned} E\beta_\infty(a,b) \triangleq E \lim_{N\longrightarrow\infty} \beta_N(a,b) &= \lim_{N\longrightarrow\infty} E\beta_N(a,b) \\ &\leq \frac{\sup_N E\xi_N^+ + |a|}{b-a} < \infty. \end{aligned} \tag{B.1.5}$$

However, (B.1.3) implies $P(\beta_\infty(a,b) = \infty) > 0$, which contradicts (B.1.5). Hence,

$$P(\xi^* = \xi_*) = 1,$$

or ξ_k converges to a limit ξ which is finite a.s.

By Fatou lemma it follows that

$$E\xi^+ = E \liminf_{k\longrightarrow\infty} \xi_k^+ \leq \liminf_{k\longrightarrow\infty} E\xi_k^+ \leq \sup_k E\xi_k^+ < \infty$$

and

$$\begin{aligned} E\xi^- &= E \liminf_{k\longrightarrow\infty} \xi_k^- \leq \liminf_{k\longrightarrow\infty} E\xi_k^- \leq \sup_k E\xi_k^- \\ &= \sup_k (E\xi_k^+ - E\xi_k) \leq \sup_k (E\xi_k^+ - E\xi_1^-) < \infty, \end{aligned}$$

where $E\xi_1 \leq E\xi_k$, $\forall k \geq 1$ is invoked. Hence, $E|\xi| < \infty$. □

Corollary B.1.1 If $\{\xi_k, \mathcal{F}_k\}$ is a nonnegative supermartingale or nonpositive submartingale, then

$$\lim_{k\longrightarrow\infty} \xi_k = \xi \quad \text{a.s.} \quad \text{and } E|\xi| < \infty.$$

Because for nonpositive submartingales $\xi_k^+ = 0$, the corollary follows from the theorem; while for a nonnegative supermartingale $\{\xi_k, \mathcal{F}_k\}$, $\{-\xi_k, \mathcal{F}_k\}$ is a nonpositive submartingale.

Corollary B.1.2 If $\{\xi_k, \mathcal{F}_k\}$ is a martingale with $\sup_k E|\xi_k| < \infty$, then $\lim_{k\longrightarrow\infty} \xi_k = \xi$ a.s. and $E|\xi| < \infty$.

This is because for a martingale $\{\xi_k \mathcal{F}_k\}$, $E\xi_k = E\xi_1$, and $E|\xi_k| = E\xi_k^+ + E\xi_k^- = 2E\xi_k^+ - E\xi_k = 2E\xi_k^+ - E\xi_1$, and hence

$$\sup_k E\xi_k^+ = \frac{1}{2} \sup_k (E|\xi_k| + E\xi_1) = \frac{1}{2} \sup_k E|\xi_k| + \frac{1}{2} E\xi_1 < \infty.$$

B.2. Convergence Theorems for MDS I

Let $\{\xi_k, \mathcal{F}_k\}$ be an adapted process, $\xi_k \in \mathbb{R}^m$, and let G be a Borel set in $\mathbb{R}^m$. Then the first exit time τ from G defined by

$$\tau = \begin{cases} \min\{k : \xi_k \notin G\} \\ \infty, \text{ if } \xi_k \in G, \forall k \end{cases}$$

is a Markov time. This is because

$$\{\tau = k\} = \{\xi_0 \in G, \xi_1 \in G, \ldots, \xi_{k-1} \in G, \xi_k \notin G\} \in \mathcal{F}_k.$$

Lemma B.2.1. Let $\{\xi_k, \mathcal{F}_k\}$ be a martingale (supermartingale, submartingale) and τ a Markov time. Then the process $\{\xi_{\tau\wedge k}, \mathcal{F}_k\}$ stopped at τ is again a martingale (supermartingale, submartingale), where $\tau \wedge k \triangleq \min(\tau, k)$.

Proof. Note that

$$\xi_\tau I_{[\tau \le k-1]} = \xi_0 I_{[\tau=0]} + \cdots + \xi_{k-1} I_{[\tau=k-1]}$$

is $\mathcal{F}_{k-1}$-measurable.

If $\{\xi_k, \mathcal{F}_k\}$ is a martingale, then

$$\begin{aligned} E(\xi_{\tau\wedge \mathcal{F}_k} | \mathcal{F}_{k-1}) =& E[(\xi_\tau I_{[\tau \le k-1]} + \xi_k I_{[\tau > k-1]}) | \mathcal{F}_{k-1}] \\ =& \xi_\tau I_{[\tau \le k-1]} + E(\xi_k I_{[\tau > k-1]^c} | \mathcal{F}_{k-1}) \\ =& \xi_\tau I_{[\tau \le k-1]} + I_{[\tau > k-1]} E(\xi_k | \mathcal{F}_{k-1}) \\ =& \xi_\tau I_{[\tau \le k-1]} + I_{[\tau > k-1]} \xi_{k-1} = \xi_{\tau\wedge(k-1)}. \end{aligned}$$

This shows that $(\xi_{\tau\wedge k}, \mathcal{F}_k)$ is a martingale. For supermartingales and submartingales the proof is similar. □

Theorem B.2.1. Let $\{\xi_k, \mathcal{F}_k\}$ be a one-dimensional MDS. Then as $k \longrightarrow \infty$ $\eta_k = \sum_{i=1}^{k} \xi_i$ converges on

$$A \triangleq \{\omega : \sum_{i=1}^{\infty} E(\xi_i^2 | \mathcal{F}_{i-1}) < \infty\}.$$

Proof. Since $\sum_{i=1}^{k+1} E(\xi_i^2 | \mathcal{F}_{i-1})$ is $\mathcal{F}_k$-measurable, the first exit time

$$\tau_M \triangleq \begin{cases} \min\{k : k \ge 1, \sum_{i=1}^{k+1} E(\xi_i^2 | \mathcal{F}_{i-1}) > M\} \\ \infty, \text{ if } \sum_{i=1}^{\infty} E(\xi_i^2 | \mathcal{F}_{i-1}) \le M \end{cases}$$

is a Markov time and by Lemma B.2.1 $\{\eta_{k\wedge\tau_M}, \mathcal{F}_k\}$ is a martingale, where M is a positive constant.

Noticing that $\eta_{k\wedge\tau_M} = \sum_{i=1}^{k} \xi_k I_{[i \le \tau_M]}$ and that $[i \le \tau_M] = [\tau_M < i]^c = [\tau_M \le i-1]^c$ is $\mathcal{F}_{i-1}$-measurable, we find

$$(E|\eta_{k\wedge\tau_M}|)^2 \le E\eta^2_{k\wedge\tau_M} = E\sum_{i=1}^{k} \xi_i^2 I_{[i \le \tau_M]}$$

$$=E[\sum_{i=1}^{k} E(\xi_i^2 I_{[i\le\tau_M]}|\mathcal{F}_{i-1}] = E[\sum_{i=1}^{k} I_{[i\le\tau_M]}E(\xi_i^2|\mathcal{F}_{i-1})]$$
$$=E[\sum_{i=1}^{k\wedge\tau_M} E(\xi_i^2|\mathcal{F}_{i-1})] \le M.$$

By Corollary B.1.2 $\eta_{k\wedge\tau_M}$ converges as $k \longrightarrow \infty$. It is clear that $\eta_{k\wedge\tau_M} = \eta_k$ on $[\tau_M = \infty]$. Therefore, as $k \longrightarrow \infty$, η_k pathwisely converges on $[\tau_M = \infty]$. Since M is arbitrary, η_k converges on $\cup_{M=1}^{\infty}[\tau_M = \infty]$ which equals A. □

Theorem B.2.2. Let $\{\xi_k, \mathcal{F}_k\}$ be an MDS and $\eta_k = \sum_{i=1}^{k} \xi_i$. If $E(\sup_k \xi_k)^+ < \infty$, then $\{\eta_k\}$ converges on $A_1 \triangleq \{\omega : \sup_k \eta_k < \infty\}$. If $E(\inf_k \xi_k)^- < \infty$, then $\{\eta_k\}$ converges on $A_2 \triangleq \{\omega : \inf_k \eta_k > -\infty\}$.

Proof. It suffices to prove the first assertion, because the second one is reduced to the first one if ξ_k is replaced by $-\xi_k$.

Define
$$\tau_M = \begin{cases} \min(k : \eta_k > M, k \ge 1), \\ \infty, \text{ if } \eta_k < M, \forall k. \end{cases}$$

By Lemma B.2.1 $(\eta_{k\wedge\tau_M}, \mathcal{F}_k)$ is a martingale. It is clear that

$$\eta_{k\wedge\tau_M} = \begin{cases} \le M, \text{ if } k < \tau_M, \\ = \eta_{\tau_M - 1} + \xi_{\tau_M} \le M + \sup_k \xi_k, \text{ if } k \ge \tau_M. \end{cases}$$

Consequently,

$$\sup_k E(\eta_{k\wedge\tau_M})^+ \le E \sup_k (\eta_{k\wedge\tau_M})^+ \le E(M + (\sup_k \xi_k)^+) < \infty.$$

By Theorem B.1.2 $\eta_{k\wedge\tau_M}$ converges as $k \longrightarrow \infty$.

Since $\eta_{k\wedge\tau_M} = \eta_k$ on $[\tau_M = \infty]$, as $k \longrightarrow \infty$ η_k converges on $[\tau_M = \infty]$ and consequently on $\cup_{M=1}^{\infty}[\tau_M = \infty]$ which equals A_1. □

B.3. Borel-Cantelli-Lévy Lemma

Theorem B.3.1. (Borel-Cantelli-Lévy Lemma) Let $\{B_i\}$ be a sequence of events, $B_i \in \mathcal{F}_i$. Then $\sum_{i=1}^{\infty} I_{B_i} < \infty$ if and only if $\sum_{i=1}^{\infty} P(B_i|\mathcal{F}_{i-1}) < \infty$, or equivalently,

$$\bigcap_{k=1}^{\infty}\bigcup_{i=k}^{\infty} B_i = \{\omega : \sum_{i=1}^{\infty} P(B_i|\mathcal{F}_{i-1}) = \infty\}. \tag{B.3.1}$$

Proof. Define

$$\xi_k = \sum_{i=1}^{k}[I_{B_i} - E(I_{B_i}|\mathcal{F}_{i-1})]. \tag{B.3.2}$$

Clearly, $\{\xi_k, \mathcal{F}_k\}$ is a martingale and $(I_{B_i} - E(I_{B_i}|\mathcal{F}_{i-1}), \mathcal{F}_i)$ is an MDS.

Since $|I_{B_i} - E(I_{B_i}|\mathcal{F}_{i-1})| \le 1$, by Theorem B.2.2, ξ_k converges on

$$\{\omega : (\sup_k \xi_k < \infty) \cup (\inf_k \xi_k > -\infty)\}.$$

If $\sum_{i=1}^{\infty} I_{B_i} < \infty$, then from (B.3.2) it follows that $\sup_k \xi_k < \infty$, which implies that ξ_k converges. Then, this combining with $\sum_{i=1}^{\infty} I_{B_i} < \infty$ by (B.3.2) yields $\sum_{i=1}^{\infty} P(B_i|\mathcal{F}_{i-1}) < \infty$.

Conversely, if $\sum_{i=1}^{\infty} P(B_i|\mathcal{F}_{i-1}) < \infty$, then from (B.3.2) it follows that $\inf_k \xi_k > -\infty$. Noticing that $\{\omega : \inf_k \xi_k > -\infty\}$ is contained in the set where ξ_k converges by Theorem B.2.2, from the convergence of ξ_k by (B.3.2) it follows that $\sum_{i=1}^{\infty} I_{B_i} < \infty$. □

Theorem B.3.2 (Borel-Cantelli Lemma) Let $\{B_i\}$ be a sequence of events. If $\sum_{i=1}^{\infty} PB_i < \infty$, then the probability that B_i occur infinitely often is zero, i.e.,

$$P(\bigcap_{k=1}^{\infty} \bigcup_{i=k}^{\infty} B_i) = 0. \tag{B.3.3}$$

If B_i are mutually independent and $\sum_{i=1}^{\infty} PB_i = \infty$, then $P \cap_{k=1}^{\infty} \cup_{i=k}^{\infty} B_i = 1$.

Proof. Denote by $\mathcal{F}_k$ the σ-algebra generated by $\{B_1, B_2, \dots, B_k\}$.

If $\sum_{i=1}^{\infty} PB_i < \infty$, then

$$\infty > \sum_{i=1}^{\infty} PB_i = E(\sum_{i=1}^{\infty} E(I_{B_i}|\mathcal{F}_{i-1})),$$

and hence $\sum_{i=1}^{\infty} E(I_{B_i}|\mathcal{F}_{i-1}) < \infty$ a.s. which, by (B.3.1), implies (B.3.3).

When B_i are mutually independent, then

$$\sum_{i=1}^{\infty} P(B_i|\mathcal{F}_{i-1}) = \sum_{i=1}^{\infty} PB_i.$$

Consequently, $\sum_{i=1}^{\infty} PB_i = \infty$ implies $\sum_{i=1}^{\infty} P(B_i|\mathcal{F}_{i-1}) = \infty$, and $P \cap_{k=1}^{\infty} \cup_{i=k}^{\infty} B_i = 1$ follows from (B.3.1). □

B.4. Convergence Criteria for Adapted Sequences

Let $\{y_k, \mathcal{F}_k\}$ be an adapted process.

Theorem B.4.1 Let $\{b_i\}$ be a sequence of positive numbers. Then

$$\{\omega : \sum_{i=1}^{\infty} y_i \text{ converges }\} \cap A = \{\omega : \sum_{i=1}^{\infty} y_i I_{[|y_i| \le b_i]} \text{ converges }\} \cap A,$$

where

$$A = \{\omega : \sum_{i=1}^{\infty} P(|y_i| > b_i|\mathcal{F}_{i-1}) < \infty\}. \tag{B.4.1}$$

Proof. Set

$$B_i = \{\omega : |y_i| > b_i\}.$$

By Theorem B.3.1

$$(\omega : \sum_{i=1}^{\infty} P(B_i|\mathcal{F}_{i-1}) < \infty) = (\omega : \sum_{i=1}^{\infty} I_{B_i} < \infty),$$

or $A = (\omega : \sum_{i=1}^{\infty} I_{B_i} < \infty)$.

This means that A is the set where events B_i may occur only finitely many times. Therefore, on A the series $\sum_{i=1}^{\infty} y_i$ converges if and only if $\sum_{i=1}^{\infty} y_i I_{[|y_i| \le b_i]}$ converges. □

Theorem B.4.2 (Three Series Criterion) Denote by S the ω-set where the following three series converge:

$$\sum_{i=1}^{\infty} P(|y_i| \ge c|\mathcal{F}_{i-1}), \quad \sum_{i=1}^{\infty} E[y_i I_{[|y_i| \le c]}|\mathcal{F}_{i-1}],$$

and

$$\sum_{i=1}^{\infty} \{E[y_i^2 I_{[|y_i| \le c]}||\mathcal{F}_{i-1}] - (E[y_i I_{[|y_i| \le c]}|\mathcal{F}_{i-1})^2\},$$

where c is a positive constant.

Then $\eta_k = \sum_{i=1}^{k} y_i$ converges on S as $k \longrightarrow \infty$.

Proof. Taking $b_i = c$ in (B.4.1), we have $S \subset A$ and

$$(\omega : \sum_{i=1}^{\infty} y_i \text{ converges }) \cap S = (\omega : \sum_{i=1}^{\infty} y_i I_{[|y_i| \le c]} \text{ converges }) \cap S \tag{B.4.2}$$

by Theorem B.4.1.

Define

$$\xi_i = y_i I_{[|y_i| \le c]} - E[y_i I_{[|y_i| \le c]}|\mathcal{F}_{i-1}].$$

Since $\sum_{i=1}^{\infty} E[y_i I_{[|y_i| \le c]}|\mathcal{F}_{i-1}]$ converges on S, from (B.4.2) it follows that

$$(\omega : \sum_{i=1}^{\infty} y_i \text{ converges }) \cap S = (\omega : \sum_{i=1}^{\infty} \xi_i \text{ converges }) \cap S. \tag{B.4.3}$$

Noticing that $(\xi_i, \mathcal{F}_i)$ is an MDS and

$$E(\xi_i^2|\mathcal{F}_{i-1}) = E(y_i^2 I_{[|y_i| \le c]}|\mathcal{F}_{i-1}) - (E[y_i I_{[|y_i| \le c]}|\mathcal{F}_{i-1}])^2,$$

we see

$$S \subset (\omega : \sum_{i=1}^{\infty} E(\xi_c^2|\mathcal{F}_{i-1}) < \infty).$$

By Theorem B.2.1 $\sum_{i=1}^{k} \xi_i$ converges on S, or

$$(\omega : \sum_{i=1}^{\infty} \xi_i \text{ converges }) \cap S = S.$$

Then from (B.4.3) it follows that

$$(\omega : \sum_{i=1}^{\infty} y_i \text{ converges }) \cap S = S$$

or $S \subset (\omega : \sum_{i=1}^{\infty} y_i$ converges). □

B.5. Convergence Theorems for MDS II

Let $\{\xi_i, \mathcal{F}_i\}$ be an MDS.

Theorem B.5.1 (Y. S. Chow) $\eta_k \triangleq \sum_{i=1}^{k} \xi_i$ converges on

$$A \triangleq (\omega : \sum_{i=1}^{\infty} E(|\xi_i|^p|\mathcal{F}_{i-1}) < \infty), \quad 1 \le p \le 2. \tag{B.5.1}$$

Proof. By Theorem B.4.2 it suffices to prove $A \subset S$, where S is defined in Theorem B.4.2 with y_i replaced by ξ_i considered in the present theorem.

We now verify that three series defined in Theorem B.4.2 are convergent on A if y_i is replaced by ξ_i.

For convergence of the first series it suffices to note

$$\begin{aligned} P(|\xi_i| \le c|\mathcal{F}_{i-1}) =& E(I_{[|\xi_i| \ge c]}|\mathcal{F}_{i-1}) \\ \le& \frac{1}{c^p} E[|\xi_i|^p I_{[|\xi_i| \ge c]}|\mathcal{F}_{i-1}] \le \frac{1}{c^p} E(|\xi_i|^p|\mathcal{F}_{i-1}). \end{aligned}$$

For convergence of the second series, taking into account $E(\xi_i|\mathcal{F}_{i-1}) = 0$ we find

$$\begin{aligned} \frac{1}{c}\sum_{i=1}^{\infty} |E(\xi_i I_{[|\xi_i| \le c]}|\mathcal{F}_{i-1})| =& \frac{1}{c}\sum_{i=1}^{\infty} |E(\xi_i I_{[|\xi_i| > c]}|\mathcal{F}_{i-1})| \\ \le& \sum_{i=1}^{\infty} E(\frac{|\xi_i|}{c} I_{[|\xi_i| > c]}|\mathcal{F}_{i-1}] \le \frac{1}{c^p}\sum_{i=1}^{\infty} E(|\xi_i|^p|\mathcal{F}_{i-1}). \end{aligned}$$

Finally, for convergence of the last series it suffices to note

$$\frac{1}{c^2}\sum_{i=1}^{\infty} E(\xi_i^2 I_{[|\xi_i| \le c]}|\mathcal{F}_{i-1}) \le \frac{1}{c^p}\sum_{i=1}^{\infty} E(|\xi_i|^p|\mathcal{F}_{i-1})$$

and

$$E(\xi_i I_{[|\xi_i| \le c]}|\mathcal{F}_{i-1})^2 \le E(\xi_i^2 I_{[|\xi_i| \le c]}|\mathcal{F}_{i-1})$$

by the conditional Schwarz inequality. □

Theorem B.5.2. The conclusion of Theorem B.5.1 is valid also for $0 < p < 1$.

Proof. Define

$$\zeta_i = |\xi_i|^p - E(|\xi_i|^p|\mathcal{F}_{i-1}), \quad 0 < p < 1.$$

Then we have

$$\begin{aligned} \sum_{i=1}^{\infty} E(|\zeta_i|\,|\mathcal{F}_{i-1}) \le& \sum_{i=1}^{\infty} E[(|\xi_i|^p + E(|\xi_i|^p|\mathcal{F}_{i-1}))|\mathcal{F}_{i-1}] \\ \le& 2\sum_{i=1}^{\infty} E(|\xi_i|^p\,|\mathcal{F}_{i-1}) < \infty \end{aligned}$$

on A where A is still defined by (B-5-1) but with $p: 0 < p < 1$.

Applying Theorem B.5.1 with $p = 1$ to the MDS $\{\zeta_i, \mathcal{F}_i\}$ leads to that $\sum_{i=1}^{\infty} \zeta_i$ converges on A, i.e.,

$$(\omega : \sum_{i=1}^{\infty}(|\xi_i|^p - E(|\xi_i|^p|\mathcal{F}_{i-1})) \text{ converges }) \supset (\omega : \sum_{i=1}^{\infty} E(|\xi_i|^p|\mathcal{F}_{i-1}) < \infty).$$

This is equivalent to

$$(\omega : \sum_{i=1}^{\infty} |\xi_i|^p \text{ converges }) \supset (\omega : \sum_{i=1}^{\infty} E(|\xi_i|^p|\mathcal{F}_{i-1}) < \infty). \tag{B.5.2}$$

Notice that convergence of $\sum_{i=1}^{\infty} |\xi_i|^p$ implies convergence of $\sum_{i=1}^{\infty} |\xi_i|$ since $|\xi_i| < 1$ for sufficiently large i.

Consequently, from (B.5.2) it follows that

$$(\omega : \sum_{i=1}^{\infty} \xi_i \text{ coonverges }) \supset (\omega : \sum_{i=1}^{\infty} E(|\xi_i|^p|\mathcal{F}_{i-1}) < \infty).$$

□

B.6. Weighted Sum of MDS

Theorem B.6.1 Let $\{\xi_k, \mathcal{F}_k\}$ be an l-dimensional MDS and let $\{M_k, \mathcal{F}_k\}$ be a matrix adapted process. If

$$\sup_k E(\|\xi_{k+1}\|^\alpha|\mathcal{F}_k) \triangleq \sigma < \infty \text{ a.s.}$$

for some $\alpha \in (0, 2]$, then as $k \longrightarrow \infty$:

$$\sum_{i=0}^{k} M_i\xi_{i+1} = O(s_k(\alpha)(\log(s_k^\alpha(\alpha) + e))^{\frac{1}{\alpha}+\eta}) \quad a.s. \quad \forall \eta > 0,$$

where

$$s_k(\alpha) = (\sum_{i=0}^{k} \|M_i\|^\alpha)^{\frac{1}{\alpha}}.$$

Proof. Without loss of generality, assume $M_0 \neq 0$.

We have the following estimate:

$$\begin{aligned}
&\sum_{i=1}^{\infty} E[\|[s_i(\alpha)(\log(s_i^\alpha(\alpha)+e))^{\frac{1}{\alpha}+\eta}]^{-1}M_i\xi_{i+1}\|^\alpha|\mathcal{F}_i] \\
&\qquad \leq \sigma\sum_{i=1}^{\infty}[s_i^\alpha(\alpha)(\log(s_i^\alpha(\alpha)+e)^{1+\alpha\eta}]^{-1}\|M_i\|^\alpha \\
&\qquad = \sigma\sum_{i=1}^{\infty}[s_i^\alpha(\alpha)(\log(s_i^\alpha(\alpha)+e))^{1+\alpha\eta}]^{-1}\int_{s_{i-1}^\alpha(\alpha)}^{s_i^\alpha(\alpha)}dx \\
&\qquad \leq \sigma\sum_{i=1}^{\infty}\int_{s_{i-1}^\alpha(\alpha)}^{s_i^\alpha(\alpha)}\frac{dx}{x(\log(x+e))^{1+\eta\alpha}} \\
&\qquad \leq \sigma\int_{s_0^\alpha(\alpha)}^{\infty}\frac{dx}{x(\log(x+e))^{1+\eta\alpha}}<\infty.
\end{aligned}$$

By Theorems B.5.1 and B.5.2 it follows that

$$\sum_{i=1}^{\infty} a_i^{-1}M_i\xi_{i+1}<\infty \quad \text{a.s.,} \tag{B.6.1}$$

where

$$a_i = s_i(\alpha)(\log(s_i^\alpha(\alpha)+e))^{\frac{1}{\alpha}+\eta}.$$

Notice that a_i is nondecreasing as $i \longrightarrow \infty$. If $\{a_i\}$ is bounded, then the conclusion of the theorem follows from (B.6.1). If $a_i \xrightarrow[i\longrightarrow\infty]{} \infty$, then by the Kronecker lemma (see Section 3.4) the conclusion of the theorem also follows from (B.6.1). □

References

[1] B. D. O. Anderson and T. B. Moore, Optimal Control: Linear Quadratic Methods, Prentice-Hall, N. J., 1990.

[2] K. J. Åström, Introduction to Stochastic Control, Academic Press, New York, 1970.

[3] M. Benaim, A dynamical systems approach to stochastic approximation, SIAM J. Control & Optimization, 34:437–472, 1996.

[4] A. Benveniste, M. Metivier and P. Priouret, Adaptive Algorithms and Stochastic Approximation, Springer-Verlag, New York, 1990.

[5] B. Bercu, Weighted estimation and tracking for ARMAX models, SIAM J. Control & Optimization, 33:89–106, 1995.

[6] P. Billingsley, Convergence of Probability Measures, Wiley, New York, 1968.

[7] J. R. Blum, Multidimensional stochastic approximation, Ann. Math. Statist., 9:737–744, 1954.

[8] V. S. Borkar, Asynchronous stochastic approximations, SIAM J. Control and Optimization, 36:840–851, 1998.

[9] O. Brandière and M. Duflo, Les algorithmes stochastiques contournents-ils les pièges? Ann. Inst. Henri Poincaré, 32:395–427, 1996.

[10] P. E. Caines, Linear Stochastic Systems, Wiley, New York, 1988.

[11] H. F. Chen, Recursive algorithms for adaptive beam-formers, Kexue Tongbao (Science Bulletin), 26:490–493, 1981.

[12] H. F. Chen, Recursive Estimation and Control for Stochastic Systems, Wiley, New York, 1985.

[13] H. F. Chen, Asymptotic efficient stochastic approximation, Stochastics and Stochastics Reports, 45:1–16, 1993.

[14] H. F. Chen, Stochastic approximation and its new applications, Proceedings of 1994 Hong Kong International Workshop on New Directions of Control and Manufacturing, 1994, 2–12.

[15] H. F. Chen, Convergence rate of stochastic approximation algorithms in the degenerate case, SIAM J. Control & Optimization, 36:100–114, 1998.

[16] H. F. Chen, Stochastic approximation with non-additive measurement noise, J. of Applied Probability, 35:407–417, 1998.

[17] H. F. Chen, Convergence of SA algorithms in multi-root or multi-extreme cases, Stochastics and Stochastics Reports, 64: 255–266, 1998.

[18] H. F. Chen, Stochastic approximation with state-dependent noise, Science in China (Series E), 43:531–541, 2000.

[19] H. F. Chen and X. R. Cao, Controllability is not necassry for adaptive pole placement control, IEEE Trans. Autom. Control, AC-42:1222–1229, 1997.

[20] H. F. Chen and X. R. Cao, Pole assignment for stochastic systems with unknown coefficients, Science in China (Series E), 43:313–323, 2000.

[21] H. F. Chen, T. Duncan, and B. Pasik-Duncan, A Kiefer-Wolfowitz algorithm with randomized differences, IEEE Trans. Autom. Control, AC-44:442–453, 1999.

[22] H. F. Chen and H. T. Fang, Nonconvex stochastic optimization for model reduction, Global Optimization, 2002.

[23] H. F. Chen and L. Guo, Identification and Stochastic Adaptive Control, Birkhäuser, Boston, 1991.

[24] H. F. Chen, L. Guo, and A. J. Gao, Convergence and robustness of the Robbins-Monro algorithm truncated at randomly varying bounds, Stochastic Processes and Their Applications, 27:217–231, 1988.

[25] H. F. Chen and R. Uosaki, Convergence analysis of dynamic stochastic approximation, Systems and Control Letters, 35:309–315, 1998.

[26] H. F. Chen and Q. Wang, Adaptive regulator for discrete-time nonlinear nonparametric systems, IEEE Trans. Autom. Control, AC-46: , 2001.

[27] H. F. Chen and Y. M. Zhu, Stochastic approximation procedures with randomly varying truncations, Scientia Sinica (Series A), 29:914–926, 1986.

[28] H. F. Chen and Y. M. Zhu, Stochastic Approximation (in Chinese), Shanghai Scientific and Technological Publishers, Shanghai, 1996.

[29] E. K. P. Chong and P. J. Ramadge, Optimization of queues using an infinitesimal perturbation analysis-based stochastic algorithm with general update times, SIAM J. Control & Optimization, 31:698–732, 1993.

[30] Y. S. Chow, Local convergence of martingales and the law of large numbers, Ann. Math. Statst. 36:552–558, 1965.

[31] Y. S. Chow and H. Teicher, Probablility Theory: Independence, Interchangeability, Martingales, Springer Verlag, New York, 1978.

[32] K. L. Chung, A Course in Probability Theory, (second edition), Academic Press, New York, 1974.

[33] M. H. A. Davis, Linear Estimation and Stochastic Control, Chapman and Hall, New York, 1977.

[34] K. Deimling, Nonlinear Functional Analysis, Springer, Berlin, 1985.

[35] B. Delyon and A. Juditsky, Stochastic optimization with averaging of trajectories, Stochastics and Stochastics Reports, 39:107–118, 1992.

[36] E. F. Deprettere (eds.), SVD and Signal Processing, Elsevier, Horth-Holland, 1988.

[37] N. Dunford and J. T. Schwartz, Linear Operators, Part 1: General Theory, Wiley-Interscience, New York, 1966.

[38] V. Dupač, A dynamic stochastic methods, Ann. Math. Statist. 36:1695–1702.

[39] V. Dupač, Stochastic approximation in the presense of trend, Czeshoslovak Math. J., 16:454–461, 1966.

[40] A. Dvoretzky, On stochastic approximation, Proceedings of the Third Berkeley Symposium on Mathematical Statistics and Probability, pp. 39–55, 1956.

[41] S. N. Ethier and T. G. Kurtz, Markov Processes: Characterization and Convergence, Wiley, New York, 1986.

[42] E. Eweda, Convergence of the sign algorithm for adaptive filtering with correlated data, IEEE Trans. Information Theory, IT-37:1450-1457, 1991.

[43] V. Fabian, On asymptotic normality in stochastic approximation, Ann. of Math. Statis., 39: 1327–1332, 1968.

[44] V. Fabian, On asymptotically efficient recursive estimation, Ann. Statist., 6: 854–856, 1978.

[45] V. Fabian, Simulated annealing simulated, Computers Math. Applic., 33:81–94, 1997.

[46] F. W. Fairman, Linear Control Theory, The State Space Approach, Wiley, Chichester, 1998.

[47] H. T. Fang and H. F. Chen, Sharp convergence rates of stochastic approximation for degenerate roots, Science in China (Series E), 41:383–392, 1998.

[48] H. T. Fang and H. F. Chen, Stability and instability of limit points of stochastic approximation algorithms, IEEE Trans. Autom. Control, AC-45:413–420, 2000.

[49] H. T. Fang and H. F. Chen, An a.s. convergent algorithm for global optimization with noise corrupted observations, J. Optimization and Its Applications, 104:343–376, 2000.

[50] H. T. Fang and H. F. Chen, Asymptotic behavior of asynchronous stochastic approximation, Science in China (Series F), 44:249–258, 2001.

[51] B. A. Francis, A Course in H_∞ Control Theory, Lecture Notes in Control and Information Sciences, Vol. 18, 1987.

[52] S. B. Gelfand and S. K. Mitter, Recursive stochastic algorithms for global optimization in $\mathbb{R}^d$, SIAM J. Control & Optimization, 29:999–1018, 1991.

[53] E. G. Gladyshev, On stochastic approximation (in Russian), Theory Probab. Appl., 10:275–278, 1965.

[54] G. C. Goodwin and K. S. Sin, Adaptive Filtering, Prediction and Control, Prentice-Hall, N.J., 1984.

[55] L. Guo, Self-convergence of weighted least squares with applications to stochastic adaptive control, IEEE Trans. Autom. Control, AC-41:79–89, 1996.

[56] P. Hall and C. C. Heyde, Martingale Limit Theory and Its Applications, Academic Press, New York, 1980.

[57] S. Haykin, Adaptive Filter Theory, Prentice-Hall, Englewood Cliffs, NJ, 1990.

[58] Y. C. Ho and X. R. Cao, Perturbation Analysis of Discrete Event Dynamical Systems, Kluwer, Boston, 1991.

[59] A. Juditsky, A Stochastic estimation algorithm with observation averaging, IEEE Trans. Autom. Control, 38:794–798, 1993.

[60] T. Kailath, Linear Systems, Prentice-Hall, N. J., 1980.

[61] J. Kiefer and J. Wolfowitz, Stochastic estimation of the maximum of a regression function, Ann. Math. Statist., 23:462–466, 1952.

[62] P. V. Kokotovic (Ed.), Foundations of Adaptive Control, Springer, Berlin, 1991.

[63] J. Koronaski, Random-seeking methods for the stochastic unconstrained optimization, Int. J. Control, 21:517–527, 1975.

[64] H. J. Kushner, Approximation and Weak Convergence Methods for Random Processes with Applications to Stochastic Systems Theory, MIT Press, Cambridge, MA, 1984.

[65] H. J. Kushner and D. S. Clark, Stochastic Approximation for Constrained and Unconstained Systems, Springer-Verlag, New York, 1978.

[66] H. J. Kushner and J. Yang, Stochastic approximation with averaging of the iterates: Optimal asymptotic rates of convergence for general processes, SIAM J. Control & Optimization, 31:1045–1062, 1993.

[67] H. J. Kushner and J. Yang, Stochastic approximation with averaging and feedback: Rapidly convergent "on line" algorithms, IEEE Trans. Autom. Control, AC-40:24–34, 1995.

[68] H. J. Kushner and G. Yin, Stochastic Approximation Algorithms and Applications, Springer-Verlag, New York, 1997.

[69] J. P. LaSaller and Lefchetz, Stability by Lyapunov's Direct Methods with Applications, Academic Press, New York, 1961.

[70] R. Liptser and A. N. Shiryaev, Statistics of Random Processes, Springer-Verlag, New York, 1977.

[71] R. Liu, Blind signal processing: An introduction, Proceedings 1996 Intl. Symp. Circuits and Systems, Vol. 2, 81–83, 1996.

[72] L. Ljung, Analysis of recursive stochastic algorithms, IEEE Trans. Autom. Control, AC-22:551-575, 1977.

[73] L. Ljung, On positive real transfer functions and the convergence of some recursive schemes, IEEE Trans. Autom. Control, AC-22:539–551, 1977.

[74] L. Ljung, G. Pflug, and H. Walk, Stochastic Approximation and Optimization of Random Systems, Birkhäuser, Basel, 1992.

[75] L. Ljung and T. Söderström, Theory and Practice of Recursive Identification, MIT Press, Cambridge, MA, 1983.

[76] M. Loéve, Probability Theory, Springer, New York, 1977–1978.

[77] R. Lozano and X. H. Zhao, Adaptive pole placement without excitation probing signals, IEEE Trans. Autom. Control, AC-39:47–58, 1994.

[78] M. B. Nevelson and R. Z. Khasminskii, Stochastic Approximation and Recursive Estimation, Amer. Math. Soc., Providence, RI, 1976, Translation of Math. Monographs, Vol. 47.

[79] E. Oja, Subspace Methods of Pattern Recognition, 1st ed., Letchworth, Research Studies Press Ltd., Hertfordshire, 1983.

[80] B. T. Polyak, New stochastic approximation type procedures, (in Russian) Autom. i Telemekh., 7:98–107, 1990.

[81] B. T. Polyak and A. B. Juditsky, Acceleration of stochastic approximation by averaging, SIAM J. Control & Optimization, 30:838–855, 1992.

[82] H. Robbins and S. Monro, A stochastic approximation method, Ann. Math. Statist., 22:400–407, 1951.

[83] D. Ruppert, Stochastic approximation, In B. K. Ghosh and P. K. Sen, Editors, Handbook in Sequential Analysis, 503–529, Marcel Dekker, New York, 1991.

[84] A. N. Shiryaev, Probability, Springer, New York, 1984.

[85] J. C. Spall, Multivariate stochastic approximation using a simultaneous perturbation gradient approximation, IEEE Trans. Autom. Control, AC-37:331–341, 1992.

[86] Q. Y. Tang and H. F. Chen, Convergence of perterbation analysis based optimization algorithm with fixed-number of customers period, Discrete Event Dynamic Systems, 4:359–373, 1994.

[87] Q. Y. Tang, H. F. Chen, and Z. J. Han, Convergence rates of perturbation-analysis-Robbins-Monro-Single-run algorithms, IEEE Trans. Autom. Control, AC-42:1442–1447, 1997.

[88] J. N. Tsitsiklis, Asynchronous stochastic approximation and Q-learning, Machine Learning, 16:185–202, 1994.

[89] N. J. Tsitsiklis, D. P. Bertsekas, and M. Athans, Distributed asynchronous deterministic and stochastic gradient optimization algorithms, IEEE Trans. Autom. Control, 31:803–812, 1986.

[90] Ya. Z. Tsypkin, Adaptation and Learning in Automatic Systems, Academic Press, New York, 1971.

[91] K. Uosaki, Some generalizations of dynamic stochastic approximation processes, Ann. Statist., 2:1042–1048, 1974.

[92] J. Venter, An extension of the Robbins-Monro procedure, Ann. Math. Stat., 38:181–190, 1967.

[93] G. J. Wang and H. F. Chen, Behavior of stochastic approximation algorithm in root set of regression function, Systems Science and Mathematical Sciences, 12:92–96, 1999.

[94] I. J. Wang, E. K. P. Chong and S. R. Kulkarni, Equivalent necessary and sufficient conditions on noise sequences for stochastic approximation algorithms, Adv. Appl. Probab., 28:784–801, 1996.

[95] C. Z. Wei, Multivariate adaptive stochastic approximation, Ann. Stat., 15:1115–1130, 1987.

[96] G. Xu, L. Tong, and T. Kailath, A least squares approach to blind identification, IEEE Trans. Signal Processing, SP-43:2982–2993, 1995.

[97] S. Yakowitz, A globally convergent stochastic approximation, SIAM J. Control & Optimization, 31:30–40, 1993.

[98] G. Yin, On extensions of Polyak's averaging approach to stochastic approximation, Stochastics and Stochastics Reports, 36:245–264, 1991.

[99] G. Yin and Y. M. Zhu, On w.p.l. convergence of a parallel stochastic approximation algorithm, Probability in the Eng. and Infor. Sciences, 3:55–75, 1989.

[100] R. Zeilinski, Global stochastic approximation: A review of results and some open problems. In F. Archetti and M. Cugiani (eds.), Numerical Techniques for Stochastic Systems, 379–386, Northholland Publ. Co., 1980.

[101] J. H. Zhang and H. F. Chen, Convergence of algorithms used for principal component analysis, Science in China (Series E), 40:597–604, 1997.

[102] K. Zhou, J. C. Doyle, and K. Glover, Robust Optimal Control, Prentice-Hall, New Jersey, 1996.

Index

Nonconvex Optimization and Its Applications

22. H. Tuy: *Convex Analysis and Global Optimization*. 1998 ISBN 0-7923-4818-4
23. D. Cieslik: *Steiner Minimal Trees*. 1998 ISBN 0-7923-4983-0
24. N.Z. Shor: *Nondifferentiable Optimization and Polynomial Problems*. 1998 ISBN 0-7923-4997-0
25. R. Reemtsen and J.-J. Rückmann (eds.): *Semi-Infinite Programming*. 1998 ISBN 0-7923-5054-5
26. B. Ricceri and S. Simons (eds.): *Minimax Theory and Applications*. 1998 ISBN 0-7923-5064-2
27. J.-P. Crouzeix, J.-E. Martinez-Legaz and M. Volle (eds.): *Generalized Convexitiy, Generalized Monotonicity: Recent Results*. 1998 ISBN 0-7923-5088-X
28. J. Outrata, M. Kočvara and J. Zowe: *Nonsmooth Approach to Optimization Problems with Equilibrium Constraints*. 1998 ISBN 0-7923-5170-3
29. D. Motreanu and P.D. Panagiotopoulos: *Minimax Theorems and Qualitative Properties of the Solutions of Hemivariational Inequalities*. 1999 ISBN 0-7923-5456-7
30. J.F. Bard: *Practical Bilevel Optimization*. Algorithms and Applications. 1999 ISBN 0-7923-5458-3
31. H.D. Sherali and W.P. Adams: *A Reformulation-Linearization Technique for Solving Discrete and Continuous Nonconvex Problems*. 1999 ISBN 0-7923-5487-7
32. F. Forgó, J. Szép and F. Szidarovszky: *Introduction to the Theory of Games*. Concepts, Methods, Applications. 1999 ISBN 0-7923-5775-2
33. C.A. Floudas and P.M. Pardalos (eds.): *Handbook of Test Problems in Local and Global Optimization*. 1999 ISBN 0-7923-5801-5
34. T. Stoilov and K. Stoilova: *Noniterative Coordination in Multilevel Systems*. 1999 ISBN 0-7923-5879-1
35. J. Haslinger, M. Miettinen and P.D. Panagiotopoulos: *Finite Element Method for Hemivariational Inequalities*. Theory, Methods and Applications. 1999 ISBN 0-7923-5951-8
36. V. Korotkich: *A Mathematical Structure of Emergent Computation*. 1999 ISBN 0-7923-6010-9
37. C.A. Floudas: *Deterministic Global Optimization: Theory, Methods and Applications*. 2000 ISBN 0-7923-6014-1
38. F. Giannessi (ed.): *Vector Variational Inequalities and Vector Equilibria*. Mathematical Theories. 1999 ISBN 0-7923-6026-5
39. D.Y. Gao: *Duality Principles in Nonconvex Systems*. Theory, Methods and Applications. 2000 ISBN 0-7923-6145-3
40. C.A. Floudas and P.M. Pardalos (eds.): *Optimization in Computational Chemistry and Molecular Biology*. Local and Global Approaches. 2000 ISBN 0-7923-6155-5
41. G. Isac: *Topological Methods in Complementarity Theory*. 2000 ISBN 0-7923-6274-8
42. P.M. Pardalos (ed.): *Approximation and Complexity in Numerical Optimization: Concrete and Discrete Problems*. 2000 ISBN 0-7923-6275-6
43. V. Demyanov and A. Rubinov (eds.): *Quasidifferentiability and Related Topics*. 2000 ISBN 0-7923-6284-5

Nonconvex Optimization and Its Applications

44. A. Rubinov: *Abstract Convexity and Global Optimization*. 2000 ISBN 0-7923-6323-X
45. R.G. Strongin and Y.D. Sergeyev: *Global Optimization with Non-Convex Constraints*. 2000 ISBN 0-7923-6490-2
46. X.-S. Zhang: *Neural Networks in Optimization*. 2000 ISBN 0-7923-6515-1
47. H. Jongen, P. Jonker and F. Twilt: *Nonlinear Optimization in Finite Dimensions*. Morse Theory, Chebyshev Approximation, Transversability, Flows, Parametric Aspects. 2000 ISBN 0-7923-6561-5
48. R. Horst, P.M. Pardalos and N.V. Thoai: *Introduction to Global Optimization*. 2nd Edition. 2000 ISBN 0-7923-6574-7
49. S.P. Uryasev (ed.): *Probabilistic Constrained Optimization*. Methodology and Applications. 2000 ISBN 0-7923-6644-1
50. D.Y. Gao, R.W. Ogden and G.E. Stavroulakis (eds.): *Nonsmooth/Nonconvex Mechanics*. Modeling, Analysis and Numerical Methods. 2001 ISBN 0-7923-6786-3
51. A. Atkinson, B. Bogacka and A. Zhigljavsky (eds.): *Optimum Design 2000*. 2001 ISBN 0-7923-6798-7
52. M. do Rosário Grossinho and S.A. Tersian: *An Introduction to Minimax Theorems and Their Applications to Differential Equations*. 2001 ISBN 0-7923-6832-0
53. A. Migdalas, P.M. Pardalos and P. Värbrand (eds.): *From Local to Global Optimization*. 2001 ISBN 0-7923-6883-5
54. N. Hadjisavvas and P.M. Pardalos (eds.): *Advances in Convex Analysis and Global Optimization*. Honoring the Memory of C. Caratheodory (1873-1950). 2001 ISBN 0-7923-6942-4
55. R.P. Gilbert, P.D. Panagiotopoulos† and P.M. Pardalos (eds.): *From Convexity to Nonconvexity*. 2001 ISBN 0-7923-7144-5
56. D.-Z. Du, P.M. Pardalos and W. Wu: *Mathematical Theory of Optimization*. 2001 ISBN 1-4020-0015-4
57. M.A. Goberna and M.A. López (eds.): *Semi-Infinite Programming. Recent Advances*. 2001 ISBN 1-4020-0032-4
58. F. Giannessi, A. Maugeri and P.M. Pardalos (eds.): *Equilibrium Problems: Nonsmooth Optimization and Variational Inequality Models*. 2001 ISBN 1-4020-0161-4
59. G. Dzemyda, V. Šaltenis and A. Žilinskas (eds.): *Stochastic and Global Optimization*. 2002 ISBN 1-4020-0484-2
60. D. Klatte and B. Kummer: *Nonsmooth Equations in Optimization*. Regularity, Calculus, Methods and Applications. 2002 ISBN 1-4020-0550-4
61. S. Dempe: *Foundations of Bilevel Programming*. 2002 ISBN 1-4020-0631-4
62. P.M. Pardalos and H.E. Romeijn (eds.): *Handbook of Global Optimization, Volume 2*. 2002 ISBN 1-4020-0632-2
63. G. Isac, V.A. Bulavsky and V.V. Kalashnikov: *Complementarity, Equilibrium, Efficiency and Economics*. 2002 ISBN 1-4020-0688-8

KLUWER ACADEMIC PUBLISHERS – DORDRECHT / BOSTON / LONDON